Asphalt Pavements

Asphalt Pavements

A practical guide to design, production, and maintenance for engineers and architects

Patrick G. Lavin

LONDON AND NEW YORK

First published 2003
by Spon Press
11 New Fetter Lane, London EC4P 4EE

Simultaneously published in the USA and Canada
by Spon Press
29 West 35th Street, New York, NY 10001

Spon Press is an imprint of the Taylor & Francis Group

Typeset in Times New Roman by
Newgen Imaging Systems (P) Ltd, Chennai, India
Printed and bound in Great Britain by
TJ International Ltd, Padstow, Cornwall

British Library Cataloguing in Publication Data
A catalogue record for this book is available
from the British Library

Library of Congress Cataloging in Publication Data
A catalog record for this book has been requested

ISBN 0–415–24733–0

Contents

Plates

Figures

Tables

Acknowledgments

I would like to thank my original commissioning editor, Mr Richard Whitby, for the foresight and faith in my ability to start and even complete this undertaking. I typed the manuscript on the one and only computer in our household. My dearest appreciation and thanks goes to my wife, Ada, who not only tolerated the endless use of the computer, but gave assistance in the review of the manuscript and provided moral support. I am sure she used it to put herself to sleep. Our pets, two St Bernard's, were always at my side while typing, giving whatever encouragement dogs can give.

One special acknowledgment is to my colleague and friend, Mr Roger Rixom. Roger gave me inspiration, advice, and encouragement throughout this project. The assistance Roger provided to me in all types of ways will always be remembered and my gratitude to him will not be forgotten. Last, but not least, I want to pass on my appreciation to my employer, the ARR-MAZ Products Division of Process Chemicals and its President, Mr Glen Varnadoe.

I would like to acknowledge Mr David Whiteoak and Shell Bitumen, UK for allowing me to use excerpts out of their outstanding handbook. The Asphalt Institute and its publications provided extensive references for this text. *Asphalt Surfacings*, edited by Mr Cliff Nicholls, an excellent reference that also provided me an insight into asphalt usage in Europe. I have compiled an extensive amount of material from many sources, including the Internet. All of the sources have been acknowledged in the references, and if I have missed any, I sincerely apologize.

Introduction

The purpose of this book is to provide a single source reference for the architect or engineer who may be a novice or not technically versed in the asphalt or bitumen industry, but has the desire to understand and practice in it. If I have accomplished that task then this endeavor was worthwhile. This book was not written to provide deep insight into the theory of asphalt technology or research. Its primary focus is to clearly demonstrate and illustrate the practical application of asphalt and asphalt paving mixtures. The novice may believe that asphalt (bitumen) and asphalt mixtures are one and the same, but after reading Chapter 1, will clearly understand they are not. Many texts, papers, and conferences attend to the needs of the designer of a major motorway or the researcher, but few address the needs of an architect designing an asphalt pavement for a parking lot (car park) or a city engineer maintaining a subdivision street. Yet, these applications account for almost half of the asphalt cement usage in the world. I hope to have made complex subjects such as the United States Strategic Highway Research Program (SHRP) developments in asphalt technology, fairly understandable and practical. This book provides methods and examples on how to adopt recent developments in the asphalt industry and apply them to the design of a parking lot, city street, or even a motorway.

My education and experience in the asphalt industry has been in the United States. Most of the text involves materials and methods commonly used in North America. The information presented is intended not only for use in North America but also in any other country. The concepts behind the successful application of asphalt technology are the same in Europe as they are in North America. I have presented the material with enough of a straightforward and practical approach that a novice designer in the United Kingdom will either be able to practice the material directly or understand it enough to modify it for use in the United Kingdom or anywhere else. The significant differences between the two continents with regard to asphalt technology are the terminology. For example, the term *asphalt cement* or *asphalt binder* in North America is *bitumen* in Europe. What is known as a *surface dressing* in the United Kingdom, is a *chip seal* in North America. *Macadam* and *hot rolled asphalt* is not commonly referenced at all in North America. *Porous Asphalt* is known as *open graded friction course* or a *popcorn mixture* in the United States. In an attempt to keep the text from becoming too cumbersome, instead of using cross references throughout the text, I grouped them together in Appendix A. There is also some asphalt mixture differences between North America and Europe. Various types of proprietary and specialty mixtures are more commonly used in Europe than North America. Some mixtures such as hot rolled asphalt (HRA) and proprietary thin surfacings are not used in the United States but have extensive use in the United Kingdom. Fundamentally, these mixtures are not that unique and are easily understood and can be used by almost anyone, anywhere. Asia, especially Japan, uses many of the terms and mixture types found in the United Kingdom, especially porous asphalt. South Korea, on the other hand, is mixed, using some United States terminology and methods, including the SHRP SuperPave system.

With regard to specifications and testing methods, I used an extensive amount of references to the American Society for Testing and Materials (ASTM) Standards. ASTM is well known and commonly referenced in specifications in North America and many other parts of the world. ASTM standards are the most referenced standards in the construction industry throughout the world. ASTM standards are commonly referenced by architects and engineers in building specifications, so it would be a common fit it include ASTM references for the parking facility. For that reason, I have referenced ASTM standards in this text. I have explained the salient points of each test method and specification cited. ASTM standards can be easily compared to local or regional standards to determine the equivalency.

The harmonization of specification and test methods in Europe was still underway during the preparation of this book. The European Committee for Normalization or Comité Europées de Normalization (CEN) is developing standards and methods that can be used across Europe. The harmonization includes terminology in the asphalt industry. For example, the term *wearing* course becomes *surface* course. In North America, the trend on that example is actually the reverse; the term *wearing course* is becoming more common and is used interchangeably. The harmonization terminology also replaced the term *base course*, with *binder course*. These terms are also used interchangeably in North America.

Harmonization was not in practice at the time of preparation of this text, and this book is not exclusive to European audiences. I have used the terms *wearing course* to describe a surface course and a combination of *base* or *binder course* to describe a bound base course containing an asphalt binder. If the readers understand the concepts, they will have no trouble understanding the various and equivalent terminology.

One unique subject that I have covered is the operation and proportioning of a hot mix asphalt mixing plant. This subject is rarely addressed thoroughly in other texts and I thought it was time to do so. Example problems with solutions are provided. This subject may have limited value to the architect or designer of a parking facility, but engineers and others in the public works industry and in the asphalt industry will find it beneficial. The same approach is taken on the chapter on asphalt mixture design, including example problems with solutions. The information provided on construction techniques is practical and straightforward and detailed enough to be completely familiar with the entire operation of constructing the typical asphalt pavement.

I have included two model specifications at the end of the book, one for a parking lot and one for a residential street. Both are similar, with differences in terms of the performance expectations of the two types of pavements. The model specifications are for the asphalt pavement portion of a project and can be used as is or used to create a custom specification. European terminology can be directly substituted for the North American terminology, if necessary. It is important to remember that good materials, design, and construction practice know no geographical or language barriers.

Chapter 1

Binders

Binder is a general description for the adhesive or glue that is used in asphalt pavements. These liquid binders can be defined as tars and asphalt binders. Asphalt binders are either petroleum derived or naturally occurring. The American Society for Testing and Materials (ASTM) defines bitumen as a class of black or dark colored (solid, semisolid, or viscous) cementitious substances, natural or manufactured, composed principally of high molecular weight hydrocarbons of which asphalt, tars, pitches and asphaltite are typical (ASTM 2001b). The asphalt binder is what gives an asphalt pavement its flexibility, binds the aggregate together, and gives waterproofing properties to the pavement. In North America, the binder is generally known as asphalt cement while in Europe it is known as "bitumen." Binder, or more specifically, asphalt binder has been developed in more recent terminology under the auspices of the United States Strategic Highway Research Program (SHRP) to include modified asphalt cements, unmodified asphalt cements, asphalt emulsions, and asphalt cutbacks.

The term "asphalt binder" has been selected to more specifically describe the asphalt material and any other modifiers or ingredients. The terms "asphalt," "asphalt cement," "bitumen," and "asphalt binder" may be used interchangeably, with asphalt cement and bitumen referring more specifically to their petroleum origins and asphalt binder referring to the asphalt cement and any other added ingredient that provides the engineering adhesive used in asphalt pavements. In this text, asphalt, asphalt cement, and asphalt binder may be interchanged depending on the context of the sentence and all will be referring to the same material.

Tars

Tars are obtained by the condensation of distillates, resulting from the destructive distillation of organic substances such as coal and wood. ASTM defines tar as a brown or black bituminous material, liquid or semi-solid in consistency in which the predominating constituents are bitumen obtained as condensate in the destructive distillation of coal, petroleum, oil-shale, wood, or other organic materials, and which yields substantial quantities of pitch when distilled (ASTM 2001b). Coal tar is the most common form of tar used in the paving industry. It has had extensive use in the past partly due large quantities of it being made available as the by-product of the distillation or carbonization of coal for the production of lamp gas. Crude coal tar is currently being produced in coke ovens during the carbonization of bituminous coal when manufacturing metallurgical coke.

Another crude coal tar is low temperature tar that is produced at temperatures of 600–700 °C during the manufacture of smokeless solid fuel, such as cooking briquettes. This type of coal tar is less viscous than coke oven tar, is paraffinic and has a pitch content of about 35 percent (Whiteoak 1991). The carbonization process occurs when coal is subjected to temperatures in the range of 600 °C (low temperature tar) to 1200 °C (coke oven tar) in the absence of air. Crude coal tar is a dark viscous liquid that is condensed from the coke oven effluent gases produced during carbonization. The crude coal tar is further distilled to remove water and light oils. Dehydrated tar after preheating to 300 °C is fed into a primary flash unit where under vacuum conditions light oils are separated. The residue is further preheated to 370 °C and fed to a secondary flash unit where heavier oils are vaporized. The remaining coal tar pitch may be used for roofing, waterproofing, and other industrial applications. It may also be blended with coal tar derived flux oils to yield a road tar of the required viscosity for a particular grade or application (The Asphalt Institute 1975).

Coal tar is not used in pavements nearly as extensively as it has in the past. Its use now is almost exclusively in pavement sealers. These pavement sealers usually are emulsions consisting of coal tar, water, clay or betonite, sand, and other additives. The emulsions are also referred as clay-stabilized emulsions. Tars are more susceptible to temperature changes than asphalt binders and may cause a "chicken wire" cracking pattern in the sealer, especially if a heavy coat is placed over an asphalt pavement. A more significant reason for the decline in the use of coal tar is the health hazards due to coal tar's high proportions of polycyclic aromatic compounds, which are known to be carcinogenic (Nicholls 1998). Coal tar sealers do have advantages in that they are highly resistant to fuel and lubrication oil spillage damage and give an aesthetically pleasing jet-black appearance to a parking lot or driveway. A jet-black coating gives an excellent contrast with pavement markings. These two properties are highly desired in parking lot construction and maintenance. Coal tar sealers are usually specified in aircraft refueling areas to protect the asphalt pavement.

Asphalt binder composition

Asphalt cement or bitumen is a complex mixture of hydrocarbons of various molecular weights. ASTM defines asphalt as a dark brown to black cementitious material in which the predominating constituents are bitumen, which occur in nature or are obtained in petroleum processing. ASTM further defines asphalt cement as a fluxed or unfluxed asphalt specially prepared as to quality and consistency for direct use in the manufacture of bituminous pavements, and having a penetration at 25 °C of between 5 and 300 dmm under a load of 5 grams applied for 5 s (ASTM 2001b). Asphalt is composed of asphaltenes and maltenes or petrolenes. Asphaltenes are high molecular weight amorphous solids that make up 5–25 percent of asphalt cement. The low molecular weight maltenes or petrolenes are made up of aromatics, resins, and saturates. Aromatics and saturates are viscous liquids while resins are semi-solid polar materials that give the adhesive properties to asphalt. Asphalt can also be described as a colloidal system consisting of the asphaltenes surrounded by resins being dispersed in the aromatics and saturates. Typical chemical analytical analysis has consisted of acid number, molecular weight, trace metal content, and elemental analysis. The acid number of the asphalt has impacts on the emulsibility and in the interaction with anti-stripping or adhesion additives. High acid numbers can also be corrosive to equipment. The trace metal content can assist in determining the crude oil origin or source of asphalt. The rheological and physical properties of a binder can be easily measured and specified.

These are the properties that are most useful when designing and maintaining an asphalt pavement. Chemical analysis of a binder is not routinely done for quality control or acceptance, mainly due to the complexity and variability of asphalt or bitumen. Asphalt binders are viscoelastic materials with rheological properties and tests providing the most useful information for the design and maintenance of asphalt pavements. The architect or engineer will be specifying and designing using the viscoelastic parameters of an asphalt binder and not its chemical composition.

Native asphalt

Prior to 1907 the major source of asphalt cements occurred naturally. The best-known naturally occurring asphalt is on the island of Trinidad and is in the form of a pitch lake about 87 meters deep and four-tenths of a square kilometer in area. Trinidad Lake asphalt is approximately 63 percent asphalt or bitumen and the remainder being mineral and organic material. For all practical purposes Trinidad Lake asphalt is too hard or stiff to be used exclusively for asphalt pavements, but can be blended with softer asphalt cements. Bermudez asphalt is another lake occurring asphalt from Bermudez Lake in Venezuela. Bermudez asphalt has not been available since the 1940s. Rock asphalt deposits can contain from 5 to 30 percent asphalt and the remainder either sandstone or limestone. The asphalt itself can vary from very hard to very soft consistency. Rock asphalt has been found in the United States in the states of Alabama, California, Colorado, Kentucky, Oklahoma, Texas, and Utah. Rock asphalt deposits found in Europe are located in Seyssel, France; Ragusa Sicily; Val-de-Travers, Switzerland; and Vorwhole, Germany (Kirk-Othmer 1978). Gilsonite is a naturally occurring resinous hydrocarbon named by its discoverer, Samuel Gilson, that is better described as a rock asphalt that is mined in the Unitah Basin in the northeastern portion of the state of Utah. This natural asphalt is similar to hard petroleum asphalt and is often called "asphaltite," "uintaite," or "asphaltum." Gilsonite is soluble in aromatic and aliphatic solvents as well as refined petroleum asphalt. It is a shiny black substance similar in appearance to the mineral obsidian. It is brittle and can be easily crushed into a dark brown powder (Ziegler 2001). Gilsonite is generally used as an additive for asphalt cements to increase their hardness. Native asphalt is not commonly used as a binder for asphalt pavements. An architect or engineer designing an asphalt pavement will be generally specifying a petroleum derived asphalt binder.

Petroleum asphalt

Petroleum asphalt cements are colloidal dispersed hydrocarbons derived from the refining and processing of crude petroleum. Large scale refining of petroleum crude oil to manufacture fuel and lubricants has made native asphalt less economical to use. Almost all paving asphalt binders used today are petroleum derived. Asphalt cement is manufactured by refining a crude oil through atmospheric distillation followed by vacuum distillation (Figure 1.1). The atmospheric distillation is completed at temperatures in the range of 300–350 °C. The atmospheric residue is then sent to the vacuum distillation unit that operates around 350 °C with 40 mm Hg of vacuum. The vacuum distillation residue or "resid" is the basic asphalt cement produced.

Different crude oils contain various amounts of asphalt ranging from Australian Gippsland with less than 1 percent asphalt cement to Sicilian Vittorio with 90 percent.

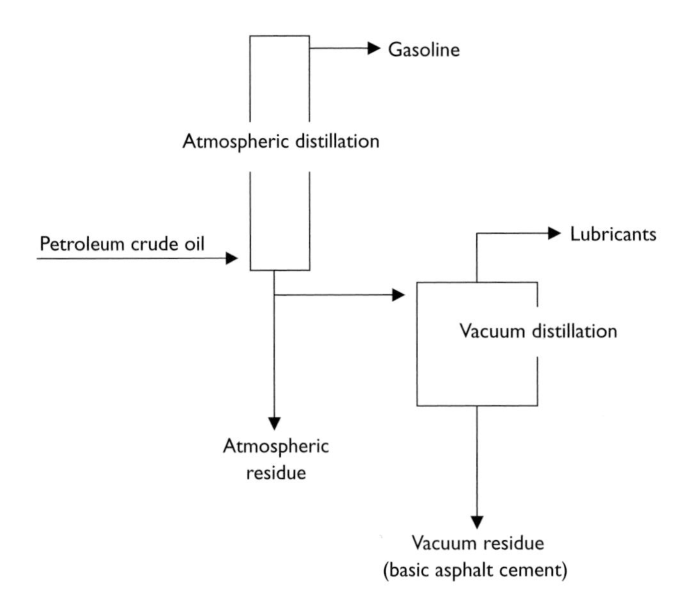

Figure 1.1 Asphalt refining example.

For some select crude oils, asphalt paving grades can be made by varying the temperature and/or vacuum inside the vacuum distillation unit and separating out more or less distillate from the residue. The vacuum residue is run out of the vacuum unit at 350 °C and passed through a heat exchanger to cool to 180 °C. The maximum equivalent distillation temperature or cut point that can be achieved by the vacuum distillation unit is about 560 °C. Other crude oils may require additional steps to make paving grade asphalt cements (Nicholls 1998). These other steps can include propane or solvent deasphalting, residue oil supercritical extraction and air blowing.

Solvent deasphalting involves exposing vacuum distillation residue to solvents such as propane. These solvents extract high boiling fractions for use as lubricants. The remaining residue gives a harder asphalt cement. Residue oil supercritical extraction (ROSE) is a more advanced process of solvent deasphalting, allowing individual components of the vacuum residue, such as asphaltenes and resins, to be extracted and separated (Sanders 1984). These individual components can then be blended back into the vacuum residue to give asphalt cement meeting specific requirements.

Air blowing involves passing air through vacuum residue that has been heated to 180 °C. The exothermic reaction further raises the residue to 250 °C (Figure 1.2). The purpose of air blowing is to produce an asphalt cement with improved high temperature performance and harder physical properties. The most common use for this type of asphalt cement is in roofing applications such as shingles, paper, and coatings. Recently this technique has been applied to select asphalt crudes to assist an asphalt refiner in meeting the SHRP binder performance grading system. The air blowing process will generally increase the upper temperature range of a performance graded (PG) asphalt binder.

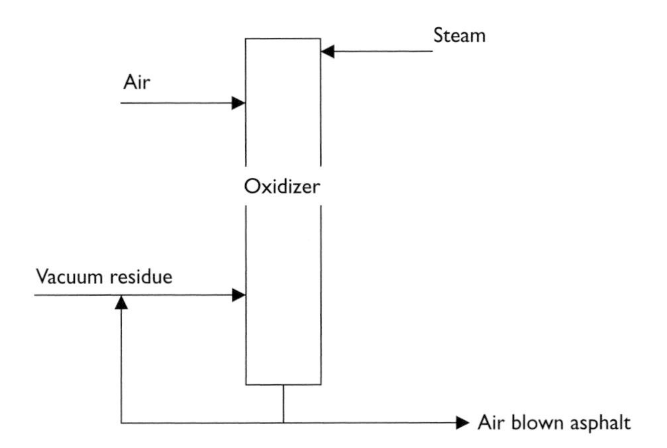

Figure 1.2 Asphalt air blowing.

Air blown asphalt may be produced in a shorter period of time through the addition of catalysts such as ferric chloride, zinc chloride, copper sulfate, aluminum chloride, and phosphorus pentoxide. The addition rate of these catalysts varies from one-tenth to 3 percent depending on the crude oil source. The addition of catalysts during air blowing can also produce asphalt with harder physical properties than an air blown asphalt alone. Cracking is a refining process that includes thermal cracking, catalytic cracking, or hydrocracking. Cracking refers to the process that reduces the weight of hydrocarbons by breaking their molecular bonds. Thermal cracking is an obsolete method once used in the production of gasoline (Bohacz 2001). Fluid catalytic cracking involves preheating a crude oil to 480–600 °C and then discharging it into a reaction vessel with aluminum silicate catalyst or zeolite in a continuous fluid bed.

Hydrocracking is catalytic cracking that takes place in the presence of hydrogen and high pressure. Pressures in a hydrocracker can reach 15,000 KPa. The cracking decomposition in all processes takes place to form both lighter and heavier petroleum products.

Distillation is then used to separate the overall products into gas, gasoline, middle distillate, and asphalt residue. The asphalt residue is sometimes known as cracked asphalt (Kirk-Othmer 1978). Cracked asphalt can then be air blown with or without catalysts to further alter its physical properties.

Asphalt cement can also have its physical properties altered through the addition of modifiers. These modifiers can have an affect on all types of binder grading systems, including penetration, viscosity, and PG. The most common modifiers added to asphalt cement are polymers at an addition level of 2–6 percent. Polymers will usually increase the asphalt binder's viscosity and increase the upper temperature range as measured by the PG binder grading system. As previously discussed, vacuum residue is the basic asphalt cement and the major asphalt product produced at a petroleum refinery. The vacuum residue can be sold as is or further modified at the refinery. In most instances the vacuum residue is shipped by ship (Plate 1.1) or rail to asphalt terminals to be sold as is or made into specific products such as asphalt emulsions or PG paving asphalt binders.

Plate 1.1 Coastal freighter delivering asphalt cement.

Asphalt terminals do not have any refining capability, but can generally blend several products to make specific grades or do so through the addition of modifiers. Terminals can be as simple as a single tank (Plate 1.2) to complex multi-product facility (Plate 1.3) that is also capable of producing asphalt emulsions and polymer modified binders.

Cutback asphalt

Asphalt or bitumen is a viscoelastic material that depending on temperature can be either a solid or a liquid. Asphalt can become liquid and easier to handle through heat, solvents, or water. The solvents or water will evaporate, leaving the asphalt cement to perform its intended function. Cutback asphalt is an asphalt cement that has a solvent or distillate such as gasoline, diesel fuel, kerosene, or naphtha added to make the asphalt liquid at ambient temperatures and improve its ability to coat aggregates. Two letters followed by a numerical digit designate or name a cutback asphalt. The two letters signify whether the cutback asphalt is slow curing (SC), medium curing (MC), or rapid curing (RC). The following numbers signify the minimum kinematic viscosity the cutback asphalt meets at 60 °C. For example, an MC-250 is a MC cutback asphalt meeting a minimum kinematic viscosity of 250 centistokes at 60 °C. Cutback asphalt may also be termed "fluxed asphalt." When gasoline or naphtha is used to flux asphalt cement, the final product is called "rapid curing" (RC) cutback asphalt. Specifications permit up to 45 percent distillate in manufacturing a RC-70 cutback asphalt. When diesel fuel or kerosene is used, MC cutback asphalt is produced. A SC liquid asphalt may be obtained by fluxing an asphalt cement with a less

Plate 1.2 Single tank asphalt terminal.

volatile distillate such as gas oil. The SC-70 and SC-250 grade cutback asphalts are very similar to residual refinery products that are used as heavy fuel oils such as bunker C or Number six fuel oils (NCHRP 1975). These SC cutback materials, whether straight run or fluxed with a relatively non-volatile material such as gas oil, are also known as "road oils" (Wright and Paquette 1987). The amount of distillate determines the viscosity and the grade of the cutback asphalt (Table 1.1). Cutback asphalt can contain between 12 and 40 percent distillate. The high demand for the distillates to be used in energy applications and ever-increasing air quality regulations has caused a steady decline in the use of cutback asphalt. Current common uses are in penetrating prime coats and in producing patching or stockpile mixtures. Some further examples are given in Table 1.2. Cutback asphalt used in mixing with aggregate will usually contain an adhesion agent to assist in the coating of the aggregate surface.

Plate 1.3 Multi-product asphalt terminal.

Table 1.1 Cutback asphalt viscosity

Cutback asphalt grade	*Kinematic viscosity, 60 °C, centistokes*
MC-30	30–60
SC-70, MC-70, RC-70	70–140
SC-250, MC-250, RC-250	250–500
SC-800, MC-800, RC-800	800–1600
SC-3000, MC-3000, RC-3000	3,000–6,000

Asphalt emulsions

Asphalt emulsions are two-phase systems consisting of two immiscible liquids, asphalt cement and water. The asphalt cement is dispersed in the continuous water phase, thus making the whole system liquid at ambient temperatures. Asphalt emulsions have little or no hydrocarbon emissions, allowing their use in geographical areas that may restrict the use of cutback asphalt in pavement construction and maintenance. Emulsions also have the ability to coat damp aggregate surfaces, thus reducing the fuel requirements for heat for drying aggregates. An asphalt emulsion is about 60 percent asphalt cement or bitumen and 40 percent soap solution. The soap solution is water and a small amount of chemical emulsifier, usually between 0.2 and 3 percent by weight of emulsion. These two ingredients are

Table 1.2 Some cutback asphalt applications

Typical applications	*Cutback asphalt grade*
Dust control	SC-70, MC-70, RC-70
Tack coat	RC-70
Prime coat	MC-30, MC-70, MC-250, RC-70, RC-250
Penetration prime	MC-30, MC-70
Chip seal, single, or multiple treatment	MC-800, MC-3000, RC-250, RC-800, RC-3000
Sand seal	MC-250, MC-800, RC-250
Structural and surface plant mixture	SC-250, SC-800, SC-3000, MC-250, MC-800, MC-3000, RC-250
Stockpiled patching mixture	SC-250, SC-800, MC-250, MC-800
In-place road mixing	SC-250, SC-800, MC-250, MC-800
Crack, joint coating	RC-70

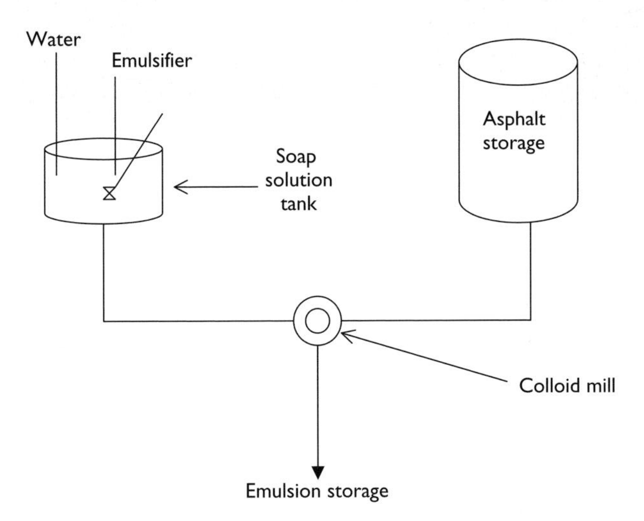

Figure 1.3 Asphalt emulsion plant.

sheared together in a colloid mill (Figure 1.3) allowing the asphalt cement to be dispersed as tiny droplets in the water and held in suspension through the assistance of the chemical emulsifier (Figure 1.4).

The chemical emulsifier, which is also known as a "surface-active agent" or "surfactant," controls how quickly the water evaporates and leaves the asphalt cement. The emulsifier consists of a long hydrocarbon chain, which terminates with either a cationic or an anionic functional group. It is an important requirement in all asphalt binder applications that the asphalt binder "wets" the surface to create the maximum contact area to ensure good adhesion. A similar expectation is of asphalt emulsions. With dry aggregates, the critical surface

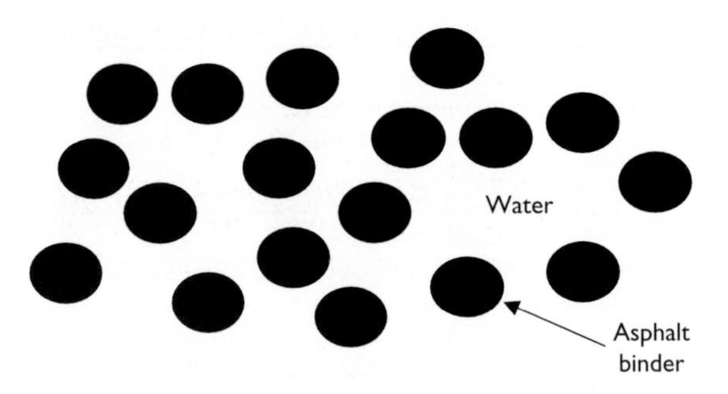

Figure 1.4 Asphalt binder dispersed in water or continuous phase.

tension of wetting of the aggregate must be high enough to ensure that the asphalt emulsion spreads easily over the surface such that the resultant adhesion exceeds the cohesion of the asphalt cement droplet. When an aggregate surface is covered with water, the wetting of the aggregate becomes a three-phase phenomenon, which can only occur if the balance of the energies favors wetting by the asphalt cement droplet. Cationic emulsifiers reduce the free surface energy of a polar aggregate, forming a thermodynamically stable condition of minimum surface energy by the emulsifier being attracted to the aggregate surface.

The chemical emulsifier also controls what class an asphalt emulsion belongs to. There are two classes of asphalt emulsions in common pavement use. Cationic emulsions have positively charged asphalt particles and anionic emulsions have negatively charged asphalt particles. This identification system comes from basic laws of electricity in that like charges repel one another and unlike charges attract. There has also been a development of non-ionic asphalt emulsions that have seen a larger usage in industrial applications rather than pavement usage. In non-ionic emulsions, the asphalt particles are neutral or do not possess any strong charge, anionic or cationic. Cationic emulsions were first developed in Europe to give emulsions better adhesive and coating characteristics when used with gravel and siliceous aggregates. Cationic emulsion's positively charged asphalt particles adhere better to electronegative aggregates. Anionic emulsions generally have better adhesion on positively charged aggregates that can include limestone. However, as there is such a large variety of aggregates used in pavement construction and maintenance, particle charge and ionic characterization may be of secondary importance to such properties as proper design, formulation, and good construction practice. Cationic and anionic asphalt emulsions have been interchanged with their respective aggregate ionic properties with some success. Emulsions are divided into two classes, anionic and cationic, and subdivided into three designations of how fast they set or "break." These designations are slow set, medium set, and rapid set. A further subset of slow set is known as quick set. The capital letter "C" designates a cationic emulsion, and the absence of such letter implies an anionic emulsion. In some European countries, a cationic emulsion is designated by the capital letter "K" and an anionic emulsion is designated by the capital letter "A." The letters "HF" designate a high float emulsion of which by current specifications are only anionic and meet a specific test criteria known as "the float test." High float emulsions tend to give a slightly higher level of aggregate coating and improved aggregate retention when used in chip seals, especially in higher temperature climates. Some of these same benefits are also derived from polymer modified emulsions. The numbers one or two following an emulsion name indicates the amount of asphalt binder and any letter following this number more specifically describes the emulsion as whether it contains a hard asphalt binder, polymer, and solvent. For example, CRS-2P is a cationic rapid set emulsion containing polymer and a minimum asphalt binder content of 65 percent. SS-1h is an anionic slow set emulsion made from a minimum of 57 percent of a hard asphalt binder. Table 1.3 gives some typical emulsion designations and shows how to decipher them.

The asphalt emulsifier determines whether an emulsion is anionic, cationic, or nonionic. The emulsifier also controls how quickly an emulsion sets or "breaks." The emulsifier also controls the emulsion's storage stability, ease of emulsification, and to a limited extent, the final emulsion viscosity. Asphalt emulsifiers possess a long lipophilic hydrocarbon tail and a polar hydrophilic head. Emulsifiers concentrate at the interface between the asphalt cement phase and the water phase. The hydrophilic head orients itself towards the more polar water phase and lipophilic tail towards the less polar asphalt phase. The emulsifier acts as a bridge between the two phases.

Anionic emulsifiers are usually wood or paper processing derivatives such as tall oil fatty acids, rosin acids, hydroxystearic acid, and lignin sulfonates. The electrovalent and polar

Table 1.3 Typical emulsion designations

Emulsion grade	*Cationic*	*Anionic*	*Set*	*Minimum asphalt content*	*Notes*
SS-1		X	Slow	57	
SS-1h		X	Slow	57	
QS-1h		X	Slow	57	Meet ASTM D3910
CSS-1	X		Slow	57	
CSS-1h	X		Slow	57	
CSS-1hP	X		Slow	57	With 3% polymer
CQS-1h	X		Slow	57	Meet ASTM D3910
MS-1		X	Medium	55	
MS-2		X	Medium	65	
MS-2h		X	Medium	65	
CMS-2	X		Medium	65	
CMS-2h	X		Medium	65	
HFMS-1		X	Medium	55	High float
HFMS-2		X	Medium	65	High float
HFMS-2s		X	Medium	65	With 1–7% solvent
HFMS-2h		X	Medium	65	High float
RS-1		X	Rapid	55	
RS-2		X	Rapid	63	
CRS-1	X		Rapid	60	
CRS-2	X		Rapid	65	
CRS-2P	X		Rapid	65	With 3% polymer
HFRS-2		X	Rapid	63	High float

Notes
h designates a harder emulsion residue penetration between 40 and 90 dmm at 25 °C.
s designates a *requirement* that the emulsion also contain solvent or distillate.

hydrocarbon group is part of the negatively charged ion when combined with sodium hydroxide (NaOH) when making the soap solution (equation 1.1).

$$RCOOH + NaOH = RCOO^- + Na^+ + H_2O \quad (1.1)$$

Cationic emulsifiers are fatty amines, fatty quartenary ammonium salts, amidoamines, and imidazolines. The fatty amines are usually used to make CMS and CRS emulsions; the fatty quartenary ammonium salts make CSS and CMS emulsions and the amidoamines and imidazolines generally make CQS emulsions. The electrovalent and polar hydrocarbon group is part of the positively charged ion when combined and hydrochloric acid (HCl) when making the soap solution (equation 1.2). The positively charge ion orientates itself on the surface of the asphalt cement droplet. The negatively charged chloride ions are then attracted to the positively charged surface and together with the water, create an electrical double-layer. The thickness of this layer has an impact on the stability and viscosity of the asphalt emulsion. The quartenary ammonium salts, however, are water-soluble salts and do not require the addition of any acid.

$$RNH_2 + HCl = RNH_3^+ + Cl^- \quad (1.2)$$

Emulsifiers are added to the water phase to make what is called a "soap solution." Emulsifiers are measured in percentage as weight of total emulsion. As a general rule, the higher the emulsifier content, the more stable the emulsion is. Rapid set emulsions may contain emulsifier contents as low as 0.2 percent by weight of total emulsion, while a slow set may contain up to 3 percent emulsifier.

The asphalt phase and the water phase or soap phase are combined together in a colloid mill. A colloid mill has a high-speed rotor or disc that revolves at a speed of 2,000–6,000 rpm with rotor clearances or gaps of 0.2–0.5 mm. These mills will produce asphalt droplets in the size of 0.001–0.01 mm. The asphalt binder used to make the emulsion must be preheated to allow it to be pumped and emulsified. The temperature will depend on the grade of asphalt binder used, but is typically in the range of 115–135 °C to give a viscosity of 0.2 Pa-s. The soap solution is usually about 50 °C. The temperatures are critical in that they are controlled to the extent that the final discharge temperature of the emulsion as it leaves the colloid mill is less than the boiling point of water or 100 °C. If the final emulsion temperature exceeds the boiling point of water, the water will flash off from the emulsion. A calculation evolving the specific heat value of asphalt and the percentage of asphalt binder and soap solution will give an estimate of the final discharge temperature (equation 1.3). This calculation is also useful in proportioning an emulsion mill, but may not be as accurate as flow meters.

$$(M_1ST_1 + M_2T_2)/(M_2 + M_1S) = T_3 \quad (1.3)$$

where M_1 is the percent asphalt binder; M_2, the percent soap solution or water; T_1, the asphalt binder temperature; T_2, the soap solution or water temperature; T_3, the colloid mill discharge temperature; S, the specific heat of asphalt (usually around 0.4–0.43).

For example, if the emulsion being produced contains 65 percent asphalt binder at 125 °C, 35 percent soap solution at 50 °C, and an asphalt binder specific heat value of 0.42; the final mill discharge temperature is given by:

$$[(65 * 0.42 * 125) + (35 * 50)]/[35 + (65 * 0.42)] = 83\,°C$$

The emulsion mill would produce a correctly proportioned emulsion when the discharge temperature reaches 83 °C.

The two most important properties to consider when manufacturing an asphalt emulsion are its stability and viscosity. Formulating an asphalt emulsion is a complex stability problem. A stable emulsion is required for storage and transport, but at the final application point, the asphalt emulsion should break quickly.

Asphalt emulsion formulation and adjustments

Formulating an asphalt emulsion consists of considering the following emulsion attributes:

- The asphalt emulsion class, anionic, cationic, or nonionic.
- The grade of the asphalt emulsion, slow set, quick set, medium set, or rapid set.
- The setting or breaking rate of the asphalt emulsion.
- The asphalt emulsion adhesion characteristics.
- The final asphalt emulsion viscosity.
- The storage stability requirement.
- The particle size distribution of the asphalt emulsion.

Asphalt emulsion storage stability and setting or breaking rate are two conflicting stability requirements. An inadequacy in the storage stability of an asphalt emulsion is initially indicated by a settlement of the emulsion. Settlement appears as a high asphalt cement content or solids on the bottom layer of a cylinder of asphalt emulsion and a low asphalt cement content or solids in the top layer. Settlement is the result of gravity acting on the denser discontinuous phase of asphalt cement droplets. The velocity of the downward movement of these particles can be estimated by using Stoke's Law (equation 1.4).

$$V = [0.222\, gr^2\, (d_1 - d_2)]/\eta \qquad (1.4)$$

where V is the asphalt emulsion settlement velocity; g, the gravitational force; r, the asphalt cement particle radius; d_1, the asphalt cement specific gravity; d_2, the soap solution specific gravity; η, the soap solution viscosity.

Stoke's Law applies to particles that are free to move through a solution, such as fine grained soil particles in water. In an asphalt emulsion, the droplets are so tightly packed that Stoke's Law will overestimate the asphalt emulsion settlement velocity. In addition to gravity, repulsive and attractive forces are acting on an emulsion. The repulsive force originates from repulsion between the electrostatic double layers on the asphalt droplets created by the ionized emulsifier. The repulsive force can be increased by increasing the emulsifier concentration and the force can be decreased by the addition of an excessive amount of negatively charged ions. The attractive force is associated with the mass of the asphalt cement droplets. If the droplets are large, or if the distribution of the particle sizes is wide, the attractive force will become the repulsive force.

Asphalt emulsion coalescence follows settlement in two stages. First, the asphalt droplets agglomerate into clumps creating a flock. This phenomenon is called "flocculation," which is also reversible. Flocculation is followed by coalescence, which is irreversible, in which the flocks fuse together to form larger globules. This process can be spontaneous or it can be induced by mechanical action (Whiteoak 1991).

An asphalt emulsion contains emulsifier molecules in both the water phase and on the surface of the asphalt cement droplets. Some of the emulsifier ions have also formed micelles in the water phase and in a stable emulsion there is equilibrium. If some of the emulsifier ions are removed from the water phase, the equilibrium will be restored by ions from the micelles and from the surface of the asphalt cement droplets. This same phenomenon occurs when an asphalt emulsion comes into contact with aggregate or a pavement surface. The negatively charged aggregate surface rapidly absorbs some of the ions from the water phase, which results in the weakening of the charge on the asphalt cement droplet, thus initiating the overall setting or breaking of the asphalt emulsion. A point is reached where the charge on the surface of the asphalt cement droplets is so depleted, that rapid coalescence now takes place. The aggregate is now covered in hydrocarbon chains and the liberated asphalt cement adheres strongly to the aggregate surface (Whiteoak 1991).

There are a number of adjustments that can be made to modify the basic properties of an asphalt emulsion. These properties include viscosity, storage stability, setting or breaking rate, and particle size distribution. Basic changes to an emulsion formulation include changing the grade, use level or source of asphalt cement and changing the emulsifier or its use level. An asphalt binder that provides good emulsification properties will consist of:

- Low sulfur content, less than 4 percent.
- Acid number greater than 0.5.
- pH less than 7.

Increasing the asphalt emulsion viscosity

- *Increasing the asphalt cement content.* Limitations to this include specification, cost, and if the asphalt cement content is already high, a small increase in use level can induce a dramatic viscosity increase.
- *Modification of the water or soap phase.* The viscosity of an asphalt emulsion is highly dependent on water phase composition. Asphalt emulsion viscosity can be increased by decreasing the water phase acid content, increasing the emulsifier content and by increasing the neutralization ratio between the acid and amine emulsifier content.
- *Increasing the flow rate through the emulsion mill.* By increasing the flow rate through the emulsion mill, the particle size distribution will be changed. At asphalt cement contents less than 65 percent, the viscosity of the asphalt emulsion is virtually independent of flow rate. However at asphalt cement contents greater than 65 percent, the globules of asphalt cement are packed relatively closely together, such that inducing a change in particle size distribution by changing the flow rate has a marked affect on viscosity.

- *Decreasing the viscosity of the asphalt cement.* This statement sounds contradictory to the purpose of increasing the asphalt emulsion viscosity. How it works is that if the viscosity of the asphalt cement entering the colloid mill, is reduced, the particle size will be reduced which will tend to increase the final viscosity of the asphalt emulsion.

Decreasing the asphalt emulsion viscosity

- *Reducing the asphalt cement content.* This practice is controlled by specification for the minimum asphalt cement content. The effect of this practice on emulsions containing less than 60 percent asphalt cement is small.
- *Modification of the water or soap phase.* To decrease the asphalt emulsion viscosity, it is necessary to either increase the acid content or to decrease the amine emulsifier content.
- *Decrease the flow rate through the emulsion mill.* Changes the particle size distribution that is converse of the practice for increasing asphalt emulsion viscosity.
- *The addition of calcium or sodium chloride.* Small addition levels of calcium or sodium chloride added to the completed asphalt emulsion can control or reduce viscosity rise in the emulsion. Addition levels are usually 0.05–0.2 percent by weight of asphalt emulsion.

Changing the setting or breaking rate of an asphalt emulsion

- *Aggregate size.* The aggregate size in the application affects the setting rate in that the smaller aggregates will break an asphalt emulsion faster than larger aggregates. This is a direct function of surface area.
- *Modification of the water or soap phase.* An asphalt emulsion can set or break faster by reducing the water phase acid content, increasing the amine emulsifier content or by decreasing the ratio between the ratio between the acid and amine emulsifier content. Emulsifiers have also been designed to increase or decrease setting times.
- *Increasing or decreasing the asphalt cement content.* Increasing the asphalt content will break an emulsion faster while decreasing the content will slow the set down.
- *Ambient and emulsion temperatures.* Higher temperatures will cause an asphalt emulsion to break faster.
- *Asphalt emulsion particle size and distribution.* The finer the asphalt cement droplets size and the smaller the dispersion will increase the setting time of an emulsion.
- *Additives.* The use of additives can increase or decrease the setting or breaking of an asphalt cement. In applications such as slurry seals, hydrated lime or Portland cement is added to set the mixture faster. Likewise aluminum chloride or aluminum sulfate can retard the setting of a slurry seal mixture.

Asphalt emulsion storage stability

- *Asphalt cement specific gravity.* Asphalt cement with a high specific gravity will tend to settle when emulsified. The specific gravity of the asphalt cement can be reduced by the addition of kerosene prior to emulsification. However, this will result in an increased emulsion viscosity and a reduced asphalt cement viscosity on the aggregate and pavement surface. The water or soap phase specific gravity can be increased through the addition of sodium or calcium chloride. However this will also reduce the asphalt emulsion viscosity.

- *Low viscosity asphalt emulsions.* Low viscosity emulsions are more prone to settlement than high viscosity emulsions because the particles have more freedom to move. The storage stability can be improved by increasing the asphalt emulsion viscosity. Increasing the emulsifier use level will also decrease the settlement rate.
- *Electrolyte content of the asphalt cement.* The presence of cations in the asphalt cement can reduce the storage stability of the asphalt emulsion. In cationic emulsions, a high sodium content in the asphalt cement may induce settlement and breaking during storage. This can be counteracted by the addition of salt to the water phase.
- *Particle size distribution of the asphalt emulsion.* Asphalt emulsions with a wide range of particle or droplet sizes are more prone to settlement than an emulsion with a relatively narrow size distribution. Large particles settle more rapidly due to the repulsion forces between the particles.

Asphalt emulsion particle size distribution

The distribution of the asphalt emulsion droplet size is dependent on the interfacial tension between the asphalt cement and the water or soap phase. The lower the interfacial tension, the easier the asphalt cement disperses. The droplet size is also dependent on the energy used in dispersing the asphalt cement. For a given mechanical energy input, harder asphalt cements will produce coarser emulsions, while softer asphalt cements produce finer emulsions. It is possible to influence the particle size and distribution to achieve a finely divided emulsion with a narrow range of particle sizes.

- *The addition of acid to the asphalt cement.* The addition of naphthenic acids to a non-acidic asphalt cement is important for the production of anionic asphalt emulsions. An asphalt cement that has a pH of greater than seven can cause storage stability problems. The acids react with the alkaline water or soap phase to form additional soaps which are surface active and which help stabilize the asphalt emulsion. The addition of naphthenic acids causes a decrease in the mean particle size of emulsion while changing its particle distribution. In addition, the specific surface area of the asphalt emulsion is increased, and consequently the amount of emulsifier absorbed at the asphalt cement particle surface is increased. This results in a reduction of emulsifier concentration in the water phase causing the asphalt emulsion to break or set faster.
- *Temperature.* Increasing the temperature of either the asphalt cement or the water phase during manufacture reduces the viscosity, causing an increase in the mean particle size of the asphalt emulsion.
- *Asphalt cement content.* Increasing the asphalt cement content increases the mean particle size and tends to reduce the range of particle sizes.
- *Composition of the water or soap phase.* For cationic emulsions manufactured with an amine emulsifier and hydrochloric acid, increasing either the emulsifier or the acid content can decrease the particle size. If the ratio of acid to amine emulsifier is kept constant, increasing both the amine and acid content can reduce the particle size. The size distribution alone is not related to the use level these two components.
- *Colloid mill operation.* The gap and rotational speed of the colloid mill significantly control the particle size and distribution of the emulsion. A small gap will result in a small particle size with a relatively narrow range of sizes. A high rotational speed will also produce a small particle size (Whiteoak 1991).

Table 1.4 Emulsion storage and spray temperatures

Emulsion grade	*Storage temperature (°C)*	*Spray temperature (°C)*
CSS-1, CSS-1h, CSS-1hp, CQS-1h, SS-1, SS-1h, MS-1, HFMS-1, RS-1	10–60	20–70
CMS-2, CMS-2h, MS-2, MS-2h, HFMS-2, HFMS-2h, HFMS-2s	50–85	20–70
CRS-1, CRS-2, CRS-2P, RS-2, HFRS-2	50–85	50–85

It is important to point out that in an asphalt emulsion, it is almost impossible to change one property without affecting another. There are strong relationships between emulsifier level, particle size and distribution, breaking rate, and viscosity.

Emulsion storage and handling

Asphalt emulsions are combinations of two unlike materials, asphalt cement and water, and do need proper storage and handling.

- Store the emulsion above freezing temperatures and below 90 °C.
- Store the emulsion at the required temperature for the particular grade (Table 1.4).
- Spray or apply the emulsion at the required temperature for the particular grade (Table 1.4).
- Store emulsions in vertical tanks in order to minimize the effects of skimming.
- Do not allow the emulsion to freeze, this will break the emulsion.
- Do not mix cationic emulsions and anionic emulsions together, this will break the emulsion.
- Clean tanks and equipment with water before changing classes of emulsions.
- Do not use air mixers or spargers to agitate an emulsion, this will break the emulsion.
- Do not over agitate or over pump an emulsion, which can cause a drop in viscosity or stability.
- Do use clean warm water for diluting an emulsion and always add the water slowly to the emulsion and not the emulsion to the water.
- Haul emulsions in tankers equipped with baffles to prevent sloshing.
- Do not dilute rapid set emulsions. Medium and slow set emulsions may be diluted with water by adding it to the finished emulsion (The Asphalt Institute 1998a).

Asphalt emulsion uses

The grade of an asphalt emulsion is selected by the desired application or end result. The class (anionic or cationic) is generally given by specification or local availability. Specifically a cationic emulsion would be selected for an aggregate or application possessing a negative charge such as limestone or other calcareous stone, and an anionic emulsion would be selected for a positively charged aggregate such as the siliceous granite. However, the aggregate's charge is of secondary importance to proper design and construction techniques and emulsions have been interchanged with success with each type of aggregate charge.

Table 1.5 Some asphalt emulsion applications

Typical applications	*Asphalt emulsion grade*
Dust control	SS-1, CSS-1
Tack coat	SS-1, SS-1h, CSS-1, CSS-1h
Prime coat	SS-1, SS-1h, CSS-1, CSS-1h
Fog seal	SS-1, CSS-1, MS-1
Penetration prime	SS-1
Chip seal, single, or multiple treatment	RS-1, RS-2, HFMS-1, HFMS-2, HFRS-2, CRS-2, CRS-2P
Sand seal	SS-1, SS-1h, CSS-1, CSS-1h, CMS-2s
Slurry seal	SS-1h, CSS-1h, CSS-1hP
Microsurfacing	CQS-1h
Structural and surface plant mixture	SS-1, SS-1h, HFMS-2s, CSS-1, CSS-1h
Stockpiled patching mixture	HFMS-2s, CMS-2s
In-place road mixing	SS-1, SS-1h, HFMS-2s, CSS-1, CSS-1h
Crack filler, joint coating	RS-2, HFMS-2h, MS-2, CMS-2h, CRS-2P

Slurry seal applications have had the greatest success when applied with a cationic emulsion, usually a CSS-1h. A highly developed slurry seal application, known as "microsurfacing" requires the use of a cationic emulsion. The break or cure time of the emulsion will decrease when the emulsion is used with the opposite charge application. When writing a specification regarding the class of emulsions, an understanding of what is available in the area where the emulsion will be applied will reduce excessive transportation costs for an otherwise unavailable emulsion. The specification regarding the grade of the emulsion will be dictated by the application. New construction will usually require the use of slow set emulsions for tack or prime coats, while maintenance activities will encompass all grades, from slow set to rapid set. Table 1.5 gives various applications for the different grades of asphalt emulsions.

The architect or engineer will be specifying the particular class and grade of asphalt emulsions for use in the paving project. The particular performance or testing requirement specifications are best made by specifying a universally accepted standard or a well-known and used local test standard such as one used by a highway agency.

ASTM D977 is a universally recognized standard for anionic asphalt emulsions (Table 1.6) and ASTM D2397 for cationic asphalt emulsions (Table 1.7). For example, the specification may read "CSS-1 asphalt emulsion meeting the requirements of ASTM D2397 shall be used for tack coat at an application rate of 0.2 liters per square meter."

The test requirements for asphalt emulsions usually have a direct correlation to construction performance and the final performance characteristics. The following summary illustrates the significance of each criteria.

- *Viscosity at 25°C*: Measures the viscosity of mostly slow set emulsions to ensure that emulsion is capable to be pumped and sprayed or mixed.
- *Viscosity at 50°C*: Measures the viscosity at a higher temperature to simulate the actual conditions the particular grade emulsion is used at. Mostly used for rapid set emulsions to ensure that the emulsion is capable of being sprayed.

Table 1.6 Anionic emulsion requirements

Test requirement	*Emulsion grade*							
	SS-1	*SS-1h*	*QS-1h*	*MS-1*	*MS-2*	*MS-2h*	*RS-1*	*RS-2*
Viscosity, SSF, 25 °C (s)	20–100	20–100	20–100	20–100	≥100	≥100	20–100	
Viscosity, SSF, 50 °C (s)								75–400
Storage stability, 24 h (%)	≤1	≤1		≤1	≤1	≤1	≤1	≤1
Demulsibility, 0.02N $CaCl_2$ (%)							≥60	≥60
Cement mixing (%)	≤2.0	≤2.0						
Sieve (%)	≤0.1	≤0.1	≤0.1	≤0.1	≤0.1	≤0.1	≤0.1	≤0.1
Residue by distillation (%)	≥57	≥57	≥57	≥55	≥65	≥65	≥55	≥63
Residue penetration, 25 °C	100–200	40–90	40–90	100–200	100–200	40–90	100–200	100–200
Residue ductility, 25 °C, 5 cm/min (cm)	≥40	≥40	≥40	≥40	≥40	≥40	≥40	≥40
	High float emulsion grade							
	HFMS-1	*HFMS-2*	*HFMS-2h*	*HFMS-2s*	*HFRS-2*			
Viscosity, SSF, 25 °C (s)	20–100	≥100	≥100	≥50				
Viscosity, SSF, 50 °C (s)					75–400			
Storage stability, 24 hours (%)	≤1	≤1	≤1	≤1	≤1			
Demulsibility, 0.02N $CaCl_2$ (%)					≥60			
Sieve (%)	≤0.10	≤0.10	≤0.10	≤0.10	≤0.10			
Residue by distillation (%)	≥55	≥65	≥65	≥65	≥63			
Residue penetration, 25 °C	100–200	100–200	40–90	≥200	100–200			
Residue ductility, 25 °C, 5 cm/min (cm)	≥40	≥40	≥40	≥40	≥40			
Float of residue, 60 °C (s)	1,200	1,200	1,200	1,200	1,200			

Source: ASTM 2001b; © *ASTM*: reprinted with permission.

Table 1.7 Cationic emulsion requirements

Test requirement	*Emulsion grade*						
	CSS-1	*CSS-1h*	*CQS-1h*	*CMS-2*	*CMS-2h*	*CRS-1*	*CRS-2*
Viscosity, SSF, 25 °C (s)	20–100	20–100	20–100				
Viscosity, SSF, 50 °C (s)				50–450	50–450	20–100	100–400
Storage stability, 24 h (%)	≤ 1	≤ 1		≤ 1	≤ 1	≤ 1	≤ 1
Demulsibility, 0.8% dioctyl sodium sulfosuccinate (%)						≥ 40	≥ 40
Cement mixing (%)	≤ 2.0	≤ 2.0					
Sieve (%)	≤ 0.10	≤ 0.10	≤ 0.10	≤ 0.10	≤ 0.10	≤ 0.10	≤ 0.10
Residue by distillation (%)	≥ 57	≥ 57	≥ 57	≥ 65	≥ 65	≥ 60	≥ 65
Residue penetration, 25 °C	100–250	40–90	40–90	100–250	40–90	100–250	100–250
Particle charge	Positive	Positive	Positive	Positive	Positive	Positive	Positive
Residue ductility, 25 °C, 5 cm/min (cm)	≥ 40	≥ 40	≥ 40	≥ 40	≥ 40	≥ 40	≥ 40

Source: ASTM 2001b; © *ASTM*: reprinted with permission.

- *Storage stability*: ensures that the emulsion is capable of being stored and does not settle out.
- *Demusibility*: applicable to rapid set emulsions for determining how quickly they break or set. A general rule is that the higher the value, the faster the break. However, if the specification is set too high, the emulsion will not be very stable and be incapable of being stored. If an asphalt emulsion has a demusibility requirement it will not have a cement-mixing requirement.
- *Cement mixing*: The opposite requirement to demulsibility, the cement mixing test applies to slow set emulsions and determines how stable the emulsion is. A general rule is that the lower the value, the more stable the emulsion is. If an asphalt emulsion has a cement-mixing requirement it will not have a demusibility requirement.
- *Sieve test*: An asphalt emulsion is passed through a sieve to determine the amount of large asphalt globules or "shot" in it. Shot can cause emulsion instability and have the potential to clog sprayers.
- *Asphalt binder residue*: Determines the amount of asphalt binder in an asphalt emulsion. Criteria for quality control and a needed parameter used in the design of surface treatments and in determining the amount of asphalt emulsion to spray.
- *Penetration*: Determines the final grade of the asphalt binder used to manufacture an asphalt emulsion. Distinguishes the difference between emulsions made with a "soft" asphalt binder or a "hard" asphalt binder. The test is completed on the asphalt binder residue.
- *Particle charge*: Used to classify an emulsion and only for determining if an emulsion is cationic. This test is not used on anionic emulsions.
- *Ductility*: The asphalt binder controls this value with the asphalt emulsion manufacturer using the test to help select the base asphalt binder. Ductility is one indicator of how temperature susceptible an asphalt binder can be (The Asphalt Institute 1988). The ductility test measures how much an asphalt binder will "stretch" at a given temperature. Ductility can also have a small significance in the reduction of aggregate loss from a chip seal. This test is completed on the asphalt binder residue.
- *Float test*: Used to determine if an asphalt emulsion is a high float emulsion. This test is an indicator of the asphalt binder residue's ability to flow at high temperatures (The Asphalt Institute 1998a). This test is completed on the asphalt binder residue.

The emulsion tests listed are standard test methods published in ASTM D244 and any competent asphalt laboratory should be capable of completing the tests. Most asphalt emulsion manufacturers are also capable of performing the tests. The architect or engineer responsible for designing a small paving project such as a parking lot is the best person to specify a widely accepted asphalt emulsion specification and the one who requires assurance that the asphalt emulsion used meets the specified requirements. Larger projects may require an independent testing laboratory sampling the emulsion at a selected interval and performing the required tests.

Asphalt binder specifications

In pavement applications, asphalt binders are specified by their physical properties and not their chemical properties. The most important physical properties of the binder to the architect or engineer are its rheological characteristics. The rheological properties can also be described as its flow or resistance to flow behavior. Asphalt cement is a viscoelastic material by possessing viscous properties at high temperatures and elastic properties at low temperatures. A viscous property is analogous to placing your foot in the sand and leaving a footprint. An elastic property is how a spring returns to its original shape after being stretched. Elastic stiffness is important to the designer at decreasing low temperatures, an asphalt binder becomes progressively stiffer and eventually brittle. On the other hand, if an asphalt binder is too viscous, it will flow at high temperatures and cause pavement deformation. The rheological properties of an asphalt binder are expressed in both empirical and fundamental properties. Until the 1970s only empirical properties have been used to characterize and specify asphalt binders throughout the world (Nicholls 1998).

There are three significant systems available to specify or grade asphalt binders. They are penetration grading, viscosity grading, and performance grading. There is no direct correlation between the three systems, however an asphalt binder may meet all the criteria of all three systems. Other physical properties such as the softening point, the Fraass breaking point, and ductility may supplement the asphalt binder specification as dictated by local practice.

Penetration grading of asphalt cement is the oldest grading method that is currently in practice. The penetration test was originally developed as a measure of consistency of bituminous materials including asphalt and tar. Penetration defines the consistency of asphalt or tar expressed as the distance in tenths of a millimeter (dmm) that a standard test needle vertically penetrates a sample of the material under known conditions of loading, time, and temperature (ASTM 2001b). The loading is usually a 100-gram weight placed on the needle and allowed to penetrate the 25 °C sample for 5 s (Plate 1.4). Higher values of penetration

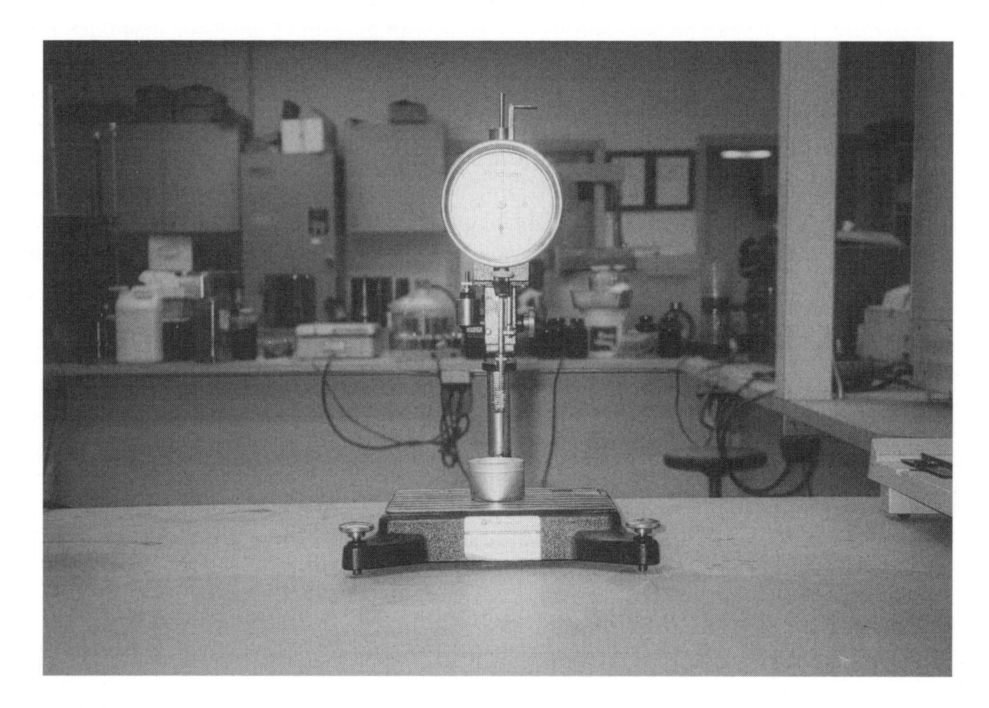

Plate 1.4 Asphalt penetration test.

indicate a softer consistency of the material. ASTM D5 is a typical test method used to penetration grade asphalt cements. The asphalt cement will be specified and referred to in terms of the minimum and maximum range of penetration. Typical standard penetration graded asphalt cements in the United States are 40–50 dmm, 60–70 dmm, 85–100 dmm, 120–150 dmm, and 200–300 dmm. The asphalt cement would be specified and named by these minimum and maximum values. For example, if specifying an asphalt binder to be used in conjunction with a chip seal or seal coat, you would ask for 120–150 "penetration" asphalt.

Penetration grading of an asphalt binder is an empirical method that gives no real indication of a binder's in-service handling characteristics or its performance characteristics. At a given temperature, the test can determine which asphalt binder is softer or harder. ASTM D946 is a standard specification for PG asphalt binder.

A fundamental grading system that at the least gives an indication of the asphalt binder's handling characteristics is the viscosity grading system. This grading system incorporates the use of a capillary tube viscometer. There are several types of capillary tube viscometers, the Asphalt Institute vacuum viscometer, the Cannon–Manning vacuum viscometer, the Modified Koppers vacuum viscometer, the Cannon–Fenske viscometer, the Lant–Zeitfuchs viscometer, the BS U-Tube modified reverse flow viscometer, and the Zeitfuchs cross arm viscometer. The asphalt binder is usually graded at two temperatures, 60 °C and 135 °C. The 60 °C temperature was chosen as it approximates the maximum temperature of asphalt pavement surfaces and the 135 °C temperature was chosen as one that approximates mixing and placement temperatures for asphalt pavements (Asphalt Institute 1988). The asphalt binder is also graded and referred to by its viscosity at 60 °C. Because asphalt cement is too viscous to flow readily at 60 °C, one of the vacuum viscometers mentioned is used with a partial vacuum to induce flow through the tube (Plate 1.5). The amount of time it takes for the material to flow through the tube is used to calculate the viscosity in poises (equation 1.5). At 135 °C, the Zeitfuchs cross-arm viscometer or one of the other kinematic viscometers mentioned is used, since asphalt cement flows readily at this temperature. All the viscometers will give satisfactory results with their intended application as the choice being the personal preference of the laboratory or specifying agency.

$$V_2 = T_2 (V_1/T_1) \tag{1.5}$$

where V_1 is the viscosity of reference standard; T_1, the time for reference standard to pass through tube; V_2, the viscosity of asphalt binder; T_2, the time for asphalt binder to pass through tube.

Among other types of viscometers used to test asphalt binders are the cone and plate and sliding viscometer. These tests are not as common as capillary tube viscometers to grade or specify asphalt binder. The cone and plate viscometer is used on asphalt binders at a wide range of temperatures, especially at temperatures less than 30 °C. The sliding plate viscometer applies the definition of dynamic viscosity by taking the shear stress applied to a film of asphalt binder sandwiched between two plates and measuring the resulting rate of strain. The sliding plate viscometer is also used to investigate the phenomenon of shear stress dependence (Whiteoak 1991). For the practice of specifying an asphalt binder by its viscosity, the architect or engineer will use the capillary tube viscosity at 60 °C. ASTM D 2171 is a useful test procedure to determine an asphalt binder's viscosity by a vacuum capillary viscometer at 60 °C.

Plate 1.5 Cannon–Manning capillary tube viscometer.

The kinematic viscosity at 135 °C ensures the ability to pump and handle the asphalt binder. ASTM D2170 can be used to determine the kinematic viscosity at 135 °C. Standard grades of viscosity graded asphalt binder are AC-2.5, AC-5, AC-10, AC-20, AC-30, and AC-40. The letters "AC" designates asphalt cement and the numerical value is one percent of the viscosity at 60 °C in poises. For example an AC-5 is 500 poises at 60 °C, while an AC-40 is 4,000 poises at 60 °C. Table 1.8 illustrates standard viscosity graded asphalt binders along with the minimum penetration values.

A further subset development of the viscosity grading system is the aged residue (AR) viscosity. This system also measures the viscosity of the asphalt binder with capillary tube vacuum viscometers at 60 °C, but artificially ages the asphalt binder before determining the viscosity. The method of artificially aging is the rolling thin film oven or RTFO. The RTFO procedure involves placing the asphalt binder in a bottle into a specially designed oven at 163 °C (Plate 1.6) that rotates the bottle for 85 min while 163 °C air is blown on

Table 1.8 Viscosity graded asphalt binder

Test requirement	*Viscosity grade*					
	AC-2.5	*AC-5*	*AC-10*	*AC-20*	*AC-30*	*AC-40*
Viscosity at 60 °C, poises	250 ± 50	500 ± 100	1000 ± 200	2000 ± 400	3000 ± 600	4000 ± 800
Penetration at 25 °C, dmm	≥200	≥120	≥70	≥60	≥50	≥40

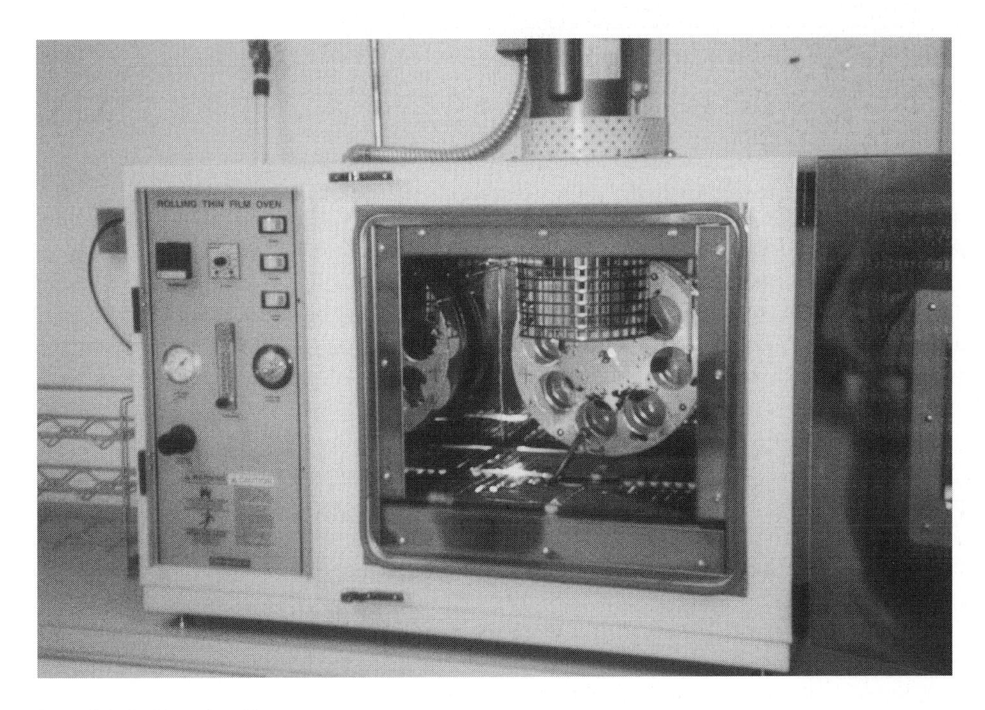

Plate 1.6 Rolling thin film oven.

the sample. The rotating bottle continuously exposes fresh films of asphalt binder to the hot air. The sample is then removed and the viscosity at 60 °C is then determined using a capillary tube vacuum viscometer. It is thought that the RTFO simulates the aging that takes place on the asphalt binder during construction or more specifically, through a hot mix plant.

The AR viscosity grading system is not commonly used with the exception of part of the western United States. The same asphalt binder can meet either system (or all three if the penetration system is included) while the AR system may more readily identify problematic asphalt binders that are highly volatile or more susceptible to aging. Typical AR grades are AR-1000, AR-2000, AR-4000, AR-8000, and AR-16000. With this system, the letters "AR"

Table 1.9 Viscosity graded on RTFO[1] aged residue asphalt binder

Test requirement	*Aged residue viscosity grade*				
	AR-1000	*AR-2000*	*AR-4000*	*AR-8000*	*AR-16000*
Approximate original asphalt binder grade[2]	AC-2.5 AC-5	AC-5 AC-10	AC-10 AC-20	AC-20 AC-30 AC-40	AC-40
Viscosity at 60 °C, poises	1000 ± 250	2000 ± 500	4000 ± 1000	8000 ± 2000	16000 ± 4000
Penetration at 25 °C, dmm	≥65	≥40	≥25	≥20	≥20

Notes
1 Rolling thin film oven.
2 Not to be used as a specification, but for general guidelines.

indicate AR and the numerical value is the viscosity at 60 °C. Table 1.9 illustrates standard AR viscosity graded asphalt binders along with the minimum penetration values.

ASTM D3381 may be used to specify both original and AR by viscosity grading. This specification is universally accepted and recognized, but based on asphalt binder availability, it may be more practical to follow local practice. Local guidelines are best obtained from a government highway agency close to the project or from a local asphalt binder supplier.

The third major system of grading or specifying asphalt binders is the PG classification system. In 1987 the United States Department of Transportation implemented SHRP. One of the outcomes of the program is the SuperPave™ Performance Grade binder specification. The American Association of State Highway and Transportation Officials (AASHTO) designation for the Performance Grade binder specification is MP-1 (AASHTO 2000a).

The PG binder specification differs from the penetration and viscosity grading systems in that the tests used measure physical properties that can be directly related to field performance by engineering principles. These tests require equipment that have been developed or modified under the SHRP program. It is called a binder specification because it is intended for both modified and unmodified asphalt cements. A unique feature of the PG binder specification is that instead of performing a test at a constant temperature and obtaining a varying test value, the specified test value is constant and the test temperature at which the value must be achieved is varied. The basis of the PG binder specification is that fundamental properties of the asphalt binder are measured at actual pavement temperatures where the critical pavement distress modes occur. Three critical pavement distresses addressed by the specification are permanent deformation such as rutting, fatigue cracking, and thermal or low temperature cracking. PG binders are graded by a designation such as PG 64-22.

In this example, the designation means that the asphalt binder was classified under the PG system, meets a seven-day average high temperature requirement of 64 °C and meets a low temperature requirement of −22 °C.

The designation means that binder possesses adequate physical properties with a seven-day average high pavement temperature of 64 °C and a low pavement temperature of −22 °C (The Asphalt Institute 1997).

The major feature of the PG binder specification is its reliance on testing asphalt binders in conditions that simulate the three critical periods during an asphalt pavement's life. Tests performed on the original asphalt binder represent its transportation, storage, and handling. The second period represents the asphalt binder aging during mixture production and pavement construction, and is simulated in the PG binder specification by aging the asphalt binder in a RTFO. The final period occurs as the asphalt binder ages over a long time as part of the pavement. This period is simulated in the PG binder specification by the pressure-aging vessel (PAV). This procedure exposes asphalt binder samples to heat and pressure conditions that simulate years of in-service aging in the pavement (The Asphalt Institute 1997).

There are currently a total of six major pieces of testing equipment used to performance grade an asphalt binder. They are the rotational viscometer, the RTFO, the PAV, the dynamic shear rheometer, bending beam rheometer, and direct tension tester. The RTFO and the PAV do not actually test the asphalt binder but rather condition and age it. An additional test incorporated in the PG binder specification, that is not actually an asphalt binder performance indicator, is the flash point test that incorporates the Cleveland open flash apparatus. The minimum flash point requirement of ≥230 °C is used to address safety and transport concerns.

The test procedure used to determine the flash point is the AASHTO test method, T48 (AASHTO 2000b). The kinematic viscosity of the asphalt binder is also determined, but with the Zeitfuchs cross-arm or other kinematic capillary tube viscometers being replaced by the rotational viscometer. The rotational viscometer is sometimes referred to by its more common name, the Brookfield viscometer. The purpose of the rotational viscosity requirement is to insure pumpability of the asphalt binder during storage, transport, and at the mixing or emulsion plant. The rotational viscosity was selected due to its ability to successfully test both unmodified and modified asphalt binders. Modified asphalt binders, especially binders containing polymer have proven to be difficult to test with standard capillary tube viscometers. The PG binder specification for rotational viscosity is a maximum of 3,000 centipoise or 3 P-s at 135 °C as measured by the ASTM procedure, D4402.

Two instruments new to the asphalt grading systems are the DSR and the bending beam rheometer (BBR). These two instruments are used to grade an asphalt binder under the PG binder specification. The dynamic shear rheometer is used to determine the asphalt binder's high temperature properties and the BBR is used to determine the low temperature properties. A third new instrument, known as the direct tension tester (DTT), is sometimes used in conjunction with the BBR, to determine the final low temperature properties of the asphalt binder.

The AASHTO specification MP-1a incorporates the DTT, while the MP-1 specification does not (AASHTO 2000a). Table 1.10 summarizes the uses of equipment used in the PG binder specification.

The DSR (Plate 1.7) measures the asphalt binder's properties at intermediate to high temperatures. The critical pavement distress to be concerned with at higher temperatures is permanent deformation, which includes rutting and shoving. One cause of rutting in an asphalt

Table 1.10 PG binder specification equipment

Equipment	*Purpose*
Open cup flash point apparatus	Safety, storage, and transportation
Rotational viscometer	Measures the asphalt binder's handling properties at high temperatures
Rolling thin film oven	Simulates short-term aging such as at a mixing plant
Pressure aging vessel	Simulates 7–10 years of pavement aging
Dynamic shear rheometer	Measures the asphalt binder's properties at high and intermediate temperatures
Bending beam rheometer Direct tension tester	Measures the asphalt binder's properties at low temperatures

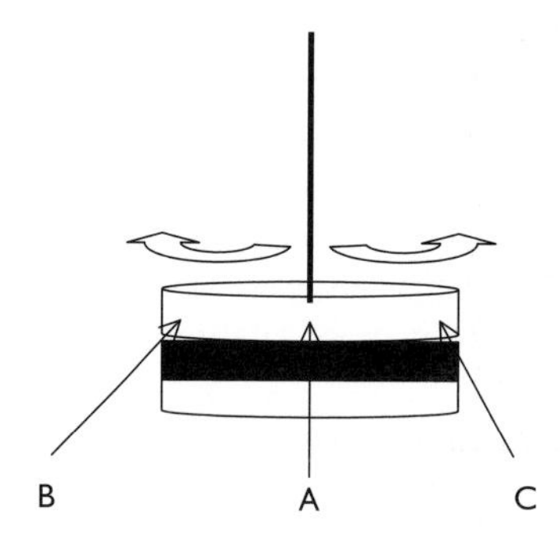

Figure 1.5 Dynamic shear rheometer plate operation.

pavement can be attributed to an accumulated non-recoverable shear strain or movement in the asphalt binder (Kriech 1994). The DSR applies a fixed torque to move an oscillating plate from point "A" to point "B" and from point "A" to point "C" (Figure 1.5). The oscillating plate is parallel to a fixed lower plate. Depending on the temperature and stiffness of the asphalt binder, the torque required moving the plate at a set oscillation or frequency will vary.

The DSR determines the shear stress of an asphalt binder dynamically through the application of a sinusoidal load. In dynamic tests, such as that measured by the DSR, the shear stress can be applied as a sinusoidally varying stress of constant amplitude and fixed frequency. The deformation of the asphalt binder between the oscillating plates also varies sinusoidally, with the same frequency as the applied stress (Figure 1.6). The shear modulus, known as the complex shear modulus, is symbolized in the PG binder specification as G*. The complex shear modulus G*, at a given frequency is given by the ratio of the amplitude of shear stress, τ, and shear strain, γ (equation 1.6).

$$G^* = \tau/\gamma \tag{1.6}$$

The DSR is used to characterize both the viscous and elastic behavior of an asphalt binder. Since an asphalt binder is viscoelastic, it deforms both as an elastic solid and as a liquid over time. The deformation is composed of instant elastic recoverable strain, delayed elastic

Plate 1.7 Dynamic shear rheometer.

recoverable strain, and permanent non-recoverable strain or viscous flow. When the applied stress is removed from the asphalt binder, the initial elastic strain is recovered instantly and the delayed elastic strain is recovered over time, but the permanent strain or viscous flow is not recovered. It is the viscous flow which after repeated loading from traffic, can eventually result in permanent deformation of the asphalt pavement (Nicholls 1998).

The complex shear modulus, G*, is used as a measure of total resistance of the asphalt binder to deformation caused by repeated pulses of shear stress, which under test conditions is caused by the oscillating plates of the DSR.

The phase lag or phase angle, δ, is a measure of the degree of elasticity of the asphalt binder. A completely elastic material will not show any difference between the shear stress and shear strain. A completely viscous material would have a phase difference or angle of

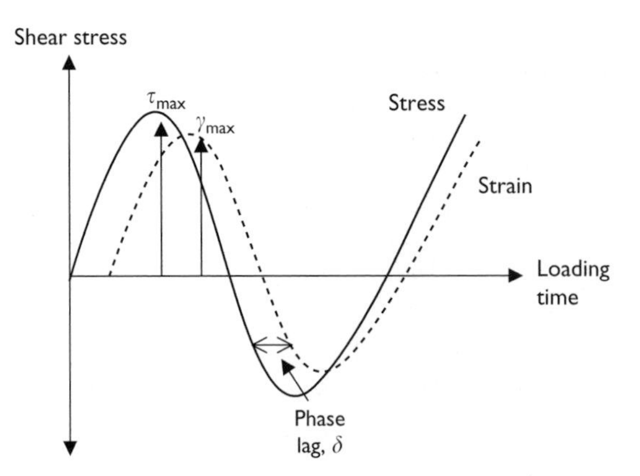

Figure 1.6 Sinusoidal loading of an asphalt binder by a DSR.

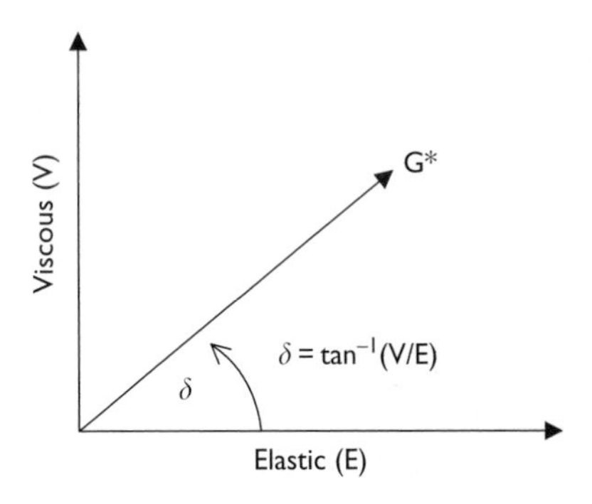

Figure 1.7 Viscoelastic relationship.

90° on the sinusoidal curve. Since asphalt binders are viscoelastic, the phase angle between the shear stress and shear strain is between 0° and 90° (Figure 1.7). The specific values of the phase angle, δ, depends on the source and grade of the asphalt binder, and on the temperature and frequency the dynamic shear test was completed. Small phase angles are determined at low temperatures and high frequencies, while phase angles closer to 90° are found at high temperatures and low frequencies (Whiteoak 1991). A phase angle of 0° would represent an elastic solid and a phase angle of 90° would be a viscous liquid.

The DSR provides the complex shear modulus, G*, and the phase angle, δ, for the PG binder specification. The specification used both G* and δ in conjunction with each other. The complex shear modulus divided by the sine of the phase angle, or G*/sin δ, is used to determine the permanent deformation resistance of an asphalt binder. To minimize deformation or rutting, G*/sin δ must be a minimum of 1.00 KPa for the original asphalt binder and a minimum of 2.20 KPa after the RTFO aging.

High values of G* and low values of δ are desirable attributes from the standpoint of rutting resistance (The Asphalt Institute 1997). The measurement of G*/sin δ after the RTFO is significant in that the stiffness of the asphalt binder immediately after construction of the pavement has the most critical influence on the overall rutting resistance of the asphalt binder. It is also important to note that an asphalt binder meeting the minimum G*/sin δ does not completely insure that an asphalt pavement will not rut. The asphalt binder contributes only 15–20 percent of the pavement rutting. Proper mixture design and the aggregate structure influence the other 80–85 percent (Kriech 1994). The DSR also measures the complex shear modulus and the phase angle on the asphalt binder after it has been aged by both the RTFO and the PAV. In this test, G* and sin δ are multiplied together to give a value that determines the fatigue cracking resistance of an asphalt binder. This factor, G*sin δ is known as the fatigue cracking factor. The PG binder specification has a G*sin δ maximum value of 5,000 KPa. Low values of G* and δ are considered desirable attributes from the standpoint of resistance to fatigue cracking.

In summary, the DSR is used in the PG binder specification to determine the resistance of an asphalt binder to two of the three pavement distresses that are addressed by the PG binder specification. Permanent deformation resistance is measured on the original and the RTFO aged asphalt binder and fatigue cracking is measured on the PAV aged asphalt binder. The temperature of the asphalt binder is increased in six-degree intervals until it meets the minimum G*/sin δ requirements and the maximum G*sin δ requirement. The corresponding G*/sin δ maximum test temperature is upper PG temperature grade designation. For example, an original asphalt binder tested has a G*/sin δ value of 1.05 KPa at 70 °C and a RTFO residue G*/sin δ value of 2.30 KPa at 70 °C. The asphalt binder upper temperature is classified as a PG70.

Aging of an asphalt binder occurs primarily due to the volatilization of light oils present in the asphalt binder and oxidation by reacting with air that is always present in the asphalt pavement environment. Oxidation causes the asphalt binder to become stiffer or brittle. Oxidative hardening occurs at a relatively slow rate in an asphalt pavement, with it occurring faster at warmer temperatures. The rolling thin film aging method in the PG binder specification is the same method used in the AR viscosity grading system that was described earlier. AASHTO T240 is the recognized test procedure for the RTFO aging in the PG binder specification; however, ASTM D2872 is a functional equivalent procedure. RTFO aging is completed on the asphalt binder to simulate the aging processes that take place during mixing at a hot mix asphalt plant and during construction. This aging process is mostly due to the volatilization of light oils that occur when the asphalt binder exists in thin films covering the aggregate. Since the volatilization of light oils occur during the RTFO aging procedure, the PG binder specification incorporates a maximum asphalt binder mass loss of 1 percent after the RTFO aging. The PAV was developed specifically for the PG binder specification. The PAV simulates oxidative hardening of the asphalt binder that may occur over seven to ten years in the life span of an asphalt pavement. The PAV accomplishes this long-term aging in the test time of 20 h with the application of 2,070 KPa of air while the asphalt binder is heated to either 90, 100, or 110 °C. The PAV aging temperature is based on simulated climatic conditions. The aging temperature is 90 °C for PG grades under 64 and 100 °C for PG grades 64 °C and higher. 110 °C is sometimes selected to simulate in extremely hot climates. It is important to remember that asphalt binder residue subjected to PAV aging is always first aged by the RTFO. This course of test aging follows the similar type of aging that actually occurs during the life of an asphalt pavement. The asphalt binder is first aged during mixing and construction (RTFO) and then slowly ages while in the environment (PAV).

The BBR and the DTT are the final two test instruments used in the PG binder specification. The BBR and the DTT are used to determine the low temperature cracking properties of an asphalt binder. Both of these instruments only use the PAV asphalt binder residue to test. Since low temperature cracking usually occurs some time after a pavement has been constructed, the PAV asphalt binder residue is the logical choice to test for low temperature cracking properties. Asphalt binders at low temperatures are too stiff to be tested by the DSR. The BBR (Plate 1.8) was developed to determine the stiffness of asphalt binder at temperatures that range from 0 to −36 °C.

The BBR determines the asphalt binder stiffness by measuring how much the asphalt binder deflects or creeps under a constant load at a constant temperature. The test temperatures are related to a pavement's lowest service temperature, when the asphalt binder acts more like an elastic solid (The Asphalt Institute 1997). The BBR applies a small creep load to a small beam made of the asphalt binder and then measures the deflection of the beam to this load (Figure 1.8). The asphalt binder's resistance to this load is the creep stiffness. The PG binder specification uses two test parameters from the BBR, creep stiffness, and m-value. The beam is tested using a three-point loading of 980 mN that is applied for 240 s. It is important to note that the load is applied for a specific length of time and not until the beam breaks or fails. The test temperatures are actually 10 °C warmer than the designated low temperature PG binder grades. This increase of temperature decreases the testing

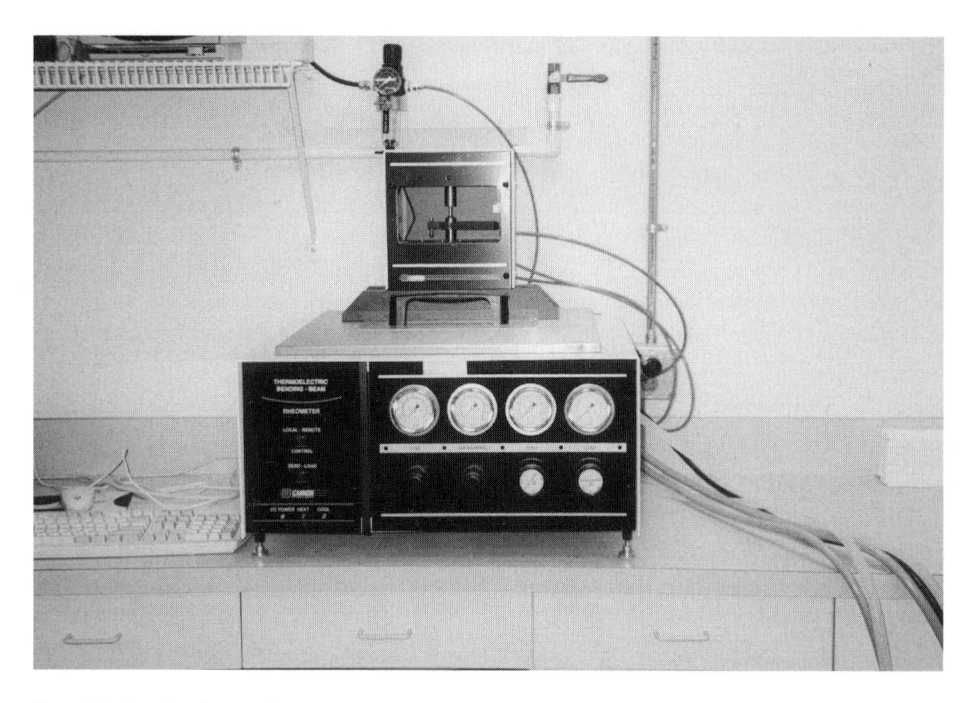

Plate 1.8 Bending beam rheometer.

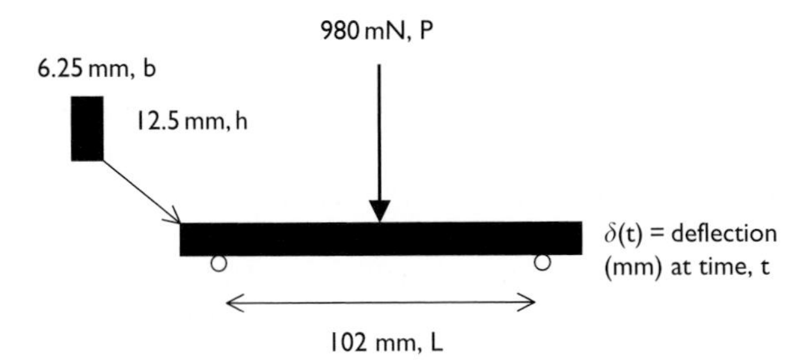

Figure 1.8 Bending beam geometry.

time and allows the creep stiffness of the asphalt binder to be determined after only a 60 s loading. The 60 s creep stiffness simulates stiffness after 2 h of loading at a 10 °C lower test temperature. For example, a PG 58-28 asphalt binder beam stiffness is actually determined by the BBR at −18 °C and 60 s of loading.

The creep stiffness of the asphalt binder beam is calculated using equation 1.6.

$$S = (PL^3)/4bh^3\delta \qquad (1.6)$$

where S is the creep stiffness, MPa; P, the applied constant load, mN; L, the distance between beam supports, mm; b, the beam width, mm; h, the beam height or thickness, mm; δ, the deflection, mm.

Low temperatures cause an asphalt binder to become very stiff with a low resistance to movement. Tensile stresses from movement can exceed the tensile strength of the asphalt pavement, leading to cracking. To prevent low temperature cracking, the PG binder specification uses a maximum creep stiffness, S, of 300 MPa. The second value determined from the BBR is the m-value. The m-value represents the rate of change in the creep stiffness versus time. The m-value is obtained by calculating the stiffness of the asphalt binder beam at several times during the 240 s loading of 980 mN to the beam. The m-value is the slope of the curve of log stiffness versus log time. The rate at which the asphalt binder stiffness changes with time at low temperatures is given by the m-value. A high m-value is desirable because as the pavement temperature decreases and pavement contraction begins to occur, the binder will respond as being less stiff. This decrease in stiffness leads to smaller tensile stresses in the asphalt binder and less likely occurrence of low temperature cracking (The Asphalt Institute 1997). The PG binder specification requires that the m-value be $\geq$0.300 at 60.

The final piece of test equipment used in the PG binder specification is the DTT. The DTT is the only PG binder test equipment that actually tests the asphalt binder until it breaks or fails. There is a strong relationship between asphalt binder stiffness and the amount of stretching they undergo before breaking. Asphalt binders that undergo considerable stretching before failure are called "ductile" and those that break without much stretching are called "brittle." It is important that an asphalt binder is capable of a minimal amount of elongation. Typically stiffer asphalt binders are more brittle and softer asphalt binders are more ductile. Creep stiffness measured by the BBR does not completely measure the ability of an asphalt binder to stretch before breaking. Some modified asphalt binders can exhibit high creep stiffness, but are more ductile than typical asphalt binders and can stretch farther before breaking. The DTT measures at very low temperatures the amount of strain in an asphalt binder. The test is performed at the temperature range of −0 to −36 °C where an asphalt binder is typically considered brittle (The Asphalt Institute 1997).

The DTT performs the test on the PAV residue for the same reasons that the BBR does. The tensile strength properties of an asphalt binder, like stiffness, also depend on both temperature and loading time. The DDT pulls a small dog bone shaped specimen (Figure 1.9) of asphalt binder at a rate of 1.0 mm/min until it breaks or fails. The elongation at the failure point is used to calculate the failure strain, which is another indication of whether an asphalt binder will be brittle or ductile at low temperatures. The failure strain, ϵ, is the change in length, ΔL, divided by the specimen or effective gauge length, L_e. Failure is defined as the load when the *stress* reaches it maximum value, not just when the specimen breaks. Failure stress, σ, is the failure load, P, divided by the original cross-section of the dog bone shaped specimen. The PG binder specification requires that the failure strain be a minimum of 1 percent at 1.0 mm/min.

Currently only the DSR and the BBR are used to designate the grade of the PG binder with one exception where the DTT is used as a referee measurement. The referee use of the DTT is "in." the ASTM Standard Specification for PG Asphalt Binder, ASTM D6373 (ASTM 2001b) and in the AASHTO counterpart, MP-1. The following describes the referee use of the DTT. If the BBR creep stiffness is below 300 MPa, the direct tension test is not required. If the creep stiffness is between 300 and 600 MPa, the direct tension failure strain requirement can be used in lieu of the creep stiffness requirement. The m-value requirement must be satisfied in both cases (AASHTO 2000a). The AASHTO specification MP-1a is an expansion of the MP-1 specification that now requires the use of the direct tension requirement in addition to the BBR to further enhance the specification. Highway agencies in the United States are requiring AASHTO MP-1, MP-1a, or their own modified version. The significance of MP-1a is that it aids in further specifying modified asphalt binders by requiring the use of the DTT. Table 1.11 is the common specification properties of the PG binder grading system.

The purpose of the PG binder specification is to give the asphalt binder purchaser additional tools to further specify all the significant properties of an asphalt binder. It is important to remember that the specification is that it is just a purchasing specification for asphalt

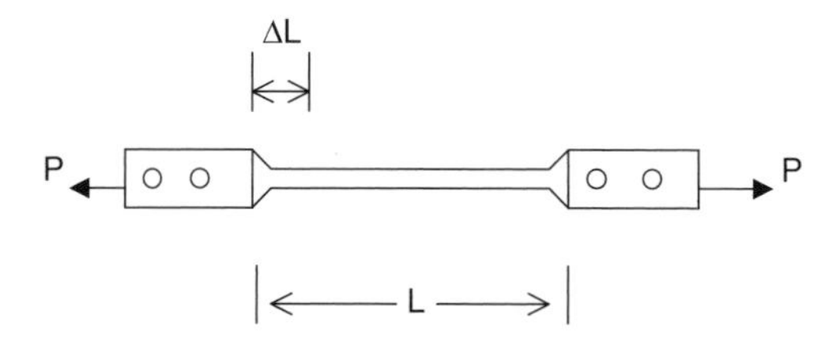

Figure 1.9 Direct tension specimen.

Table 1.11 Common performance graded asphalt binders

Performance grade	*PG 52*							*PG 58*					*PG 64*						*PG 70*						*PG 76*				
	−10	*−16*	*−22*	*−28*	*−34*	*−40*	*−46*	*−16*	*−22*	*−28*	*−34*	*−40*	*−10*	*−16*	*−22*	*−28*	*−34*	*−40*	*−10*	*−16*	*−22*	*−28*	*−34*	*−40*	*−10*	*−16*	*−22*	*−28*	*−34*
Original binder																													
Flash point, ASTM D92														≥ 230 °C															
Viscosity, ASTM D4402, ≤ 3 Pa-s														135 °C															
Dynamic Shear, AASHTO TP5 G*/sin δ ≥ 1.00 KPa	52 °C							58 °C					64 °C						70 °C						76 °C				
Rolling thin film oven, ASTM D2872, residue																													
Mass loss, maximum																1.00 %													
Dynamic Shear, AASHTO TP5 G*/sin δ ≤ 2.20 KPa	52 °C							58 °C					64 °C						70 °C						76 °C				
Pressure aging vessel, AASHTO PP1, residue																													
PAV aging temperature	90 °C							100 °C					100 °C						100(110) °C						100(110) °C				
Dynamic shear, AASHTO TP5 G*sin δ ≤ 5000 KPa	25	22	19	16	13	10	7	25	22	19	16	13	31	28	25	22	19	16	34	31	28	25	22	19	37	34	31	28	22
Creep stiffness, AASHTO TP1 S ≤ 300 MPa; m ≥ 0.300	0	−6	−12	−18	−24	−30	−36	−6	−12	−18	−24	−30	0	−6	−12	−18	−24	−30	0	−6	−12	−18	−24	−30	0	−6	−12	−18	−24
Direct tension, AASHTO TP3	0	−6	−12	−18	−24	−30	−36	−6	−12	−18	−24	−30	0	−6	−12	−18	−24	−30	0	−6	−12	−18	−24	−30	0	−6	−12	−18	−24

Table 1.12 PG asphalt binder grades

High temperature grades	*Low temperature grades*
PG 46	−34, −40, −46
PG 52	−10, −16, −22, −28, −34, −40, −46
PG 58	−16, −22, −28, −34, −40
PG 64	−10, −16, −22, −28, −34, −40
PG 70	−10, −16, −22, −28, −34, −40
PG 76	−10, −16, −22, −28, −34
PG 82	−10, −16, −22, −28, −34

Source: Reprinted with permission from Professional Publications., *Civil Engineering Reference Manual*, 7th edn, Michael R. Lindeburg, PE, © 1999, by Professional Publications, Inc.

binders. Table 1.12 illustrates the various PG asphalt binders that maybe specified or purchased.

One particular source or grade of asphalt binder may be able to meet all sited specifications, that is, penetration, viscosity, AR viscosity, and performance grade. An asphalt binder may not meet a particular grade, especially in the PG binder specification, without the addition of modifiers. It is best practice for the designer to specify the end result asphalt binder grade and not the specific modifiers that may have to be used to meet the specification. This is particularly true when designing small projects such as parking lots or local streets. The designer specifying an asphalt binder for a particular paving project should use the accepted grading method in the area of the project. For example, if the project was in Los Angeles, California you may select an AR-4000, but if it was in Tokyo, Japan, you may select the approximate equivalent asphalt binder grade as a 50 dmm penetration.

The actual selection process of a particular grade of an asphalt binder depends on climate, traffic levels, pavement location, mixture design, and several other parameters. These parameters will be discussed in the chapter on traffic analysis and material selection. The software, *LTPPBind2.1*, has a database of 2,500 weather stations from North America and is an excellent tool to easily determine the proper SHRP PG binder for the project. At the time of this text preparation, it was available from the United States Federal Highway Administration's Long-Term Pavement Performance web site (LTPP 2001). Software may be available to accomplish the same task for other parts of the world.

Chapter 2

Aggregates

Aggregates are one of the key building materials used in the construction industry and the largest portion of an asphalt pavement. Aggregates are generally naturally derived from minerals and sometimes have been further mechanically processed to be better suited for specific applications. Synthetic aggregates are also used in asphalt pavements and are most commonly blast furnace slags from the steel industry. Crushed stone, gravel, and sand are the natural types of aggregate used in the construction of an asphalt pavement. ASTM defines aggregates as a granular material of mineral composition such as sand, gravel, shell, slag, or crushed stone, used with a cementing medium to form mortars or concrete, or alone as in base course, railroad ballast, etc. (ASTM 2001a).

The United States produces nearly two billion tons of aggregate per year, which represents approximately half of the non-energy mining volume in the country. Sand and gravel

Plate 2.1 Limestone quarry.

are produced commercially in every state in the United States, and crushed stone is produced in every state except Delaware.

Crushed stone quarries range in size from small operations to those (Plate 2.1) with production of more than 10 million tons annually (Langer and Glanzman 1993).

Natural aggregates are divided into three geologic classifications of rock, sedimentary, metamorphic, and igneous (Table 2.1). Sub-classifications of sedimentary are calcareous and siliceous; metamorphic are foliated and nonfoliated; and igneous are intrusive and extrusive. Sedimentary rocks were formed by the accumulation of sediments carried by water or wind. Calcareous sedimentary rock is calcium derived and siliceous is silica derived. A key feature of a sedimentary rock structure is that it consists of layers. Sedimentary rocks make up about 70 percent of the crushed stone production. Metamorphic rock was formed

Table 2.1 Aggregate geological classifications

Class	*Type*	*Example*
Sedimentary	Calcareous	Conglomerate
		Dolomite
		Limestone
		Travertine
	Siliceous	Arenite
		Arkose
		Breccia
		Chert
		Graywacke
		Sandstone
Metamorphic	Foliated	Gneiss
		Mica Schist
		Slate
	Nonfoliated	Anatexite
		Granulite
		Marble
		Quartzite
		Serpentinite
Igneous	Intrusive	Diorite
		Granodiorite
		Granite
		Hornblendite
		Monzonite
		Pyroxenite
		Tonalite
	Extrusive	Andesite
		Basanite
		Basalt
		Dacite
		Diabase
		Obsidian
		Quartz Porphyry
		Phonolite
		Ryholite
		Tephrite
		Trachyte

from either a sedimentary or igneous rock structure put under a tremendous amount of pressure or heat. A foliated metamorphic rock contains parallel planes of minerals. These planes are weak and can delaminate and flake off. Nonfoliated rock does not contain these planes and are generally massive in structure. Metamorphic rock makes up about 8 percent of the crushed stone production. Igneous rock was formed through the cooling and solidification of molten magma that erupted from the earth. Extrusive igneous rock was formed by cooling on the earth's surface and intrusive was formed originally below the earth's surface. Igneous rock makes up about 22 percent of the crushed stone production.

The geologic derivation of natural aggregate has limited importance to the designer. Aggregate is a high-volume, low-cost commodity, with the transportation costs to the site of use being the major part of the total cost of the aggregate. Aggregate occurs where nature placed it, not where people need it (Langer and Glanzman 1993). Table 2.1 is provided to help the designer become familiar with the geologic classes of aggregates. ASTM C 294, *Standard Descriptive Nomenclature for Constituents of Concrete Aggregates*, gives standard descriptions for naturally occurring aggregate. The physical characteristics of aggregate greatly dictates the final performance of the asphalt pavement. The physical characteristics are very important to the designer and can be part of the pavement's design and specifications. Selecting an aggregate depends on the intended purpose along with the cost, availability, and quality. For example, aggregates that are suitable for use as a base course, may not be adequate for the asphalt pavement surface course.

The three types of aggregates are crushed stone, gravel, and sand. Crushed stone is mechanically crushed rocks or boulders. Most crushed stone is quarried from bedrock. Gravel is the product of the erosion of bedrock and surficial materials. Gravel can also be crushed, and that is actually recommended for its use in asphalt pavements or bases. Sand can be either the erosion of bedrock or be mechanically crushed. Sand is distinguished by its size, which according to ASTM C 125, *Standard Terminology Relating to Concrete and Concrete Aggregates*, would be particles smaller than 4.75 mm but larger than 75 μm.

Aggregates have several physical properties that are of importance to the asphalt pavement designer. These properties are:

- Grading (size)
- Shape
- Toughness
- Durability
- Surface texture
- Cleanliness (deleterious materials)
- Absorption
- Adhesion
- Skid resistance.

Grading

The grading or gradation of an aggregate is the most important property that an aggregate can contribute to the performance of an asphalt pavement. Aggregate grading is the distribution of the particle size expressed as a percentage of total weight. Grading is determined by passing the aggregate through a series of sieves stacked with progressively smaller openings (Figure 2.1). The weight of the aggregate on each sieve is determined (mass A), along with the total aggregate weight (mass B) and the percent retained, percent passing, and total percent passing can be calculated as determined by equations 2.1–2.3.

$$\textit{Percent retained} = (A/B) * 100 \quad (2.1)$$
$$\textit{Percent passing} = 100 - \text{percent retained} \quad (2.2)$$
$$\textit{Total percent passing} = 100 - \text{cumulative percent retained} \quad (2.3)$$

The weights and calculations of an example aggregate gradation are shown in Table 2.2.

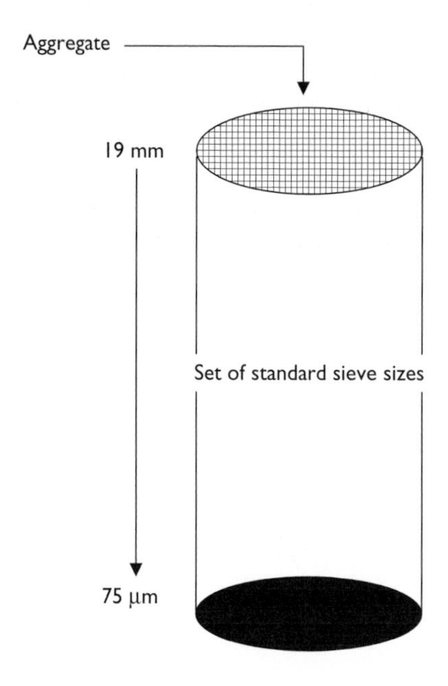

Figure 2.1 Aggregate gradation sieves.

Table 2.2 Aggregate gradation calculations

Sieve size	*Cumulative weight retained (g)*	*Retained (%)*	*Total passing (%)*
19 mm	0	0	100
9.5 mm	60.8	3	97
4.75 mm	202.6	10	90
2.36 mm	364.6	18	82
1.18 mm	709.0	35	65
600 μm	1215.4	60	40
300 μm	1701.5	84	16
150 μm	1924.3	95	5
75 μm	1972.9	97.4	2.6
Pan	2025.6	100	0
Total	2025.6		

There are two types of aggregate grading analysis a dry sieve analysis and a wet sieve analysis. The steps in a dry sieve analysis include:

1 Splitting the sample to the correct size.
2 Determining the total dry weight.
3 Pouring it over a set of sieves and shaking.
4 Weighing the material collected on each sieve size.
5 Determining the cumulative percent retained on each sieve and then converting to a cumulative or total percent passing each sieve.

When a washed sieve analysis is completed to obtain a washed gradation, the aggregates are washed over a 75 μm sieve to remove the dust or fine particles. The aggregate is then dried, weighed, and graded as a dry analysis. The amount of weight lost after washing is the passing 75 μm material or dust. In designing asphalt mixtures, an accurate determination of the amount of material finer than 75 μm has significance on the performance of the mixture. It is preferred to use the wet analysis method for determining the gradation of aggregates to be used for asphalt mixtures. Modern asphalt mixture design methods such as the SHRP SuperPave® method require the use of a washed gradation.

The maximum aggregate size in an asphalt mixture or surface treatment is important to ensure good performance. If the maximum particle size is too small, the pavement may be unstable; if it is too large segregation of the mixture may occur and the mixture may be difficult to place and work (Roberts *et al.* 1991). The maximum size of an aggregate designates the smallest sieve through which 100 percent of the particles will pass. Maximum size can also be defined as the one sieve size larger than the nominal maximum size. The nominal maximum size is one sieve size larger than the first sieve size to retain more than 10 percent by weight of the particles.

Example 2.1

Determine the maximum aggregate size of the following coarse aggregate gradation:

Sieve size	Total passing (%)
19 mm	100
12.5 mm	95
9.5 mm	88
4.75 mm	60
2.36 mm	21
1.18 mm	4
600 μm	1

Solution

First determine the nominal maximum size. This is done by converting the percent passing to total percent retained:

Sieve size	*Total retained* (%)
19 mm	$100 - 100 = 0$
12.5 mm	$100 - 95 = 5$
9.5 mm	$100 - 88 = 12$
4.75 mm	$100 - 60 = 40$
2.36 mm	$100 - 21 = 79$
1.18 mm	$100 - 4 = 96$
600 μm	$100 - 1 = 99$

The first descending sieve to retain more than 10 percent is the 9.5 mm sieve. By definition the nominal maximum size is one sieve size larger than this sieve; which is the 12.5 mm sieve. By definition the maximum size is one sieve size larger than the nominal maximum size.The maximum aggregate size is *19 mm*.

Asphalt mixture designations, including SuperPave, used the nominal maximum size of the aggregate. Paving practice can also limit the nominal maximum aggregate size to about one-half of the lift or mat thickness. The nominal maximum aggregate size is also used to designate slurry seal gradations and other surface treatment aggregates.

The grading results of aggregate are generally recorded in a table and shown graphically. When the results are graphed, the aggregate grading can be easily identified as dense graded, gap graded, or open graded (Figure 2.2).

Two types of graphs are used to plot an aggregate gradation, the semi-log chart as shown in Figure 2.2 or the 0.45 power chart (Figure 2.3). The 0.45 power chart was developed in 1962 by the United States Bureau of Public Roads (prior to the Federal Highway Administration) that uses an arithmetic scale of the sieve size raised to the 0.45 power. The vertical axis is the percent passing a sieve size and the horizontal axis is the sieve size opening raised to the 0.45 power. The horizontal axis does not contain the actual calculated numbers, but instead has marks that indicate the different sieve sizes in ascending order. The chart was developed on the assumption that the best aggregate grading to be used for an asphalt mixture is the one that gives the densest particle packing. Equation 2.4 can also be used to determine the maximum particle density prior to the addition of the asphalt binder (Nicholls 1998). The chart is still used as a tool in the design of asphalt mixtures.

$$P_d = 100\,(d/D)^{0.45} \tag{2.4}$$

where D is the the largest aggregate particle size used in the grading; P_d, the the total percent passing the sieve size, d.

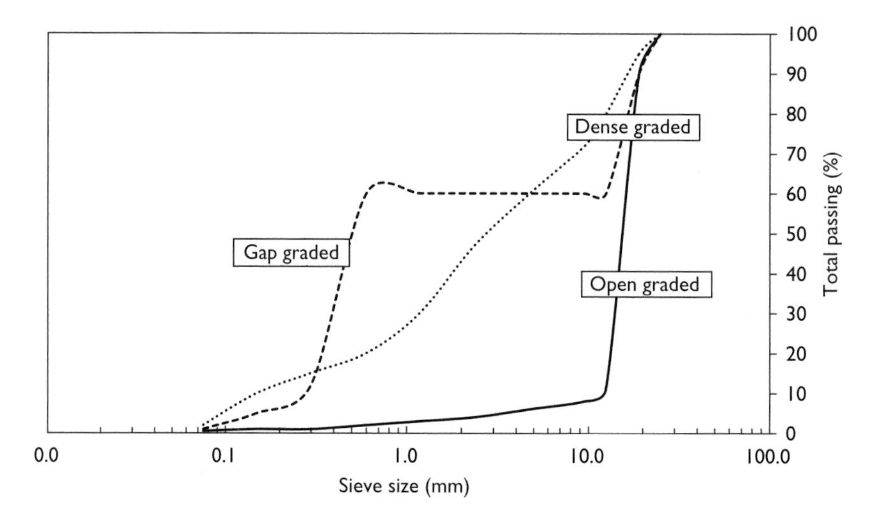

Figure 2.2 Aggregate grading.

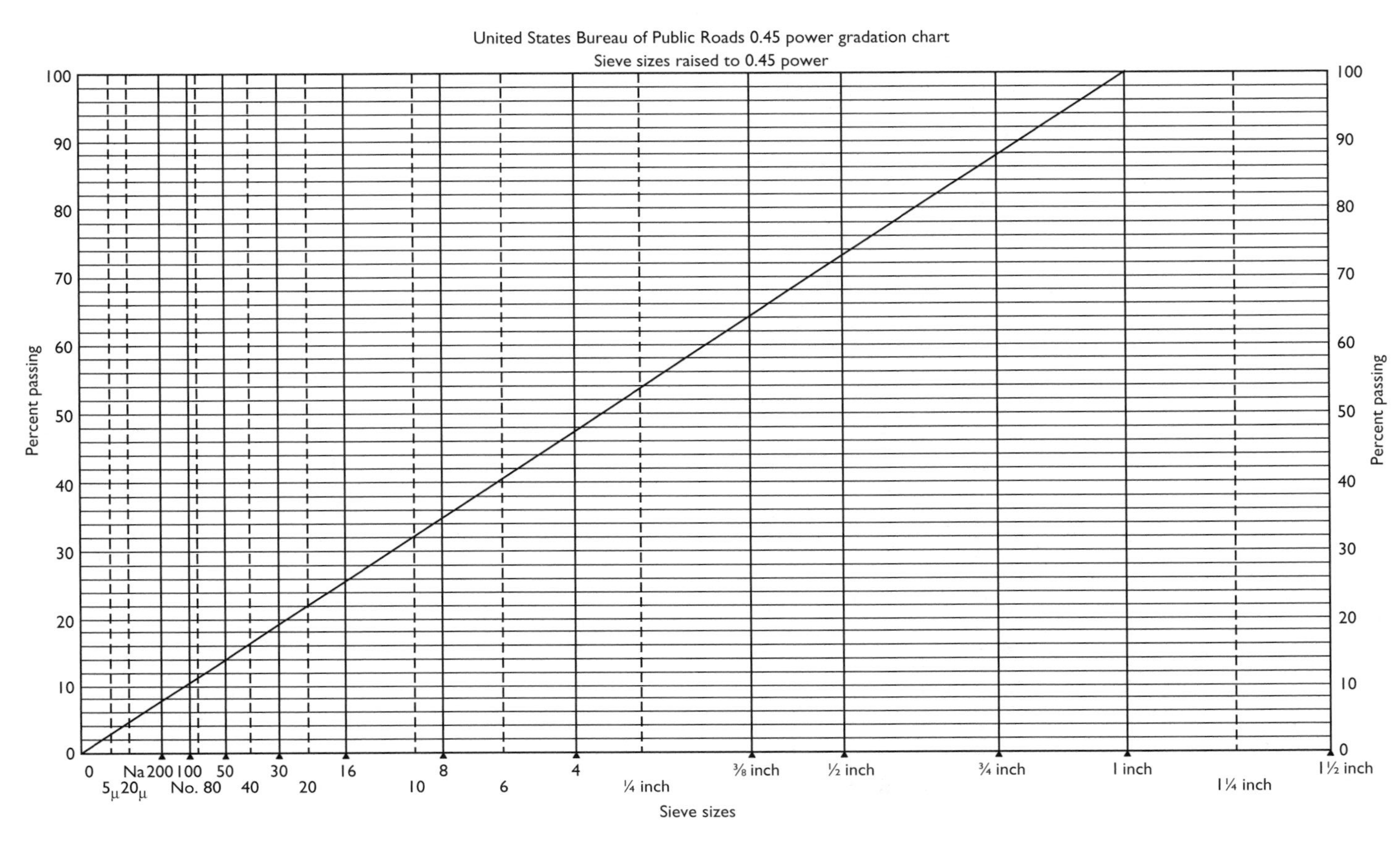

Figure 2.3 Original 0.45 power chart with maximum density line.

Table 2.3 ASTM D448 various coarse aggregate gradations

Sieve size	*Sieve number*													
	357	*4*	*467*	*5*	*56*	*57*	*67*	*68*	*7*	*78*	*8*	*89*	*9*	*10*
	Total passing (%)													
63 mm	100													
50 mm	95–100	100	100											
37.5 mm		90–100	95–100	100	100	100								
25 mm	35–70	20–55		90–100	90–100	95–100	100	100						
19 mm		0–15	35–70	20–55	40–85		90–100	90–100	100	100				
12.5 mm	10–30			0–10	10–40	25–60			90–100	90–100	100	100		
9.5 mm		0–5	10–30	0–5	0–15		20–55	30–65	40–70	40–75	85–100	90–100	100	100
4.75 mm	0–5		0–5		0–5	0–10	0–10	5–25	0–15	5–25	10–30	20–55	85–100	85–100
2.36 mm						0–5	0–5	0–10	0–5	0–10	0–10	5–30	10–40	
1.18 mm								0–5		0–5	0–5	0–10	0–10	
300 μm												0–5	0–5	
150 μm														10–30

Source: American Society of Testing Materials D448 (2001b).

The 0.45 power chart is used to determine the maximum density of an aggregate by drawing a straight line from the origin of the graph to the maximum aggregate size of the aggregate. The actual gradation of the aggregate is then plotted on the same graph and can be compared to the maximum density line. The grading will need to be finer or coarser than the gradation represented by the maximum density line in order to allow additional room for the addition of the asphalt binder and mixture air voids.

The asphalt paving mixture's gradation, also known as the "job mix formula," can also be plotted on the 0.45 power chart. By comparing the job mix formula gradation with the maximum density line, several asphalt mixture properties can be predicted. A more detailed explanation on this application is given in Chapter 5.

Aggregate gradations are generally compared to a universal specification such as ASTM D3515, *Standard Specification for Hot-Mixed, Hot-Laid Bituminous Mixtures*, or to a regional or local transportation agency specification. Coarse aggregates also have universal size numbers that reference a standard gradation band as established in ASTM C33, *Standard Specification for Concrete Aggregates*, or ASTM D448 (Table 2.3), *Standard Classification for Sizes of Aggregate for Road and Bridge Construction*. Surface treatment or surface dressing aggregate gradation specifications can be found in ASTM D1139, *Standard Specification for Aggregate for Single or Multiple Bituminous Surface Treatments*.

Table 2.4 ASTM D3515 gradation specifications for dense asphalt mixtures

Sieve size	*Dense mixtures*						
	Mix designation and nominal maximum size of aggregate (mm)						
	50	37.5	25	19	12.5	9.5	4.75
	Grading of total mixture aggregate (%) passing by weight						
63 mm	100						
50 mm	90–100	100					
37.5 mm		90–100	100				
25 mm	60–80		90–100	100			
19 mm		56–80		90–100	100		
12.5 mm	35–65		56–80		90–100	100	
9.5 mm				56–80		90–100	100
4.75 mm	17–47	23–53	29–59	35–65	44–74	55–85	80–100
2.36 mm	10–36	15–41	19–45	23–49	28–58	32–67	65–100
1.18 mm							40–80
600 μm							25–65
300 μm	3–15	4–16	5–17	5–19	5–21	7–23	7–40
150 μm							3–20
75 μm	0–5	0–6	1–7	2–8	2–10	2–10	2–10
Coarse aggregate size numbers	3,357 and 57	4 and 67 or 4 and 68	5 and 7 or 57	67 or 68 or 6 and 8	7 or 78	8	10
	Suggested asphalt cement weight (%) of total mixture						
	3–5	3–6	3–7	4–8	4–10	4–10	5–10

Source: American Society of Testing Materials D3515 (2001b).

Table 2.5 ASTM D1073 grading requirements for fine aggregates

Sieve size	*Amounts finer than each laboratory sieve, mass %*			
	Grading no. 1	*Grading no. 2*	*Grading no. 3*	*Grading no. 4*
9.5 mm	100			100
4.75 mm	95–100	100	100	80–100
2.36 mm	70–100	75–100	95–100	65–100
1.18 mm	40–80	50–74	85–100	40–80
600 μm	20–65	28–52	65–90	20–65
300 μm	7–40	8–30	30–60	7–40
150 μm	2–20	0–12	5–25	2–20
75 μm	0–10	0–5	0–5	0–10

Source: American Society of Testing Materials D1073 (2001b); © ASTM; reprinted with permission.

Due to the economical and geographical variation in aggregate sources, it is wise for the asphalt designer to select a standardized or local agency gradation specification. Local aggregate producers and hot mix asphalt producers can provide information on which aggregate gradations are available locally. Coarse aggregate sizes can be blended along with fine aggregate to give a final aggregate mixture gradation or job mix formula. Aggregate blending is discussed in detail in Chapter 5. An example gradation specification for dense graded asphalt mixtures is given in Table 2.4.

Fine aggregate or sand is defined as aggregate that entirely passes the 4.75 mm sieve but is retained on the 75 μm sieve. The 2.36 mm sieve is key fine aggregate sieve for dense graded asphalt mixtures. Fine aggregate for asphalt pavements consist of natural sand, crushed stone screenings or manufactured sand from stone, crushed gravel, or crushed blast furnace slag. Fine aggregate should not contain any clay, loam, or significant amounts of organic material. The grading of fine aggregate can be further defined into four size categories, from coarse to fine, as specified by ASTM D1073 (Table 2.5) *Standard Specification for Fine Aggregate for Bituminous Mixtures*. Mineral filler can also be another ingredient for an asphalt mixture. Mineral filler is a finely divided mineral usually consisting of rock dust, fly ash, hydraulic cement, or other inert material that has small enough particles that 100 percent pass the 600 μm sieve and 70–100 percent pass the 75 μm sieve. The size distribution of the material passing the 75 μm can influence the stiffness of the asphalt binder in the mixture. If the majority of the mineral filler is smaller than 20 μm, the asphalt binder portion of the mixture will become stiffer. Size distribution larger than 20 μm does not by itself have a stiffening effect on the asphalt binder (TRB 2000). The hot mix asphalt plant itself through the degradation of the aggregate during the drying and mixing operations generates some mineral filler. This generated mineral filler is collected through a dry dust collector known as a "baghouse." This collected dust or "baghouse fines" is introduced back into the asphalt mixture to serve two functions:

1 Serve as a required ingredient in the hot mix asphalt mixture as mineral filler. Sometimes it is necessary to supplement the baghouse fines with additional mineral filler.
2 Environmental requirements by recycling the dust back into the asphalt mixture instead of discharging it into the atmosphere or collecting it and disposing in a landfill.

Particle shape

The shape of an aggregate particle also has a significant influence on the performance of an asphalt pavement. Particle shape can be described as cubical, flat, elongated, and round. Roundness is also further divided into five categories: rounded, sub-rounded, curvilinear, sub-angular, and angular. General aggregate shape as related to geology is described in Table 2.6. Particle shape classifications are also based on the measurement of length, width, and thickness. The flatness ratio is the thickness to width ratio and the elongation ratio is the width to length ratio. Using the flatness ratio and the elongation ratio, each individual particle shape can be plotted to form what is known as a "Zingg graph" (Nicholls 1998). Flat and elongated particles are also measured for the SuperPave mixture design method. ASTM D 4791, *Standard Test Method for Flat Particles, Elongated Particles, or Flat and Elongated Particles in Coarse Aggregate*, measures the percentage by mass of coarse aggregates that have a maximum to minimum dimension ratio greater than five. This procedure uses a proportional caliper device to measure the dimensional ratio of an aggregate particle. Another procedure for determining the particle shape of the aggregate is multiple ratio analysis using a digital caliper device. This method graphically shows the shape of all the particles in a given sample by evaluating it on five different ratios instead of one.

Aggregate angularity is also defined as particles that have at least one fractured face. The SuperPave mixture design method has adopted two procedures to quantify particle shape: ASTM D 5821, *Standard Test Method for Determining the Percentage of Fractured Particles in Coarse Aggregate*, for coarse aggregate angularity and AASHTO T 304, *Compacted Void Content of Fine Aggregate*, for fine aggregate angularity. Coarse aggregate and fine aggregate angularity insures a high degree of aggregate internal friction that gives the asphalt mixture rutting and deformation resistance. Rounded aggregates, including gravel and natural sand, have better workability and compactness with less effort than mixtures containing angular aggregates. Rounded aggregates are more prone to mixture deformation and are not suited for high traffic or high loading pavements. Table 2.7 gives an example criteria on coarse and fine angularity along with flat and elongated particles for specific traffic requirements.

Chapter 3 gives a detailed analysis on traffic loading and material selection for a specific loading requirement. The architect or engineer needs to select the proper balance of crushed

Table 2.6 Classification of aggregate shape

Classification	*Description*	*Examples*
Rounded	Fully water worn or completely worn by attrition	River or ocean gravel, ocean and wind-blown sand: chert, etc.
Irregular	Naturally irregular or partly shaped by attrition with rounded edges	Pit gravel, pit sand: silica sand, etc.
Angular	Possessing a well-defined edge intersection of roughly planer faces	Crushed rock or manufactured sand of sedimentary types: limestone, dolomite, etc.
Cubical	Similar to angular, with length to thickness ratios close to one	Crushed rock or manufactured sand of metamorphic types: granite, etc.
Flake	Usually angular of which the thickness is small relative to the width and/or length	Laminated or foliated rocks: mica, etc.

Table 2.7 Aggregate shape requirements for various traffic levels

Traffic, (million ESALs)	*Mixture depth from pavement surface*				*Flat and elongated particles criteria ASTM D 4791*
	Coarse aggregate angularity criteria[1] *ASTM D 5821*		*Fine aggregate angularity criteria*[2] *AASHTO T 304*		
	<100 mm	>100 mm	<100 mm	>100 mm	*Maximum (%)*
<0.3	55/–	–/–	—	—	—
<1	65/–	–/–	40	—	—
<3	75/–	50/–	40	40	10
<10	85/80	60/–	45	40	10
<30	95/90	80/75	45	40	10
<100	100/100	95/90	45	45	10
≥100	100/100	100/100	45	45	10

Notes

1 95/90 denotes that 95 percent of the coarse aggregate has one or more fractured faces and 90 percent has two or more.

2 Percent air voids in loosely compacted fine aggregate.

aggregate for the specific loading requirement. For example, if 100 percent crushed aggregate is selected for a parking lot for a fast food restaurant, the asphalt mixture may be too harsh, being difficult to place and compact for such a low volume application. With the exception of heavy-duty applications, most parking lots would be selected at < 1 million equivalent single axle loads, giving a crushed aggregate requirement of approximately 65 percent.

Toughness

Aggregate toughness is its resistance to abrasion and degradation. Degradation can occur during the production and compaction of an asphalt mixture. Abrasive wear is due to traffic and loading. Toughness of an aggregate can be measured as the mass percent loss of material during a test known as the "Los Angeles Abrasion test." It is performed by subjecting the coarse aggregate to impact and grinding by steel spheres. ASTM C 131, *Standard Test Method for Resistance to Degradation of Small Size Coarse Aggregate by Abrasion and Impact in the Los Angeles Machine* is the procedure for aggregates smaller than 37.5 mm and ASTM C 535, *Standard Test Method for Resistance to Degradation of Large Size Coarse Aggregate by Abrasion and Impact in the Los Angeles Machine* is the procedure for aggregates larger than 19.0 mm. Maximum loss values typically range from 35 percent for surface courses and 45 percent for base courses. Surface courses receive more traffic wear than base courses. The Los Angeles (LA) Abrasion test is primarily a measure of the resistance of coarse aggregate to degradation by abrasion and impact; however, field observations have not always shown a good relationship between LA abrasion loss and performance (Roberts *et al.* 1991). The wet micro-Deval test used in France and some other countries is a test similar to the LA abrasion method, but is run with water. Other abrasion tests include the aggregate impact value test developed by Stewart of Cape Town University, South Africa, a test now part of British Standards 812 and the German Schlagversuch impact test, a test under the European Committee for Normalisation, or CEN standards (Nicholls 1998).

Durability

Durability is similar to toughness with the additional requirement that the aggregate be resistant to degradation under the actions similar to weather such as freezing and thawing. "Soundness" is another term to describe durability. Soundness is the mass percent loss of material from an aggregate during the sodium or magnesium sulfate soundness test such as ASTM C 88, *Standard Test Method for Soundness of Aggregates by Use of Sodium Sulfate or Magnesium Sulfate*. This empirical test estimates the resistance of aggregate to in-service weathering. The test is performed by exposing an aggregate sample to repeated immersions in saturated solutions of sodium or magnesium sulfate followed by oven drying. One immersion and one drying is considered to be one soundness cycle. During the oven drying phase, salts precipitate in the permeable void space of the aggregate. Upon re-immersion the salt rehydrates and exerts internal expansive forces that simulate the expansive forces of freezing water. The test result is the total percent loss over various sieve sizes for a required number of cycles. Maximum loss values typically range from 10 to 20 percent for five cycles (The Asphalt Institute 2001). ASTM D 692, *Standard Specification for Coarse Aggregate for Bituminous Paving Mixtures*, specifies a maximum of 12 percent loss after five cycles when using sodium sulfate and 18 percent loss when using magnesium sulfate. Variations of this test include British Standard 812 magnesium sulphate soundness value and the German freeze/thaw test. Some engineers do not consider the freeze/thaw durability of an aggregate as an important performance characteristic in flexible asphalt pavements as it is in rigid concrete pavements. Aggregate base courses should also specify a soundness specification that has a maximum loss of 10–20 percent for five cycles.

Surface texture

The surface texture of an aggregate particle influences the workability and durability of an asphalt mixture. Surface texture is inherent to the aggregate geology and mechanical processing. Crushed stone and most synthetic aggregates will generally have a rough texture and river gravel will have a smooth texture. The smooth texture aggregates will be easier to coat with asphalt cement and the mixture will be easier to work, but the rougher texture aggregates will form a stronger bond with the asphalt cement and increase the mixture's strength (Roberts *et al.* 1991). The surface texture of an aggregate source is indirectly specified by requiring a minimum angularity percentage and by requiring the use of specific natural or synthetic aggregates or by eliminating the use of smooth texture aggregates such as river gravel.

Cleanliness

Cleanliness is the absence of deleterious materials from aggregates. Deleterious material is vegetation, clay lumps, clay coating, shale, mica, and other objectionable material and can be measured through ASTM C 142, *Standard Test Method for Clay Lumps and Friable Particles in Aggregates*. It also refers to the removal of excessive dust that was generated during crushing of an aggregate. Washing the aggregates removes the deleterious materials including the excessive dust. When it is physically practical and economically feasible the asphalt designer should always specify washed aggregates to be used.

Clay content is the percentage of clay material contained in the sand or fine aggregate fraction. The amount of clay or plastic fines in the fine aggregate is measured by the Sand Equivalent Test as given in ASTM D 2419, *Standard Test Method for Sand Equivalent Value*

Table 2.8 Clay content criteria

Traffic (million ESALs)	*Sand equivalent, minimum (%)*
<0.3	40
0.3 to <3	40
3 to <10	45
10 to <30	45
≥ 30	50

Source: The Asphalt Institute (2001); © The Asphalt Institute; reprinted with permission.

of Soils and Fine Aggregate. A sample of fine aggregate is mixed with a flocculating solution in a graduated cylinder and is agitated and then allowed to settle. The sand then separates from the flocculated clay particles, and the heights of clay and sand in the cylinder are measured. The sand equivalent value is the ratio of the sand height to the clay height, expressed as a percentage. Cleaner aggregate will have a higher sand equivalent value. The allowable clay content values for fine aggregate expressed as a minimum percentage of the sand equivalent value is a range of 40–50 (Table 2.8).

Large amounts of plastic fines in an asphalt mixture can cause mixture deformation and contribute to asphalt/aggregate adhesion failures. Dust, in the terms of asphalt mixtures, is the aggregate grading passing the 75 μm sieve. Dust may or may not contain plastic fines. The presence of plastic fines can be verified through the methylene blue test or by the plasticity index of the dust. The plasticity index is a measure of the degree of plasticity of the dust and can indirectly indicate the amount of plastic fines. The plasticity index of aggregate uses in asphalt mixtures and surface treatments should be less than four.

Absorption

Aggregate absorption is the amount of water an aggregate will absorb when it is soaked in water. An aggregate, especially a porous one, will also absorb the asphalt binder. Due to the viscosity of the asphalt binder and the fact that most aggregates are hydrophilic (water loving), the aggregate will absorb less asphalt binder than water. Asphalt absorption of the mixture can be calculated and is discussed in Chapter 5. If an aggregate is highly absorptive, it will continue to absorb the asphalt binder after the initial mixing at the mixing plant, thus leaving less of an asphalt binder film on its surface. The absorptive capacity of the aggregate is usually accounted for in the laboratory mixture design process. The amount of asphalt binder in a mixture less than the asphalt binder absorption of the aggregate is known as the *effective* asphalt binder content. If an aggregate absorbs too much of the asphalt binder, the mixture will become less cohesive and difficult to place and compact. The mixture will also have a low asphalt film thickness on the aggregate, which can lead to mixture durability problems. These aggregates will require a higher asphalt binder content in order to provide a mixture with acceptable performance. Highly porous aggregates should contain other desirable qualities to overcome their excessive asphalt demand. Blast furnace slag and other synthetic or manufactured aggregates are lightweight materials that are also highly porous. Their lightweight and wear resistant properties frequently outweigh their high asphalt demands make them especially useful for friction courses, bridge decks, and parking structures. Absorption is not specified by the designer, but needs to be accounted for during the mixture design and during production.

Adhesion

Adhesion is the aggregate's affinity for asphalt. Asphalt cement must be able to coat the aggregate surface and not degrade in the presence of water. Certain aggregates may be unsuitable for use in an asphalt pavement or surface treatment because of the chemical composition of the aggregate particles. Aggregates that have an excessive affinity for water, or are hydrophilic, allow the removal of the asphalt film from their surface to become what is known as "stripping" or "moisture damage." These aggregates can also be termed "lipophobic" (oil hating). These aggregates may also be called "moisture sensitive." Stripping leads to the disintegration of an asphalt pavement, usually in the form of raveling. An aggregate that is hydrophobic may also be said to be one that exhibits a high degree of resistance to stripping (Wright and Paquette 1987).

The adhesion of asphalt to aggregate and stripping is a complex phenomenon involving physical and chemical interactions. Several theories have been developed on the mechanism of adhesion and stripping, and so far none can fully explain the phenomenon and the fact that more than one mechanism may be occurring at the same time (Roberts *et al.* 1991).

Several test methods have been developed to help the designer identify asphalt and aggregate combinations that may have a tendency to have moisture sensitivity problems. Tests have been as simple as visual rating to the more recent methods such as the net absorption test developed under the SHRP in the United States. The most commonly used method is the Modified Lottman mixture test that is usually performed according to the ASTM D4867 *Standard Test Method for Effect of Moisture on Asphalt Concrete Paving Mixtures*. This test is very similar to the test that is known under the American Association of State Highway and Transportation Officials as (AASHTO) test method 283 (AASHTO T-283). The procedure is normally incorporated at the mixture design stage using the final design mixture that is compacted to an air void content (Chapter 5) of 7 percent. Some of the compacted mixture is isolated at room temperature, while the other is subject to a conditioning regime involving water saturation, freezing, and then a cycle in a 60 °C water bath. The tensile strength of the conditioned and unconditioned is computed, which finally gives a tensile strength ratio. In theory, a tensile strength ratio of 1.0 or 100 percent indicates the mixture will not have any tendency to strip. In practice, and as recommended under the SuperPave mixture design method, tensile strength ratios greater than 80 percent is considered passing. The net absorption/desorption test also allows the designer to screen aggregates and asphalt for possible stripping problems. This test method measures the chemical compatibility of the asphalt and aggregate and its disruption by water and aging. The method actually measures the initial absorption and desorption of asphalt to aggregates in the presence of water.

Antistripping or adhesion agents are generally specified to improve the performance of mixtures that are susceptible to moisture damage. An adhesion agent is usually a liquid chemical known as an amine and is added to the asphalt in a very small amount, typically 0.5 percent by weight of asphalt cement. Hydrated lime is a dry form of an adhesion agent, however, it is usually added in at a rate of 1–2 percent by weight of aggregate.

Designers responsible for small paving projects, such as parking lots, should specify a test procedure and requirement that is regionally acceptable, such as one used by a transportation agency. The designer could also blanket specify an adhesion agent to be added to the mixture, whether it is actually needed or not, as a form of pavement enhancement.

Skid resistance

The texture of the aggregate, especially the coarse aggregate, contributes to the skid resistance of the pavement. The surface texture of the pavement itself contributes significantly to the skid resistance of the pavement. The designer of wearing or surface courses for asphalt pavements considers skid resistance one of the more important properties for high-speed highways or motorways. The designer of a low-speed parking lot is mostly concerned with *slip* resistance of the mixture.

The skid resistance of aggregates is actually the polishing resistance of the aggregate. High speed and high traffic can polish the aggregate, leading to a smooth surface and a reduction in the pavement skid resistance. Parking facilities and low volume roads generally do not generate enough wear on the pavement to actually polish the aggregate. Aggregates specified for wearing courses for roadways or highways must possess some resistance to polishing. The polished stone value (PSV) is a measure of the resistance of the coarse aggregate to the polishing action of a vehicle's tire under conditions that are similar to those occurring on the pavement surface. The higher the polished stone value, the more resistant the aggregate is to polishing. There are various tests that can simulate the tire polishing of the aggregate. The tests are performed in the laboratory or can be performed full scale, on the roadway.

The British Pendulum Tester is the most common laboratory test for generating PSV. The pendulum tester is a dynamic pendulum impact type tester used to measure the energy loss when a rubber slider edge is propelled over a test surface. The resistance to movement of the rubber slider or foot by the stone is measured. The measure reading is in terms of friction. The higher the amount of friction created by the stone, the greater the skid resistance. The scale on the British Pendulum tester is a direct reading of the PSV. The higher the PSV is, the more skid resistant is the aggregate. The British Pendulum Tester would not measure the amount of friction generated by the aggregate surface and not necessarily the macro texture of the pavement surface. A laboratory scale, small wheel circular track can accelerate polishing of highly textured surfaces. The small wheel track first polishing machine polishes the aggregate prior to determining the PSV of the stone. ASTM E660, *Standard Practice for Accelerating Polishing of Aggregates or Pavement Surfaces Using a Small-Wheel, Circular Track Polishing Machine*, is one method of accelerating the polishing of the aggregate in the laboratory.

The determining of the PSV is in two parts:

1 Samples of the coarse aggregate are subjected to a polishing action in an accelerated polishing machine.
2 The state of polishing reached by each sample is determined by measuring the skid resistance using the British Pendulum Tester.

ASTM E303, *Standard Test Method for Measuring Surface Frictional Properties Using the British Pendulum Tester*, describes the operation of the pendulum tester. To determine the PSV of a coarse aggregate source, the aggregate is first subjected to accelerated polishing by a polishing machine. The pendulum tester then tests the aggregate particles. Highly skid resistance aggregates have a PSV greater than 70, while lower skid resistance aggregates have a PSV less than 40. These tests enable the engineer or designer to rank aggregate sources that would be acceptable to use on high-speed or volume wearing courses (Nicholls 1998).

Synthetic aggregates

Aggregates resulting from the modification of materials, which may involve, both physical and chemical changes, and do not occur naturally, are known as synthetic aggregates. They usually take the form of the by-product that is developed in the refining of ore, or those specially produced for the ultimate use as aggregate. Blast furnace slag is the most commonly used synthetic aggregate. It is the by-product of the smelting of iron in blast furnaces. It is non-metallic and floats on molten iron. It is drawn off at intervals and reduced in size either by quenching with water or by crushing it after it has air-cooled (The Asphalt Institute 1988). Slag is typically specified for wearing courses for high-speed pavements, since it provides excellent skid resistance. The availability of slag in the geographic region of the paving project usually determines whether or not it should be specified.

Recycled asphalt pavement (RAP) is the millings or crushing of previously laid asphalt pavements (Plate 2.2). In the United States, RAP is the largest quantity material of all items that are recycled (Plate 2.3). An asphalt mixture may contain up to 50 percent of its weight as RAP, with any higher percentages being limited by asphalt plant production abilities mixture gradation requirements, and the desirable stiffness of the inplace asphalt binder.

Plate 2.2 RAP stockpile.

Plate 2.3 Large RAP stockpile.

RAP is added to a mixture as aggregate, however its significant economic impact is the asphalt binder it contains. The hot asphalt mixing plant can recover the majority of the asphalt binder contained in the RAP during the heating process of the RAP. The recoverable asphalt cement in the RAP reduces the required amount of new asphalt cement for the asphalt mixture. Laboratory testing determines the amount of asphalt binder that is recoverable from the RAP and the required amount of new asphalt binder needed. An asphalt mixture containing 50 percent RAP could reduce its overall asphalt binder requirement by 2–3 percent.

The designer should not specifically eliminate the use of synthetic aggregates or recycled asphalt pavement. Some synthetic aggregates can possess qualities that are superior to natural aggregates. These include skid resistance and lightweight. Skid resistance aggregates, however, have limited importance to an architect designing a low-speed parking lot, while lightweight aggregates would have a very beneficial impact on a parking garage or bridge deck. The designer should specify that all synthetic aggregates and RAP also meet the same physical requirements, such as grading, that natural aggregates are required to meet. The use of RAP can lower the overall cost of an asphalt mixture, so its use may be encouraged as long as there is not a reduction in pavement performance.

Aggregate storage and handling

The quality of an asphalt mixture being produced begins with the aggregate stockpiles. The aggregate should be stored in piles based on the type and gradation of the aggregate. Generally, different sized aggregates are stockpiled individually for supplying in the asphalt mixing facility. The coarse aggregate and fine aggregate should be separated. The aggregate should be stored on a clean surface that drains moisture away from the stockpile. Where feasible, consideration should be given to storing the aggregate under a shelter. This is especially important when using RAP in hot mix asphalt that is being produced by a batch mixing plant. Excess moisture in aggregate increases drying time and reduces the production capacity of the plant. Excess moisture in aggregate used for cold mix can lead to asphalt emulsion coating difficulty and durability problems.

When constructing a stockpile, care should be given, especially with coarse aggregates, to minimize the potential for segregation. Well or continuously graded aggregates tend to segregate when placed in cone shaped stockpiles. Stockpiles should be constructed in a layer manner as opposed to dumping by a conveyer, and forming a cone shaped pile. The stockpile should be constructed in progressive horizontal layers. If building the layers on a slope, each successive layer should be constructed at a slope of three to one or less. During the construction of the stockpile the aggregates should be piled or dumped in units that are not larger than one truckload. The units should be piled so that they stay in place and do not roll down slopes. These techniques help to assure uniformity of the aggregate and significantly reduce segregation. Segregation is the separation of particle sizes from a blend of different size particles. With the cone shaped piles, the larger aggregate pieces will roll down the sides and collect at the bottom, leading to segregation. RAP should be stockpiled in the same manner as aggregate.

When handling or removing aggregate from a stockpile, care should also be given to prevent segregation. Even with carefully constructed stockpiles, the outside edge of the stockpile will generally be coarser than the inside of the pile. The final mixture gradation can be affected by how this material is removed from the stockpile and introduced into the mixing plant. The loader operator should work the entire face of the stockpile and the aggregate should be removed perpendicular to the aggregate flow into the stockpile. When loading from a stockpile built in layers, the loader operator should try to obtain each bucket by entering the lower layer at the approximate midpoint of the height of the layer and scooping up through the overlying layer. This practice results in a blend of the two layers in the bucket, reducing segregation. Removal of aggregate from the bottom of a large stockpile will often result in the coarser aggregate particles rolling down at the face of the stockpile and gathering at the bottom, increasing segregation problems (TRB 2000).

Chapter 3

Traffic and environment

The primary purpose of an asphalt pavement is to support vehicles, may be the pavement is a four-lane freeway or a parking lot. The type and amount of vehicles has a significant impact on several of the variables that are used in designing and constructing an asphalt pavement. These variables include pavement thickness, material selection, and layout.

Traffic

Traffic is measured through two parameters, traffic capacity and level of service. The traffic capacity of an asphalt pavement or any transportation facility is its ability to accommodate vehicles, whether moving or stationary. It is a measure of the supply side of a transportation facility. The level of service is a measure of the quality, not quantity, of the traffic flow.

The objective of capacity analysis is the estimation of the maximum number of vehicles that can be accommodated by a pavement or facility within a specified time period. Facilities generally operate poorly at or near their capacity, they are rarely planned to operate in this range. Accordingly, traffic capacity analysis also provides a means of estimating the maximum amount of traffic that can be accommodated by a pavement or facility while maintaining prescribed operational qualities. Capacity analysis is a set of procedures for estimating the traffic-carrying ability of a pavement over a range of defined operational conditions. Operational conditions are defined through the level of service. Levels of service are defined for each type of facility and are related to amounts of traffic that can be accommodated at each level (TRB 1994).

The vehicle capacity or volume of a facility or pavement is defined as the maximum hourly rate at which vehicles can reasonably be expected to traverse a point or uniform section of a roadway or parking area during a given time period under prevailing roadway traffic and control conditions. The design hourly volume (DHV), is a future hourly volume that is used for capacity design. For a highway facility it is the thirtieth highest hourly volume of the design year. For parking facilities, it is the tenth busiest hour of the year. Traffic volumes are much heavier during certain hours of the day or year and it is for these peak hours that the roadway is designed. It is also unreasonable to build a facility that is under-utilized 90 percent of the year. Another measurement of vehicle volume is the average daily traffic (ADT). The ADT is the number of vehicles that pass a particular point on a roadway during a period of 24 consecutive hours averaged over a period of 365 days. It is usually not feasible to make continuous traffic counts 365 days a year along a roadway or parking area. ADT values are usually derived from statistically based sampling of the traffic volume.

Table 3.1 Roadway level of service designations

Level of service letter	*Description*
A	Free flow, ideal conditions
B	Reasonably free flow, delay is not unreasonable
C	Operation stable, but becoming critical, so delays at traffic signals
D	Approaching unstable flow, delays at intersections becoming extensive
E	Unstable flow, no traffic dissipation, continuous backup on approaches to intersections
F	Forced flow, vehicles backup, stop and go operation, high approach delays

ADT is the fundamental value used by designers to help determine the proper asphalt pavement thickness and the proper materials.

The concept of levels of service uses qualitative measures that characterize operational conditions within a traffic stream and their perception by motorists. The descriptions of individual levels of service characterize these conditions in terms of such factors as speed and travel time, freedom to maneuver, traffic interruptions, and comfort and convenience. Six levels of service are defined for each type of facility for which analysis procedures are available. They are given letter designations, from A to F, with the level of service, A, representing the best operating conditions and the level of service, F, representing the worst (Table 3.1). For most design and planning purposes, level of service flow rates D or C are usually used because they ensure a more acceptable final quality of service to facility users (TRB 1994). The highway agency should strive to provide the highest level of service feasible. In heavily developed sections of metropolitan areas, conditions may necessitate the use of level of service, D, for freeways and arterials, but such use should be rare and at the least the level of service, C, should be striven for. For some urban and suburban highways, conditions may necessitate the use of level of service, D (AASHTO 1995).

The level of service for a parking area has similar qualitative needs, but are rated more specifically by the application the parking area is used for. Five levels of service are used with the level of service, A, representing very high turnover rates, to the level of service, D, representing very low turnover rates. The fifth level of service, G, represents applications such as shopping centers (Plate 3.1) that use a significant amount of shopping carts, that can cause difficulties with vehicular traffic. Table 3.2 gives descriptions for each parking area level of service.

Pavement design requires a prediction in the amount of loading a pavement will receive. This loading is usually in the form of vehicular traffic. Vehicular traffic can be any mixture of passenger cars and trucks. Primary concern in designing the asphalt pavements is the number of and weight of axle loads expected to be applied to the pavement during a given period of time. These axle loads can vary significantly depending on the type of vehicle. The method of simplifying this variability is to equate all the axle weights to one common or equivalent axle. This equivalent single axle load (ESAL) is a single axle load of 80 kN (18,000 pounds when used in United States design guides). The 18,000 pound single axle load was selected for use in the United States AASHTO pavement design guide that was initially published in 1972. Research done at the original AASHO Road Test in Ottawa, Illinois has shown that the effect on pavement performance of an axle load of any mass can be represented by the number of equivalent 80 kN single axle load (ESALs) applications.

Plate 3.1 Parking lot with shopping cart corrals.

Table 3.2 Parking application service designations

Letter	*Application*
A	Very high turnover rates such as convenience stores, or areas that have greater than 20% truck traffic
B	High turnover rates such as general retail
C	Medium turnover rates such as airport, hospital, or residential parking
D	Low turnover rates such as employee parking
G	Grocery stores and other applications that use a large amount of shopping carts

For example, one application of an 89 kN single axle is equal to 1.5 applications of an 80 kN single axle. Conversely, it takes almost four applications of a 58 kN single axle to equal one application of an 80 kN single axle (The Asphalt Institute 1999). A small car with a single axle weight of 4.5 kN would be equal to 18 applications of the 80 kN single axle or ESAL. The significant point from these examples is that the larger vehicles, such as trucks with multiple axle loads of 80 kN or greater, have the greatest impact on pavement design and performance. When designing the thickness of an asphalt pavement, most designers concentrate on the truck traffic for controlling the thickness with automobile traffic loading having just a slight impact.

For the purposes of asphalt pavement thickness design for roadways, traffic capacity will be in the terms of the present or future ADT. The ADT will have a further estimate as to the number of vehicles of different types, such as passenger cars, buses, single-unit trucks,

Table 3.3 Estimated truck distributions and axle configurations (United States)

Axle configuration	*Trucks (%)*					
	Urban			*Rural*		
	Interstate	*Freeway*	*Primary*	*Interstate*	*Primary*	*Arterial*
Single-unit trucks						
2-axle, 4-tire	52	66	67	43	60	71
2-axle, 6-tire	12	12	15	8	10	11
3-axle or more	2	4	3	2	3	4
All single units	66	82	85	53	73	86
Multiple-unit trucks						
4-axle or less	5	5	3	5	3	3
5-axle	28	13	12	41	23	11
6-axle or more	1	1	1	1	1	1
All multiple units	34	18	15	47	27	14
All trucks	100	100	100	100	100	100

and multiple-unit trucks of different types expected to use the roadway. This mixed stream of different axle loads and axle configurations derived from the ADT will be converted into ESALs to the sum over the design period of the pavement. The design period is the number of years from the initial application of traffic until the first planned reconstruction or overlay. In a parking application, the design period may be the useful life, financial, or otherwise, of the facility the parking area is associated with. For example, if the fast food restaurant only has a useful business life of 8 years before rebuilding or relocating, the design period of the parking lot should also be 8 years. Design period should not be confused with pavement life. Key considerations which influence the accuracy of traffic estimates and which can significantly influence the life cycle of the asphalt pavement are: the correctness of the load equivalency values used to estimate the relative damage induced by axle loads of different mass and configurations; the accuracy of traffic volume and weight information used to represent the actual loading projections; and the prediction of ESALs over the design period (AASHTO 1993). Determining the mixed stream of different vehicle types and axle loads is obtained from highway agency weight stations. When this information is not available, Table 3.3 can assist in estimating a percentage distribution of trucks and their axle configurations. An accurate estimation of total truck traffic will need to be determined in conjunction with the ADT.

Facility designations

A freeway is a divided corridor with at least two lanes in each direction that operates in an uninterrupted flow, without fixed elements such as traffic signals, stop signs, and at-grade intersections. Access points are limited to ramp locations. Since grade turns and other features can change along a freeway, traffic capacity is evaluated along shorter segments. Multi-lane and two-lane highways contain some fixed elements and access points from at-grade intersections, though relatively uninterrupted flow can occur if signal spacing is greater than 3 km. Where signal spacing is less than 3 km, the roadway is classified as an

Table 3.4 Roadway designation and approximate ADT

Road designations	*Approximate ADT*
Local road	≤2,000
Collector road	2,000–12,000
Arterial	12,000–40,000
Freeway	≥30,000

Table 3.5 Average percentage of total truck traffic in the design lane

Number of traffic lanes (two direction)	*Trucks (%) in design lane*
1	100
2	50
3	45
4	30–45
5	35–40
≥6	25–35

arterial, and flow is considered to be interrupted. Divided highways have separate roadbeds for the opposing directions, whereas undivided highways do not. Smaller roadways are classified as local roads and streets. All roadways can be classified as urban, suburban, or rural, depending on the surrounding population density. Urban areas are considered to have populations greater than 5,000. Rural areas are outside the boundaries of urban areas (Lindeburg 1999). Table 3.4 shows some general guidelines on ADT assigned to road classifications.

The design lane is the lane where the greatest number of ESALs is expected. This would normally be either lane of a two-lane roadway or the outside lane of a multi-lane roadway. The design lane of a parking facility could be the entrance or exit areas, loading dock areas, or any other area where the largest amount of ESALs can be expected. Under some conditions more trucks may travel in one direction than the other or loaded trucks will travel in one direction and empty trucks in the other direction. Table 3.5 may be used for determining the relative proportion of trucks to be expected for the design lane.

Traffic growth, or no growth, must be considered when determining the amount of ESALs for determining the thickness of a pavement. An example of no growth would be an employee parking lot or pay to park facility, where the capacity is expected at 100 percent and the parking lot is constrained by property boundaries and the number of parking spaces. Traffic history for comparable roadways and facilities may be available from local and regional planning programs. Historical growth data may be used as an estimate for future growth, especially when there is no knowledge of any future events that may cause a significant increase or decrease in traffic levels. For example, when designing a rural highway close to an urban area, future suburban sprawl or subdivisions should be considered in addition to any typical growth rates. Growth rates for this type of example have been in the range of 6–11 percent compounded annually. Normal traffic growth in the United States has been 3–5 percent. If possible, determine separate growth percentages for passenger vehicles and trucks. This is especially significant when knowledge of a future heavy industrial complex is planned to use the roadway. When applying growth rates or factors to future traffic estimates, also be sure that the overall traffic capacity of the roadway is not exceeded. Traffic growth rates are used by applying the growth rate percentage compounded annually to the first year traffic estimate or ESALs. When there is knowledge of a significant future event that will occur at set date in the future with a different growth rate, the design period would be broken into two separate periods – applying the second growth rate to the new second design period first year traffic estimate as shown in the example following equation 3.1. The fundamentals of uniform series compounding in economic analysis also apply in determining the traffic levels of a pavement. Equation 3.1 may be used to calculate future traffic levels or a scientific calculator may also be helpful.

$$F = P\frac{(1+i)^n - 1}{i} \tag{3.1}$$

where F is the future traffic at end of year, n; P, the first year traffic estimate; i, the traffic growth rate; n, the year or years of design period.

Example 3.1

An engineer is designing a two-lane rural roadway that has a traffic growth rate of 4 percent. The design period is 10 years and the annual equivalent 80 kN single axle load (ESALs) is 8,000 (mostly cars). It is known that 6 years later, suburban sprawl will cause the traffic growth rate to increase to 12 percent. What is the total ESALs for the design period?

Solution

Divide the problem into two parts. The first growth period is 6 years long.

$$F_1 = 8{,}000[(1+4\%)^6 - 1]/4\% \Rightarrow F_1 = 8{,}000(6.633) = 53{,}064 \text{ ESALs}$$

The second growth period is 4 years long at 12 percent,

$$F_{total} = 53{,}064[(1+12\%)^4 - 1]/12\% \Rightarrow F_{total} = 53{,}064(4.779) = 253{,}590 \text{ ESALs}$$

253,590 ESALs is the total loading for the 10-year design period.

Estimating equivalent single axle loads

The design ESAL is determined by multiplying the number of vehicles in each weight class by its appropriate truck factor and obtaining the sum of the products (equation 3.2).

$$\text{Design ESALs} = \Sigma(\text{number of vehicles in each weight class} \times \text{truck factor}) \tag{3.2}$$

The truck factor is the number of 80 kN single axle load applications contributed by one passage of a vehicle. The truck factors are determined from axle-weight distribution data using load equivalency factors (Table 3.6). Load equivalency factors are the number of 80 kN single axle load applications contributed by one passage of an *axle*. The load equivalency factor increases approximately as a function of the ratio of any given axle load to the standard

Table 3.6 Load equivalency factors

Gross axle load	*Load equivalency factors*		
kN	*Single axle*	*Tandem axle*	*Triple axle*
4.45	0.00002		
8.9	0.00018		
17.8	0.00209	0.0003	
26.7	0.01043	0.001	0.0003
35.6	0.0343	0.003	0.001
44.5	0.0877	0.007	0.002
53.4	0.189	0.014	0.003
62.3	0.36	0.027	0.006
71.2	0.623	0.047	0.011
80.0	1.0	0.077	0.017
89.0	1.51	0.121	0.027
97.9	2.18	0.18	0.04
106.8	3.03	0.26	0.057
115.6	4.09	0.364	0.08
124.5	5.39	0.495	0.109
133.4	6.97	0.658	0.145
142.3	8.88	0.857	0.191
151.2	11.18	1.095	0.246
160.0	13.93	1.38	0.313
169.0	17.20	1.70	0.393
178.0	21.08	2.08	0.487
187.0	25.64	2.51	0.597
195.7	31.0	3.0	0.723
204.5	37.24	3.55	0.868
213.5	44.50	4.17	1.033
222.4	52.88	4.86	1.22
231.3		5.63	1.43
240.0		6.47	1.66

80 kN single axle load raised to the fourth power (equation 3.3). For example, the load equivalency factor of a 53 kN single axle is given as 0.189 while the load equivalency factor for an 89 kN single axle is 1.51. Thus, the 89 kN load is eight times as damaging as the 53 kN load, that is $(89/53)^4 = 7.95$.

$$\text{Load equivalence factor} = (W/80)^4 \tag{3.3}$$

where W is the axle load in kN. The load equivalency factors are based on observations made at the AASHO Road Test in Ottawa, Illinois, and United States. In this respect, some limitations should be recognized, such as limited pavement types, load applications, age, and the environment (AASHTO 1993).

The ESAL of an axle is determined by multiplying the weight of the axle times the load equivalency factor for the corresponding axle obtained from Table 3.6. For example, a 71.2 kN single axle has a corresponding load equivalency factor of 0.623. Multiply 71.2 kN by 0.623 to give an equivalent single axle load of 44 kN. This calculation is carried out for each axle of the vehicle and then added together to give the total ESALs for the vehicle or what is known as the truck factor.

Example 3.2

A 170 kN gross weight vehicle has three axles, one single and two tandem axles. The single axle has a weight of 30 kN and each of the tandem axles has a weight of 70 kN. Determine the truck factor.

Solution

Determine the load equivalency factor for each axle from Table 3.6. It is acceptable to interpolate or approximate close values. The load equivalency factor of the 30 kN single axle is 0.02 and the load equivalency factor for each of the tandem axles is 0.047.

The truck factor in ESALs is: $\sum(0.02) + (0.047) + (0.047) = 0.114$

An average truck factor is calculated by multiplying the number of axles in each weight class by the appropriate load equivalency factors and dividing the sum of the products by the total number of vehicles (equation 3.4).

$$\text{Average truck factor} = \frac{\sum (\text{number of axles} \times \text{load equivalency factor})}{\text{number of vehicles}} \tag{3.4}$$

Truck factors can be determined on each individual vehicle that may be expected to use a facility or they can be determined for specific combinations of vehicles or truck types such as three-axle single units and then determining a truck factor or all the ESALs can be determined and then be divided by the total vehicle count to give an average truck factor.

Advance knowledge of specifically what type of vehicles, their axle configurations, and their weight is usually not known with the exception of applications such as a truck terminal or other specific use facilities. In the United States, the Federal Highway Administration (FHWA) compiles truck data from weight stations and other collection points to determine a typical distribution of truck factors. Weight station information represents only a sample of the total traffic stream with measurements taken at a limited number of locations and for limited periods of time. Such information must be carefully interpreted when applied to specific projects. This information is still useful however, when it is not feasible to determine an actual truck factor (Table 3.7).

The tire contact pressure of a vehicle, especially heavy vehicles, also has an impact on determining the thickness of an asphalt pavement and can be accounted for by modifying the initial design ESALs. Today tires pressures are significantly higher than the 480 KPa pressure used in the development of the load equivalency factors. Truck tire contact pressures typically equal about 90 percent of the tire inflation pressure. To account for the additional stress caused by high contact pressures, the initial design ESAL is multiplied by the ESAL tire adjustment factor obtained from Table 3.8 for each individual vehicle type or for the average truck. An initial estimate of the asphalt pavement thickness needs to be determined to obtain the correct tire adjustment factor.

Load equivalency factors and truck factors are based on performance equations developed from the AASHO Road Test, which may not apply directly to some urban streets,

Table 3.7 Typical United States truck factor distributions

Axle configuration	*Typical truck factor*					
	Urban			*Rural*		
	Interstate	*Arterial*	*System average*	*Interstate*	*Collector*	*System average*
Single-unit trucks						
2-axle, 4-tire	0.002	0.006	0.01	0.003	0.017	0.01
2-axle, 6-tire	0.17	0.23	0.19	0.21	0.41	0.30
3-axle or more	0.61	0.76	0.82	0.61	1.26	0.86
All single units	0.05	0.04	0.10	0.06	0.12	0.08
Multiple-unit trucks						
4-axle or less	0.98	0.46	0.69	0.62	0.37	0.64
5-axle	1.07	0.77	0.90	1.09	1.67	1.36
6-axle or more	1.05	0.64	0.92	1.23	2.21	1.63
All multiple units	1.05	0.67	0.79	1.04	1.52	1.25
All trucks	0.39	0.07	0.23	0.52	0.30	0.32

Table 3.8 ESAL tire adjustment factor

Asphalt pavement thickness (mm)	*Tire contact pressure* (KPa)		
	480	700	1,050
	Adjustment factors		
100	1.2	1.8	4.0
125	1.1	1.7	3.0
150	1.0	1.5	2.25
175	1.0	1.4	1.75
200	1.0	1.3	1.5
225	1.0	1.2	1.4
250	1.0	1.1	1.3

county or rural roads, parkways, or parking lots. For city streets, service vehicles such as refuse trucks, buses, and delivery trucks will develop the major traffic loads. Load equivalency values for such vehicles are generally not well estimated by load equivalency factors developed from truck weighing stations. In urban streets and parking facilities, a significant effort should be made to determine the actual axle loads and frequencies typical of vehicles operating on those pavements (AASHTO 1993). Residential streets and parking facilities have relatively high volumes of automobile traffic and few trucks. An asphalt pavement thickness design using load equivalency factors given in Table 3.6 can result in a pavement that is too thin to withstand occasional truck traffic such as snowplows, refuse trucks, buses, and moving vans. In many instances, it may be more practical to group traffic capacity by similar ESALs with a general character or description of the traffic loading (Table 3.9). The designer may not have the actual traffic projections or the ability to determine a specific ESAL count, and these general groupings can be used to get a general estimate of traffic capacity. Vehicles are also grouped into three generic categories (Table 3.10). This method of ESAL estimation is so common, that thickness design methods have been developed

Table 3.9 Traffic design index categories

Design index	*General description*	*Daily ESAL*
DI-1	Light traffic, few vehicles heavier than passenger cars, no regular use by group 2 or 3 vehicles	≤5
DI-2	Medium-light traffic, maximum of 1,000 ADT, including not over 10% group 2, no regular use by group 3 vehicles	6–20
DI-3	Medium traffic, maximum of 3,000 ADT, including not over 10% group 2 and 3 with a maximum of 1% group 3 vehicles	21–75
DI-4	Medium-heavy traffic, maximum of 6,000 ADT, including not over 15% group 2 and 3 with a maximum of 1% group 3 vehicles	76–250
DI-5	Heavy traffic, maximum of 6,000 ADT, may include 25% group 2 and 3 with a maximum of 10% group 3 vehicles	251–900
DI-6	Very heavy traffic, over 6,000 ADT, may include over 25% group 2 and 3 vehicles	901–3,000

Table 3.10 Vehicle groupings

Vehicle group	*Description*
1	Passenger cars, pick-up trucks, panel trucks
2	Loaded 2-axle trucks such as refuse trucks, school buses, fire trucks, delivery trucks
3	Trucks or combination vehicles having 3-axle, 4-axle, or more loaded axles

Table 3.11 Light traffic groupings

Traffic class	*Description*	*Maximum heavy trucks per month*
I	Residential driveways, residential parking stalls, parking lots for automobiles and pick-up trucks	<1
II	Residential streets without regular truck traffic, scheduled bus service; traffic consisting of automobiles, home delivery trucks, refuse trucks, etc.	60
III	Collector streets, delivery lanes for retail facilities such as shopping centers, up to 10 single-unit or 3-axle semi-trailer trucks	250
IV	Heavy trucks, up to 75 5-axle semi-trailer trucks per day, equivalent trucks may include heavy loaded 3-axle and 4-axle trucks such as dump trucks	2,200

around these traffic classifications. These methods, however, are usually used for relatively low capacity pavements and parking facilities.

Parking facilities and local roads can be further defined into four classes, I through IV (Table 3.11). This information is especially useful for parking lots and residential streets and in further defining design indexes DI-1 and DI-2. Some design procedures also distinguish these four classes. Table 3.11 allows the designer to ignore the number of automobiles and just determine the loading by the limiting truck traffic.

Table 3.12 Recommended residential parking stalls

Type dwelling	*Number of vehicle stalls per dwelling*
Town or garden home	2
Multi-family – studio	1.25
Multi-family – one bedroom	1.5
Multi-family – two bedroom	1.75
Multi-family – ≥ three bedrooms	2

There are several approaches to estimating the amount of traffic that uses a parking facility. The overall use and the type of business the parking lot serves dictate the amount of traffic and how the traffic is determined. In some applications, the number of vehicle stalls in the parking lot can be used to determine the ultimate amount of traffic. This type of estimation is best suited for facilities that have a parking service level designation of letter D or possibly C. These parking facilities have a low turnover rate, such as employee parking lots or residential parking lots. The traffic estimate may be completed by first determining the amount of vehicle stalls required, account for the amount of vehicles arriving in the morning and leaving in the afternoon (employee parking) with an estimate on the percentage of those vehicles leaving for the noon meal period. Any truck traffic can also be included. In employee parking lots, the number of vehicle stalls is usually determined by the number of employees working any given day with consideration given to car-pooling. In residential parking facilities, which are usually used, for apartments or condominiums, Table 3.12 may be used to determine the number of vehicle stalls. When determining total vehicle use, consideration must also be given for any refuse or moving trucks that may also travel through the parking lot.

When there is significant mix of truck traffic and employee parking, such as in a manufacturing plant, it is more practical and economical to evaluate and design the parking facilities in sections. In this manner, the designer would not over design the thickness of the employee parking areas or under design the thickness of the truck staging or loading areas. Joint use areas such as driveways and approach aprons (Plate 3.2) would require the estimate of the total vehicle traffic, both passenger vehicles and trucks.

Parking facilities servicing shopping centers or convenience stores are more difficult to estimate the number of vehicles that may use them at any one time during the course of the year. Shopping centers may be at capacity during holiday periods, such as Christmas and have a relatively light amount of traffic during the day of week, such as at 2 p.m. on a Wednesday during the summer. The number of vehicle stalls necessary to meet the peak demand may greatly exceed those that are necessary for a typical day. The design hour being the 10 busiest hours of the year eliminates some of the absolute peaks of demand. A shopping center vehicle stall requirement is usually set at a number of stalls per square meter or square foot of the shopping center. A minimum number of vehicle stalls per square meter of rental space can also be part of a business plan in marketing the easy accessibility to the shopping center. An overestimate may be an aid in marketing, but empty vehicle stalls should not be accounted for in thickness design, which may end up in an over-engineered pavement.

Plate 3.2 Common approach driveway.

The maximum convenient parking capacity is usually the primary goal in determining parking capacity. The following criteria will assist in making the best use of the available space:

- Use rectangular parking areas when possible.
- Make the longer sides of the area for parking parallel.
- Line the perimeter of the area with parking stalls.
- Design the traffic lanes to serve two rows of parking stalls.

A convenience store's high traffic periods generally occur in the early morning such as 6–8 a.m. and in the early evenings, such as 4–7 p.m. The amount of vehicle stalls in any high turnover facility does not give an indication of the amount of the vehicles that may actually use the parking lot during the course of the day. The vehicle traffic in high turnover facilities can be estimated by actually counting the vehicles at a similar facility or using the business plan estimates on the amount of business traffic that is expected during the course of the day, along with any growth plans, for estimating the vehicle traffic.

Example 3.3

A condominium parking lot is being designed for a building containing 20 three-bedroom units, 55 two-bedroom units, and 35 one-bedroom units. A separate drive will be built for use by moving vans and refuse truck access. Determine the traffic design index.

Solution

This is a three-section problem, since a separate drive is being built for any heavy truck use. One section is just the resident or passenger vehicle parking area, another section is the separate drive for the refuse and moving van use and the third section is the common use approach entrance.

Section one, resident parking: Table 3.12 indicates the number of vehicle stalls required and thus the expected passenger vehicle use.

$$\sum \text{passenger stalls} = (20 \times 2) + (55 \times 1.75) + (35 \times 1.5) = 188.75 \sim 190 \text{ stalls}$$

Assuming every vehicle leaves and returns each day, a maximum ADT for parking section is 360 passenger vehicles. The 360 passenger vehicle ADT in ESALs is:

$$\sum \text{ADT ESALs} = 2 \text{ axles} \times 0.00018 \text{ (from Table 3.6)} \times 360 = 0.13$$

From Table 3.9, the resident parking traffic design index is DI-1.

Section two, refuse truck drive: an assumption of twice per week of refuse truck service and twice per month of moving van visits is made. From Table 3.10 both vehicles fit group 2. From Table 3.11 the refuse truck drive meets traffic class II. The traffic design index is either DI-1 or DI-2. Table 3.7 gives a typical truck factor for these types of vehicles as 0.19. If total truck visits per month is 10, the monthly ESALs is $10 \times 0.19 = 1.9$ or a daily ESAL of 0.06.

The refuse truck drive design index is DI-1.

Plate 3.3 Refuse container pad.

Section three, common approach entrance: the approach entrance is used by both section one and two, so the daily ADT or ESALs is the sum of the two.

$$\text{Daily ESAL} = 0.13 + 0.06 = 0.19$$

The common approach entrance design index is DI-1.

A design index of DI-1 may be used for the entire parking facility. If there were more truck traffic, both the refuse truck drive and the common approach entrance would have become a traffic design index of DI-2.

The preceding example of breaking the design project into several distinguishable parts can be applied to any type of parking facility or also to streets and highways. The overall goal is that the pavement is not under or over designed to serve for its intended traffic loading or purpose. A parking facility with variable traffic design indexes will also have a variable thickness of asphalt pavement depending on where it is required. For example, the main parking area for passenger vehicles may be 75 mm thick, the common entrance may be 125 mm, and the pad for refuse containers may be 150 mm thick (Plate 3.3)

The ability to correctly select traffic loading whether the asphalt pavement be a highway or a parking facility has consequences on thickness design, mixture design, material selection, geometric design, and maintenance activities. The designer should not be overly conservative in that the asphalt pavement is uneconomical or underestimate the traffic loading and use a design that has a low pavement life.

Environment

In addition to traffic loading, the environment is the other criterion that has a great effect on the life and design of an asphalt pavement. The environment has an impact on material selection, thickness design, and the geometric design of a pavement. The two critical areas of the environment that the designer is concerned with are temperature and precipitation. Temperature influences the selection of which grade asphalt binder that would be used in the asphalt pavement. Temperature can also influence the structural design of the asphalt pavement, which will be addressed in Chapter 4. The SHRP performance grade (PG) binder specification is based on specifying an asphalt binder depending on which climatic temperature range the asphalt pavement is placed in service. Climatic temperatures can also be

Table 3.13 Asphalt grade selection based on ambient temperature

Ambient annual air temperature (°C)	*Asphalt grade[1] selection*		
	Penetration	*Viscosity*	*Aged residue*
<7	120–150	AC-5	AR-2000
7–16	85–100	AC-10	AR-4000
16–22	60–70	AC-20	AR-8000
<22	40–50	AC-40	AR-16000

Note

1 Excluding SHRP PG binder system.

used in selecting the proper asphalt binder that is graded by other systems such as penetration, viscosity, and aged residue. However, these systems are not as temperature specific as the performance grade system nor do they address low service temperatures, generally below 0 °C. Table 3.13 is a general guide for selecting asphalt binders that have been graded or specified by the penetration, viscosity, or aged viscosity systems. This table should be used when specifying that a performance grade asphalt binder is not practical or available.

The advantages of the PG asphalt binder system is that the designer can specifically select an asphalt binder that has met test requirements at both the high and low temperatures that will occur during the asphalt pavement's service life. The PG asphalt binder is designated and specified at the average seven-day maximum pavement temperature (°C) and the minimum pavement temperature (°C). For example, theoretically, a PG64-22 would be used in an asphalt pavement that has an average seven-day maximum pavement temperature of 64 °C and a low pavement temperature of −22 °C.

Reliability

One point to consider is the reliability of the ambient or pavement temperatures. For example, a statement is made that the average daily high air temperature during the hottest week next year will not exceed 32 °C. In the hottest seven-day period of a typical year, the average is 32 °C. So the previous statement can only be made with 50 percent reliability. Now, if in one year in 50 years, the hottest week of the year will be 34 °C, we can then say with 98 percent reliability that the temperature will not get above 34 °C. The same reliability statements can be made concerning the low temperatures. If the coldest day of the year is typically −26 °C and one in 50 years the air temperature gets down to −32 °C, that would be the coldest temperature in 50 years. Or in other words there is only a 2 percent chance the temperature will get below −30 °C this year (Kriech 1994). In another example, consider summer air temperatures in Chicago, Illinois, which has a mean seven-day maximum of 32 °C and a standard deviation of 2 °C. In an average year, there is a 50 percent chance that the seven-day maximum air temperature will exceed 32 °C. However, assuming a normal statistical frequency distribution, there is only a 2 percent chance that the seven-day maximum will exceed 36 °C (numerical mean plus two standard deviations), therefore a design air temperature will provide 98 percent reliability. Reliability measurements allow the designer to assign a degree of risk to the high and low pavement temperatures used in selecting the asphalt binder grade. Reliability used in selecting asphalt binders is defined as the percent probability in a single year that the actual temperature (one-day low or seven-day high) will not exceed the design temperatures. A higher reliability means lower risk (The Asphalt Institute 1997). A minimum of 98 percent reliability should be used for all grade selections when the pavement's 20-year design life exceeds a traffic level of 300,000 ESALs.

Pavement temperature derivation

In the PG asphalt binder system, design temperatures to be used for selecting an asphalt binder grade are the pavement temperatures, not the air or ambient temperatures. For the surface course or layer of an asphalt pavement, the pavement design high temperature is actually defined at a depth 20 mm below the pavement surface. The pavement design low temperature can be at any part of the pavement thickness. Typically the designer does not have access to actual pavement temperatures, but does have access to weather station data, etc. to establish a seven-day average high air temperature and a maximum low air temperature. For United States and Canadian projects, the SHRP SuperPave® design software has a database of 6,092 reporting weather stations to assist the designer in selecting the proper asphalt binder grade. Since air temperature data is generally attainable, equations have been developed to convert air temperatures to pavement surface or subsurface temperatures. These equations were developed using theoretical analyses of actual conditions performed with models for net heat flow and energy balance and assuming typical values for solar absorption (0.90), radiation transmission through air (0.81), atmospheric radiation (0.70), and wind speed (4.5 m/sec) (The Asphalt Institute 1997). The high temperature pavement design temperature (20 mm below the surface) is given in equation 3.5

$$T_{20\,mm} = (T_{air} - 0.00618\ Lat^2 + 0.2289\ Lat + 42.2)(0.9545) - 17.78 \tag{3.5}$$

where $T_{20\,mm}$ is the pavement design high temperature, °C at 20 mm depth; T_{air}, the seven-day average high air temperature, °C; Lat, the geographical latitude of the project in degrees.

The pavement design low temperature at any depth of the pavement is calculated as a function of the low temperature using equation 3.6.

$$T_{pav} = -1.56 + 0.72\ T_{air} - 0.004\ Lat^2 + 6.26\ \log_{10}(H + 25) - Z(4.4 + 0.52\,\sigma_{air}{}^2)^{1/2} \tag{3.6}$$

where T_{pav} is the pavement design low temperature °C, at depth, H; T_{air}, the low air temperature, °C; Lat, the geographical latitude of the project in degrees; H, the depth to pavement surface, mm; σ_{air}, the standard deviation of the mean low air temperature °C; Z, from a standard normal distribution table, Z = 2.055 for 98% reliability.

Equation 3.6 can be used for any depth of the asphalt pavement layer including the surface by allowing H to equal zero. Allowing variable depths to be entered into the equation assists the designer to properly select the desired asphalt binder for the specific pavement layer used in the application. The significant impact is in designing heavy-duty pavements where total structural thickness may be 200 mm or more. An asphalt binder selected for the surface course may be an over designed asphalt binder for a base course. Entering the latitude of the pavement project in both equations 3.5 and 3.6 gives some accuracy in predicting pavement temperatures at various latitudes.

Asphalt binder selection

Generally, two reliability values are used in selecting asphalt binders, 50 percent and 98 percent. Normal distribution curves are used when determining the reliability of air temperatures. To achieve a reliability of at least 50 percent and provide for an average maximum pavement design temperature of, for example, at least 58 °C, the high temperature asphalt binder grade of PG 58 would match the design temperature. If the pavement design low temperature, happens to be −22 °C, for the same reliability of 50 percent, the low temperature asphalt binder grade would be −22 °C. If the desired reliability is changed to 98 percent for this example, it is necessary to select a high temperature asphalt binder grade of PG 64 in order to protect the asphalt pavement on the 2 percent chance when the pavement temperature exceeds 58 °C. Likewise the low temperature asphalt binder grade of −28 °C would be selected to protect the pavement on the 2 percent chance when pavement temperatures occur below −22 °C. For this example, if 50 percent reliability is desired, an asphalt binder grade of PG 58-22 would be selected and for 98 percent reliability, an asphalt binder grade of PG 64-28. In the case of the PG 64-28, the actual reliability exceeds 99 percent, because of the rounding up to standard PG grades in six-degree increments. Rounding up due to standard PG grade selection can introduce some overdesign or conservatism into the asphalt binder selection process. Another source of overdesign occurs during the actual asphalt binder PG testing and classification. For example, a specific asphalt binder may pass all of the criteria when tested at lower or higher temperatures, it will still be classified and named by conservatively rounding down or up to the next six-degree increment of the grading system. The net result of both testing and reliability factor selection allows a significant factor of safety to be included in the asphalt binder selection process. For example, the previous PG 58-22 selected for a 50 percent reliability may actually have been graded as a PG 60-23, had such a grade existed. Designers using the PG asphalt binder system should recognize that the considerable factors of safety are already included in the selection process. It may not be necessary or cost effective to require indiscriminately high values of reliability or abnormally conservative high or low temperature grades (The Asphalt Institute 1997).

In summary, selecting a PG asphalt binder starts with determining the air temperatures where the asphalt pavement is located, converting those to pavement temperatures followed by selecting the proper asphalt binder. Reliability selection is mostly based on the designer's comfort level with the risk that pavement temperatures may exceed the asphalt binder grade selected. In moderate climates or where an asphalt pavement application does not require a significant margin of safety, a reliability of 50 percent may be perfectly adequate. An example might be a subdivision cul-de-sac in Seattle, Washington. A reliability of 98 percent or higher may be desirable for an expressway application in areas with significant temperature extremes or high traffic loading such as Mexico City, Phoenix, or Boston. The designer may also wish to specify asphalt binders that are prevalent in the geographical area that the project is located. For example, after calculating the pavement design temperatures, it is determined that a PG 64-10 is best suited for the project. However, only PG 64-22 and PG 58-28 are available within 500 km of the project. The designer changes the required asphalt binder to a PG 64-22. In all cases, good practical engineering judgment should govern these types of decisions.

Asphalt binder grade selection for traffic

The asphalt binder selection procedures given in Table 3.13 and as described for the temperature based PG system are the basic procedures for typical asphalt pavement loading conditions. Under these conditions it is assumed that the pavement is subjected to a design number of fast, repeated, and transient loads. For the high temperature design procedure, which is controlled by specified properties related to permanent deformation, the speed of traffic loading has an additional affect on performance. Pavement rutting studies using the French (PRT) rut tester, the SST Repeated-Shear Constant-Height test and the United States FHWA Accelerated Loading Facility (ALF) all support the conclusion that additional asphalt binder stiffness can significantly reduce rutting under heavy, slow traffic. All three of these rut prediction tools suggest that the high temperature PG grading should increase one grade for significant increases in traffic ESALs. An additional shift in the selected high temperature binder grade is used to account for slow and standing load applications. Higher maximum temperature grades are used to offset the effect of slow or standing loading. For slow-moving design loads, generally traveling less than 70 km/h, the PG asphalt binder should be selected one high temperature grade higher or warmer, such as a PG 64 instead of a PG 58. For standing design loads, the PG asphalt binder should be selected two high temperature grades higher or warmer, such as a PG 70 instead of a PG 58. An additional shift should also be performed for extraordinary high numbers of heavy traffic loads such as locations where the design lane is expected to exceed 30,000,000 ESALs. If the design loading is expected to be greater than 30,000,000 ESALs, the PG asphalt binder should be selected one high temperature grade higher or warmer than the climate dictates, such as a PG 64 instead of a PG 58. Table 3.14 is an accumulation of shifting up SHRP PG grades based on traffic. Bumping of the high temperature portion of the PG grades can lead to grades such as a PG 82 that may not be available locally. The designer should consult with local binder suppliers and contractors to confirm availability and constructability when requesting grades of PG 82 or higher. Although PG 82 is available and design dictated, local constraints can limit the use of such binders. For example, air quality permits in non-attainment areas may not allow hot-mix plant temperatures to exceed 165 °C, which may not be hot enough for the effective mixture compaction of these very stiff binders. Similar grade "bumping" can also be applied to the penetration and viscosity asphalt binder systems shown in Table 3.13. As previously shown with the PG asphalt binder selection, slow moving traffic or very heavy loading requires a stiffer or more viscous asphalt binder. In the penetration grading system, stiffer asphalt binders are those generally with lower penetration values, while in the viscosity grading system higher viscosity values. Similar to the PG asphalt binder grading system, the asphalt binder will also be "bumped" to the next standard grade, such as an AC-10 becomes an AC-20 and a 85–100 dmm penetration becomes a 60–70 dmm penetration asphalt binder. Table 3.15 illustrates the "bumping" method used for penetration, viscosity, or aged residue asphalt binders.

Table 3.15 also illustrates the advantages of the PG binder grading system from the penetration or viscosity type systems. The PG binder system allows greater flexibility and the ability to select an asphalt binder that can specifically meet the climate and traffic conditions of the asphalt pavement being designed. It should also be emphasized that proper or conservative asphalt binder selection does not guarantee total or satisfactory pavement performance. The pavement structure and traffic loading also influence fatigue cracking performance. Permanent deformation or rutting is directly a function of the shear strength of the total asphalt mixture, which is greatly influenced by aggregate properties. Pavement low temperature cracking correlates most significantly to the asphalt binder properties. Designers should try to achieve a balance among the many factors when selecting an asphalt binder.

Table 3.14 SHRP PG grade shifting or bumping based on traffic

Design ESALs	*Number of PG high temperature grade increases based on traffic*		
	Standing (<20 km/h)	*Slow moving* (20–70 km/h)	*Normal* (>70 km/h)
<300,000	0	0	0
300,000 to <3,000,000	2	1	0
3,000,000 to <10,000,000	2	1	0
10,000,000 to <30,000,000	2	1	0
≥30,000,000	2	1	0

Table 3.15 Increasing grades due to standing or very heavy traffic

Original selected grade	*New or "bumped grade"*
Penetration grading	
120–150	85–100
85–100	60–70
60–70	40–50
40–50	None
Viscosity grading	
AC-5	AC-10
AC-10	AC-20
AC-20	AC-40
AC-40	None
Aged residue grading	
AR-2000	AR-4000
AR-4000	AR-8000
AR-8000	AR-16000
AR-16000	None

Precipitation

Asphalt pavements are susceptible to damage by water. The major source of the water is through precipitation, whether it is rainfall or snowfall. The designer's first line of defense to moisture-induced damage is proper drainage. If the asphalt pavement is properly drained, moisture damage is unlikely. Asphalt binder modifiers, such as anti-stripping agents or adhesion promoters can reduce the occurrence of moisture damage. Selecting materials that are not as susceptible to damage, mixture design, and quality construction can also reduce the occurrence of moisture damage. Water can also saturate the subgrade or base of an asphalt pavement and cause structural damage to a pavement in climates that have extensive freezing and thawing. The best prevention to structural damage from freezing and thawing is also proper drainage.

Chapter 4

Pavement structural design

The proper structural design of an asphalt pavement is an integral part of the pavement's life span and its ability to serve the pavement's intended function. The thickness of the pavement layer is the most significant part of the structural integrity of the asphalt pavement. Compared with a building design, the structural design of an at-grade parking lot or other low speed pavements involves a somewhat lesser risk to public safety. Pavement design is generally concerned with the economic risks of investing too little or too much in materials and construction. The term "conservative" in the context of pavement design usually refers to risks in the economic sense.

Variables that are evaluated when determining an asphalt pavement's thickness are subgrade properties, base properties, material properties, traffic loading, and environmental factors. Thickness determination of a pavement is a structural evaluation process, ensuring that the traffic loads are so distributed that the stresses and strains developed at all levels in the pavement structure and in the subgrade are within the capabilities of the materials used at those levels. It involves the selection of materials for the different pavement layers and the calculation of the required thickness. The traffic or load carrying ability of an asphalt pavement is a function of both the thickness of the material and its stiffness. Environmental conditions influence the performance of the entire pavement structure, including the subgrade. Moisture affects the subgrade, sub-base, or granular base, while temperature affects the asphalt mixtures. Most design methods take into account climatic conditions (Whiteoak 1991).

Pavements and their bases are divided into two major classifications or types: rigid and flexible. A rigid pavement is generally a pavement that has been constructed of Portland cement or a pozzolonic material. A pavement constructed of concrete is assumed to possess considerable flexural strength that will permit it to act as a beam and allow it to bridge over minor irregularities that may occur in the sub-base or subgrade on which it rests; hence the name rigid. All other types of pavements, especially asphalt binder based, are termed flexible. A common definition is that a flexible pavement is a structure that maintains contact with and distributes loads to the subgrade and depends on aggregate interlock, particle friction, and cohesion for stability. Flexible pavements include Full-Depth™ asphalt pavements or "perpetual pavements" and those pavements that are composed of a series of granular layers or courses covered by a relatively thin asphalt binder based wearing course (Wright and Paquette 1987).

The foundation of an asphalt pavement is an integral part of the overall pavement performance. Its importance is similar to the importance of proper foundation design for a building or other structures. The foundation of an asphalt pavement is composed of two layers or lifts (Figure 4.1). The upper layer is termed the sub-base, while the lower layer is the subgrade. The term "lift" in highway terminology refers to a single layer or thickness of material. It is used somewhat interchangeably with the term layer, except that a layer can consist of several lifts. Generally the sub-base is composed of granular material such as crushed stone, however it may also be a dense graded asphalt mixture as part of a Full-Depth asphalt pavement. The subgrade is usually uncompacted or compacted in-place soil or fill. The subgrade supports the pavement structure. The sub-base performs four main functions:

- It provides a structural layer that distributes loads to the subgrade.
- It provides a working platform for construction traffic and a paving platform on to which the asphalt materials can be placed and compacted.
- It acts as an insulating layer or lift protecting the subgrade from frost.
- It provides a drainage layer to remove some water from the pavement.

The base course or binder course is the main structural component of an asphalt pavement. The term "binder" is the terminology referring to the base course that "binds" the surface course to the usually granular sub-base. "Binder course" is the preferred term by the European CEN standards. The function of the base course is to distribute the traffic or loading so that the sub-base and subgrade is not overstressed. It must resist permanent deformation and fatigue cracking caused by repeated loading. It must also be capable of resisting stresses induced by temperature gradients throughout the structure. The base course is generally a dense graded asphalt mixture. For lighter loading applications, such as a parking lot or local road, a base course may also consist of granular material such as crushed stone (Figure 4.2).

The surface course is usually known as the "wearing course." "Surface course" is the preferred terminology in the CEN standards. Sometimes the surface course may actually

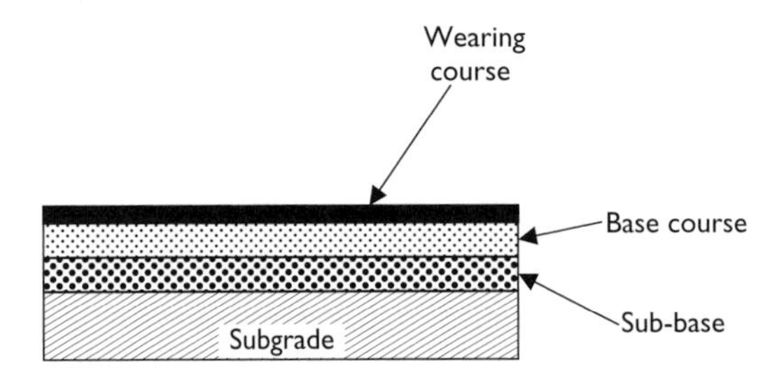

Figure 4.1 Typical asphalt pavement structure.

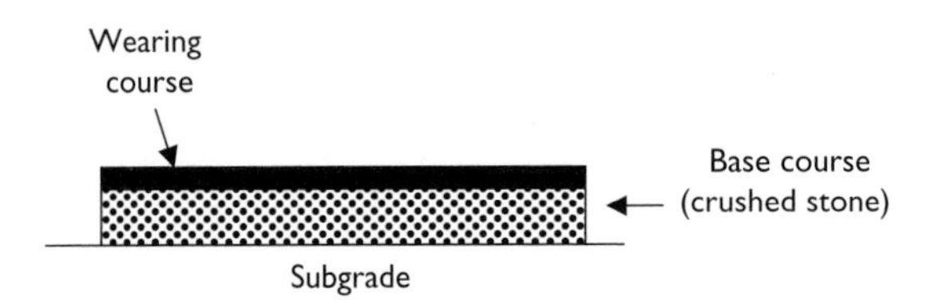

Figure 4.2 Example of a low volume design, such as a residential driveway or parking lot.

Plate 4.1 Wearing course over base or binder course.

consist of two layers – the lower layer being a thin leveling binder or base course, followed by the upper layer or lift being the actual wearing course (Plate 4.1). The purpose of the leveling base course is to provide a smooth, imperfection free surface on which to construct the wearing course. Its use is usually required on pavements with high smoothness requirements such as aircraft runways and racecar tracks. It also placed on top of asphalt pavements that have been previously milled or ground, in order to smooth out the "teeth" marks. The wearing course has several functions:

- To resist permanent deformation caused by traffic.
- Be impervious, protecting the lower layers.
- To resist the effects of weather, traffic abrasion or wearing, and fatigue.
- To provide a skid-resistant surface.
- To provide additional structural thickness or strength to the pavement.
- To provide for acceptable ride quality.

In most asphalt pavements, the stiffness in each layer or lift is greater than that in the layer below and less than that in the layer above (Whiteoak 1991).

Over the course of the past several years, various methods have been developed to determine the thickness of an asphalt pavement. No single approach is wrong, but each having its advantages or disadvantages, with each design method being correct when applied within its select variables or criteria. Most of these design methods are empirical; being based on experience and specially constructed test sections, such as the AASHO road test during 1958 to 1960. An empirical approach to the structural design of an asphalt pavement is somewhat removed from engineering principles and cannot accurately cope with variables or factors beyond those included in the trial section on which the design method is based. Some analytical methods are being used in more recent years. The Asphalt Institute multi-layered elastic system is the most common, but even it relies on some empirical functions. These methods are based on determining the actual loading stress and strain relationships, the engineering properties of the asphalt material, and any specific climatic conditions. A mechanistic design procedure is a form of an analytical design method. A mechanistic design method computes the stresses and strains in pavement in response to vehicle loads. The stresses and strains are used to estimate the amount of damage done to the pavement by vehicle loads which eventually determines how long a pavement will last. The AASHTO 2002 Pavement Design Guide designs use layered elastic models for its mechanistic-empirical design of asphalt pavements (Newcomb 2002). The AASTHO 2002 Pavement Design Guide was not available at the time of preparation of this text.

All design methods have three principle factors in common in determining the thickness of an asphalt pavement:

- Anticipated traffic in terms of both loading and volume.
- Subgrade support or strength.
- Properties of the asphalt materials selected for the pavement.

Most design methods use as their fundamental basis, the California Bearing Ratio (CBR) method of determining the strength of the soil in the supporting subgrade and of any granular base materials such as crushed stone. The CBR test is performed by compacting a sample of the subgrade soil to given moisture content, density, and compactive effort that are expected to be applied under field conditions. The CBR test may also be performed on in-place soil samples. After the sample has been compacted, a surcharge weight equivalent to the estimated weight of the pavement and sub-base is placed on the sample and the entire assembly is immersed in water for 4 days. After the completion of this soaking period, the sample is allowed to drain for 15 min. The sample, with the same surcharge imposed on it, is immediately subjected to a penetration by a piston that is 49.5 mm in diameter moving at a speed of 1.27 mm/min. The total loads corresponding to various penetration depths are recorded. A load-penetration curve is drawn, with the unit load that corresponds to a 2.5 mm penetration determined. This value is then compared to a value of 1,360 kg required to effect the same penetration in standard crushed stone. This ratio is the CBR as given in equation 4.1.

$$\text{CBR (\%)} = \frac{\text{unit load at 2.5 mm penetration}}{\text{1,360 kg}} \times 100 \tag{4.1}$$

ASTM D1883, or AASHTO T 193, *Standard Specification of the California Bearing Ratio*, gives a complete procedure for determining a CBR value. Asphalt pavement design methods that utilize or have utilized in the past the CBR method of evaluating the subgrade and relating it to the thickness of the asphalt pavement are:

- California Department of Transportation (United States).
- National Stone Association (NSA) (United States).
- AASHO or AASHTO (United States).

- Transport and Road Research Laboratory Road Note 29 (United Kingdom).
- Department of Transport Standard HD 14/87 (United Kingdom).

This list is not all-inclusive, but more to demonstrate that there are several approaches to the empirical method of determining the thickness of an asphalt pavement. All the methods with a few exceptions will give similar magnitudes of thickness for a pavement structure. For the purpose of this text, two empirical methods will be discussed, the present day AASHTO method and the NSA method. The Asphalt Institute analytical design method will also be discussed. Various thickness tables that are a combination of several methods will also be presented. The design thickness of an asphalt pavement and granular base layers determined by these methods is always the compacted thickness. Specifications should always specify the desired end thickness to be constructed in *compacted* or *finished* thickness (Plate 4.2).

AASHTO method of flexible pavement design

The 1993 *AASHTO Guide for the Design of Pavement Structures* is the basis for the AASHTO method of flexible pavement design. This design method utilizes four main design variables:

- Time
- Traffic and loading
- Reliability
- Climate.

Plate 4.2 Compacted thickness.

Material properties and pavement structural values are also evaluated through the use of the resilient modulus of the subgrade and structural number coefficients for the asphalt pavement layers and any granular sub-bases or bases. The AASHTO method is a conservative design method (in other words, pavements can be over-designed), therefore average values for the design variables can be used for all applications.

The term, "AASHO," or American Association of State Highway Officials is the previous name of "AASHTO," or American Association of State Highway *and Transportation* Officials. The AASHO road test was a $27 million project that was built in 1956 in Ottawa, Illinois and was undertaken by the various public road departments in the United States. Traffic loading began in October 1958 and ended in November 1960. Major portions of the test included flexible pavements and rigid pavements. Only one subgrade soil was used, an AASHTO classification of A-6, which can be described as a silt-clay soil that has a minimum of 36 percent passing the 75-μm sieve and a plasticity index of greater than 11. A general subgrade rating of this type of soil would be fair to poor. In the principal flexible pavement sections, the wearing or surface course was dense graded hot mix asphalt, the base course was a well-graded crushed limestone, the sub-base was a uniformly graded sand–gravel mixture and the subgrade was an A-6 soil. Major design factors in the principal experiments were variable wearing, base, and sub-base thicknesses. Three levels of wearing course thicknesses existed in combination with three levels of base thicknesses, and these combinations also existed with three levels of sub-base thickness. The wearing course varied from 25 to 150 mm, the base varied from 0 to 225 mm and the sub-base from 0 to 400 mm. Test traffic included both single and tandem axle military vehicles with 10 different axle arrangement–axle load combinations. Single axle loads ranged from 9 to 133 kN, while the tandem axle loads ranged from 107 to 214 kN. Each pavement section was tested with one of the 10 combinations and to thousands of load repetitions. Many observations were made on each of the pavement sections performance, including cracking, permanent deformation, pavement deflections, and subgrade performance.

One of the significant outcomes of the AASHO road test was the pavement serviceability index, ρ, and terminal serviceability index, ρ_t. The index involves the measurement in quantitative terms of the behavior of the pavement under traffic and its ability to serve traffic at some selected point during its service life. The road test determined, at the time evaluated, that the serviceability index at any time of a flexible or asphalt pavement is a function of roughness or slope variance in the wheel paths, the extent and type of cracking and repairs of a pavement and its permanent deformation. The scale for the pavement serviceability index is from 0 to 5, with a value of 5 representing the highest or best index of serviceability. In the beginning of the AASHO road test, researchers determined the initial pavement serviceability index, ρ, to be 4.2 for flexible pavements. In general, the value of the index declined gradually under traffic. When the serviceability index dropped to 1.5 for any pavement section, that section was taken out of service. A value of ρ of 2.5, for example, is an intermediate one between initial construction and failure of the asphalt pavement to render adequate traffic service (Wright and Paquette 1987). The terminal serviceability index, ρ_t, represents the lowest pavement serviceability index that can be experienced before rehabilitation, resurfacing, or reconstruction is required. The values are between 2.0 and 3.0 with 2.5 recommended for major highways, 2.0 for lower volume roads, and 1.5 for low use facilities such as residential parking lots. Terminal serviceability is a direct correlation to the motorist's satisfaction with the quality and ride of the existing pavement. The actual initial serviceability, ρ_0, represents the actual ride quality of the new asphalt pavement immediately after it is put into service. Obviously, this value is not known during the inital design of the asphalt pavement, but is usually assumed to be between 4.2 and 4.5. The change in serviceability, ΔPSI, is calculated as the difference between the initial serviceability, ρ_o, and the terminal serviceability, ρ_t (equation 4.2).

$$\Delta PSI = \rho_o - \rho_t \tag{4.2}$$

Serviceability

The serviceability–performance concept is based on five fundamental assumptions:

- Roadways are for the comfort and convenience of the traveling user.
- Comfort, or ride quality, is a matter of subjective response or opinion of the user.
- Serviceability can be expressed by the mean of the ratings given by all roadway users and is termed the "serviceability rating."
- There are physical characteristics of a pavement that can be measured objectively, and that can be related to subjective evaluations. This procedure produces an objective serviceability index.
- Performance can be represented by the serviceability index.

The serviceability of a pavement is expressed in terms of the present serviceability index (PSI). The PSI is obtained from measurements of roughness and distress at a particular time during the service life of a pavement. Roughness is the dominant factor in estimating the PSI of a pavement. A reliable method of determining roughness is important in monitoring the performance history of a pavement (AASHTO 1993).

Time

The pavement analysis period is the length of time that a given design strategy covers. The analysis period may also be known as the design life or design period. The pavement performance period or design period is the time that the initial pavement structure is expected to perform adequately before needing major rehabilitation or its terminal serviceability. A typical or common analysis period is 20 years only because the original United States Interstate Highway Act of 1956 required that traffic be considered through 1976. Table 4.1 is one set of suggested design value periods, but the designer may select any value that may be specifically required. This is especially true when designing parking facilities. For example, the analysis or design period on a major expressway may be selected as 30 years which would include the initial performance period and several rehabilitation periods which could include overlays or maintenance operations (Lindeburg 1999). A parking facility may have a design period of 10 years and no rehabilitation periods. The length of time that is selected for the design period could be for engineering reasons or for economic reasons. The parking facility may have served its intended marketing purpose at the end of 10 years and still have suitable performance life left.

Traffic and loading

The AASHTO method of flexible pavement design requires all traffic and loading to be converted to 80 kN (18,000 lbs) equivalent single axle loads, or ESALs (w_{18}). The numbers of

Table 4.1 Suggested asphalt pavement design period values

Roadway designation	*Suggested design or analysis period (years)*
Freeway	20–30
Primary	20–30
Arterial	15–25
Collector	10–25
Local	10–20
Heavy loading parking	10–20
High volume passenger parking	10–20
Low volume or residential parking	10–15

ESALs are added together over the pavement design period. Chapter 3 above gives a detailed explanation and procedure on converting traffic and loading to ESALs. There are four key considerations that can influence the accuracy of traffic estimates and the life cycle or design period of an asphalt pavement. These four points are:

- The accuracy of the load equivalency values used to estimate the relative damage induced by axle loads.
- The accuracy of traffic volume and weight information used to represent the actual loading estimates.
- The estimate of total ESALs over the pavement design period.
- The interaction of pavement age and traffic as it affects changes in PSI.

The load equivalency factors given in Chapter 3 are the best available at the time and are originally derived from information from the AASHO road test. Chapter 3 also provides information on estimating traffic volume and loading information. The ESALs for the performance period represent the cumulative number from when the roadway was opened to traffic until the time when the serviceability is reduced to a terminal value such as 2.5. If the traffic is underestimated the actual time to terminal serviceability will be less than the predicted performance period, resulting in increased maintenance and rehabilitation (AASHTO 1993).

Reliability

Reliability considerations ensure that the asphalt pavement will last for the designated design period. They take into account variations in traffic and performance predictions. Reliability in terms of pavements, is the probability that a pavement design will perform satisfactorily under traffic and climatic conditions for pavement design period. Pavement facilities that are designated critical are designed using higher reliability factors. Depending on the type of pavement facility, reliability values can range from 50 to 99.9 percent (Table 4.2). The 50 percent reliability may apply to a residential street or a light duty parking lot, while the 99.9 percent reliability values would apply to heavy usage expressways. It is also necessary to select an overall standard deviation, S_o, for reliability to account for traffic and loading variances that are representative of local or regional conditions. The AASHO road test provided loading standard deviations of 0.4–0.5 for asphalt pavements. These standard deviations are still adequate today.

Table 4.2 Reliability values

Roadway designation	*Suggested reliability values (%)*	
	Urban	*Rural*
Freeway	90–99.9	85–99.9
Primary	85–99	80–95
Arterial	80–99	75–95
Collector	80–95	75–95
Local	50–80	50–80
Heavy loading parking	80–95	75–95
High volume passenger parking	75–95	75–90
Low volume or residential parking	50–80	50–75

The following steps are required when determining the reliability considerations for the asphalt pavement:

1. Define the functional classification of the asphalt pavement, such as an urban arterial or a rural freeway, etc.
2. Select a reliability level from the range given in Table 4.2. The greater the reliability value, the greater the pavement structure or thickness required.
3. A standard deviation, S_o, should be selected that is representative of local conditions. A value of 0.45 may be used if there is difficulty in determining a local S_o (AASHTO 1993).

Subgrade and soil strength

The AASHTO method of flexible or asphalt pavement design requires that the strength of each pavement layer and the underlying soil subgrade be determined or assigned an assumed value. This procedure must be done prior to determining the thickness of each pavement layer. For this design procedure, the effective roadbed soil or subgrade soil resilient modulus, M_R, must be determined. The resilient modulus can either be measured in the laboratory using a test procedure such as AASHTO T274 or ASTM D4123 or it can be predicted from a correlation with nondestructive deflection measurements. For subgrade materials, laboratory resilient modulus tests should be performed on representative soils under different seasonal moisture conditions. The effective subgrade resilient modulus represents the combined effect of all the seasonal modulus values. The resilient modulus is the same as the modulus of elasticity, E, of the soil. The resilient modulus can also be estimated or converted from CBR values (equation 4.3).

$$M_R\ (\text{MPa}) = 10.3\ (\text{CBR})$$

$$M_R\ (\text{psi}) = 1{,}500\ (\text{CBR}) \qquad (4.3)$$

The correlation given in equation 4.3 is not applicable to granular, untreated base or sub-base materials. It is also valid only for soils that have an estimated resilient modulus of 207 MPa

Table 4.3 Unified soil classification definitions

Prefix	Soil type	Suffix	Soil subgroup
G	Gravel	W	Well graded
S	Sand	P	Poor graded
M	Silt	M	Silt
C	Clay	L	Clay, liquid limit $< 50\%$
O	Organic	H	Clay, liquid limit $> 50\%$

or less and are classified under the Unified Soil Classification, ASTM D2487, as CL, CH, ML, SC, SM, and SP or the equivalent soils classified under other systems. Table 4.3 gives definitions of the prefixes and suffixes used in the Unified Soil Classification system.

Soils are described as gravel, sand, silt, or clay depending on the range of particle sizes present in the greatest abundance. Gravel and sand are predominantly coarse grained. Gravel usually refers to soils having a predominant size larger than 4.75 mm and sand usually refers to soil that are finer than 4.75 mm. Silt and clay soils are composed of very fine particles. Over 50 percent of the grains are finer than 75 μm, with the individual particle not being visible without magnification (NAPA 1991).

The soil resistance value or R-value is another method used in determining subgrade or soil strength. It can be determined through laboratory testing following ASTM D2844. The resilient modulus can also be estimated from the R-value (equation 4.4).

$$M_R\,(\text{MPa}) = 8.0 + 3.8\,(\text{R-value})$$

$$M_R\,(\text{psi}) = 1{,}155 + 555\,(\text{R-value}) \qquad (4.4)$$

The relative damage value, u_f, is used to adjust the resilient modulus for seasonal damage. The relative damage value is calculated for each resilient modulus that is determined for each season (equation 4.5). All the u_f values are then summed and divided by the number of seasons to give an average seasonal u_f value.

$$u_f = (1.18 \times 10^8) M_R^{-2.32} \qquad (4.5)$$

The effective or average M_R (equation 4.6) is now solved for in terms of the average seasonal u_f value.

$$\textit{Effective } M_R = \sqrt[2.32]{\frac{1.18 \times 10^8}{\textit{seasonal } u_f}} \qquad (4.6)$$

In climates that do not have significant variations in precipitation or in areas that the subgrade soil moisture content does not vary from season to season, it may not be necessary to calculate an effective (seasonal) M_R.

Layer coefficients

The AASHTO method of flexible pavement design combines asphalt pavement layer properties and thickness into one variable, called the structural number, SN. Each pavement structural layer is assigned a layer coefficient, a_n, in order to convert the actual layer thickness, D_n, into a structural number. The layer coefficient or strength coefficient, expresses the empirical relationship between a structural number and thickness and is a measure of the ability of the material to function as an integral part of the pavement structure. Once the structural number for a pavement is determined, a set of layer thickness and the corresponding layer coefficients are chosen. When combined, the sum of each layer thickness must provide the load carrying capacity corresponding to the structural number (equation 4.7).

$$SN = \Sigma D_n a_n \tag{4.7}$$

The AASHTO design method has adopted the resilient modulus for determining the strength or quality of the subgrade, it is also necessary to identify corresponding layer coefficients. These layer coefficients are empirical in nature and are based on observations at the original AASHO road test and have been verified through more recent research and field studies. Table 4.4 gives some typical values for layer strength coefficients. These layer coefficients were developed originally in English units along with all the other information that was collected from the AASHO road test. Table 4.5 gives the layer strength coefficients converted to SI (metric) units. It may be simpler for the asphalt pavement designer to work an AASHTO design method first in English units and then convert the final layer thickness values to SI units, since the structural number nomograph (Figure 4.3) is only available in English units.

Table 4.4 Typical pavement layer strength coefficients (English units)

Material type	*Layer coefficient* (1/inch)
Sub-base course coefficients, a_3	
Sandy gravel	0.11
Sandy clay	0.08 (0.05–0.10)
Lime treated soil	0.11
Lime treated clay	0.16 (0.14–0.18)
Crushed stone	0.14 (0.08–0.14)
Base course coefficients, a_2	
Crushed stone	0.14 (0.08–0.14)
Sandy gravel	0.07
Pozzolonic base	0.28 (0.25–0.30)
Lime treated base	0.22 (0.15–0.30)
Cement treated base	0.27
Soil cement	0.20
Asphalt treated base, coarse graded	0.34
Asphalt treated base, sand graded	0.30
In-place recycled mixture	0.20
Mixing plant recycle mixture	0.40 (0.40–0.44)
Dense graded hot mix asphalt	0.44
Wearing course coefficients, a_1	
Dense graded hot mix asphalt	0.44
Sand asphalt	0.40
In-place recycled mixture	0.20
Mixing plant recycle mixture	0.40 (0.40–0.44)

Table 4.5 Typical pavement layer strength coefficients (SI units)

Material type	*Layer coefficient* (1/mm)
Sub-base course coefficients, a_3	
Sandy gravel	0.0043
Sandy clay	0.003 (0.002–0.0039)
Lime treated soil	0.0043
Lime treated clay	0.0063 (0.0055–0.0071)
Crushed stone	0.0055 (0.0031–0.0055)
Base course coefficients, a_2	
Crushed stone	0.0055 (0.0031–0.0055)
Sandy gravel	0.0028
Pozzolonic base	0.011 (0.010–0.012)
Lime treated base	0.009 (0.006–0.012)
Cement treated base	0.011
Soil cement	0.008
Asphalt treated base, coarse graded	0.013
Asphalt treated base, sand graded	0.012
In-place recycled mixture	0.008
Mixing plant recycle mixture	0.016 (0.016–0.017)
Dense graded hot mix asphalt	0.017
Wearing course coefficients, a_1	
Dense graded hot mix asphalt	0.017
Sand asphalt	0.016
In-place recycled mixture	0.008
Mixing plant recycle mixture	0.016 (0.016–0.017)

The layer design method used in the AASHTO design method for flexible pavements has been developed around a three-layer pavement composition consisting of the wearing or surface course (layer one), the base course (layer two), and the sub-base course (layer three). This design also permits omitting the sub-base course or layer three. By utilizing the three-layer composition into the design method, equation 4.7 is further developed into equation 4.8:

$$SN = D_1a_1 + D_2a_2m_2 + D_3a_3m_3 \quad (4.8)$$

where D_1 is the wearing course thickness (inches); a_1, the wearing course layer coefficient; D_2, the base course thickness (inches); a_2, the base course layer coefficient; m_2, the base course drainage coefficient; D_3, the sub-base course thickness (inches); a_3, the sub-base course layer coefficient; m_3, the sub-base course drainage coefficient.

The base layer strength coefficient, a_2 and the sub-base coefficient, a_3 can be also calculated if they are composed of granular or unbound materials. The equations 4.9 and 4.10 utilize the resilient modulus, M_R (psi) of the granular material. The resilient modulus can be determined in the laboratory or based on typical values.

$$a_2 = 0.279\ (\log_{10}M_R) - 0.977 \quad (4.9)$$

$$a_3 = 0.277\ (\log_{10}M_R) - 0.839 \quad (4.10)$$

where M_R is the resilient modulus (psi).

Equation 4.8 has also introduced the concept of drainage coefficients, m_n. Typical values of drainage coefficients range from 0.40 to 1.40. The effects that water and drainage have on base and sub-base courses modify the layer strength coefficient by using a drainage coefficient. Drainage is not considered when determining the layer thickness of the wearing course. The effect is that poor drainage can weaken a pavement and excellent drainage can actually reduce the required thickness by increasing the layer strength coefficients that were developed around the original drainage conditions at the AASHO road test. Values greater than 1.0 are assigned to base courses and sub-base courses with good or excellent drainage and are seldom saturated with water. The quality of the drainage at the AASHO road test was considered fair or gives an m value of 1.0. Thus, the pavement strength layer coefficients are not adjusted under fair drainage conditions, which are the same conditions they were developed under. Table 4.6 gives some definitions of the quality of drainage in a rating from very poor to excellent. These definitions are defined in the time it takes for water to be removed from the pavement layer, which is generally estimated by using "good engineering judgment."

Table 4.7 gives the recommended m_n values as a function of the quality of drainage and the percent of time during the year the asphalt pavement structure would normally be exposed to moisture levels approaching saturation. The moisture levels are dependent on the average yearly rainfall and the prevailing drainage conditions. These m_n values apply only to the effects that drainage has on untreated base or sub-base layers. For example, if using

Table 4.6 Drainage quality definitions

Drainage quality term	*Time water is removed*
Excellent	2 h
Good	24 h
Fair	1 week
Poor	4 weeks
Very poor	No drainage

Table 4.7 Recommended m_n values for modifying layer strength coefficients

Quality of drainage	*Percent of time pavement structure is exposed to moisture levels approaching saturation (%)*			
	<1	1–5	5–25	>25
Excellent	1.40–1.35	1.35–1.30	1.30–1.20	1.20
Good	1.35–1.25	1.25–1.15	1.15–1.00	1.00
Fair	1.25–1.15	1.15–1.05	1.00–0.80	0.80
Poor	1.15–1.05	1.05–0.80	0.80–0.60	0.60
Very poor	1.05–0.95	0.95–0.75	0.75–0.40	0.40

Source: *Guide for the Design of Pavement Structures*; © 1991, by the American Association of State Highway Transportation Officials, Washington, DC. Used by permission.

a dense graded asphalt mixture as a base layer, its layer strength coefficient would not be adjusted by m_2. Improved drainage is important to all types of pavement materials, the effects that water have on the strength of a pavement layer are not nearly as significant as they are on untreated materials such as a granular base.

AASHTO asphalt pavement design procedure

In the AASHTO design procedure for asphalt pavements, equation 4.8 and the structural number are the key focus of the procedure. The following steps summarize the procedure:

1 Determine the required reliability and overall standard deviation for the pavement.
2 Determine the ESALs (w_{18}) for the life of the pavement.
3 Determine the subgrade soil resilient modulus, M_R.
4 Determine the design serviceability loss, ΔPSI.
5 Using the four values selected above and the AASHTO design nomograph (Figure 4.3), determine the required structural number (SN) for the asphalt pavement.
6 The selected structural number and equation 4.8 and its required values are then computed to determine the thickness of each layer.

A trial and error selection of thickness values are inputted into equation 4.8 until the required structural number is arrived at. Theoretically, any combination of thickness values that satisfy equation 4.8 will work. However, minimum layer thickness results from construction techniques and strength requirements. The thickness of the flexible pavement layers is rounded to the capabilities of the paving equipment, which is usually 5–6 mm. When selecting layer thickness values, cost effectiveness as well as placement and compaction issues must be considered to avoid an impractical design (Lindeburg 1999). Table 4.8 gives some typical minimum thickness values for dense-graded asphalt mixtures and aggregate bases. The structural number concept and equation 4.8 should not be used to design the thickness for a wearing course that is above a sub-base or base course that has a M_R greater than 275 MPa. The wearing course thickness for these applications should be based on cost effective, prior experience or minimum construction thickness requirements when using the AASHTO method of flexible pavement design.

Once the design structural number (SN) is selected from the AASHTO design nomograph, it is necessary to identify a set of pavement layer thickness values which, when combined, will provide the structural capacity corresponding to the design structural number. Equation 4.8 does not have a unique solution; there are many combinations of thickness values that are satisfactory solutions. These thickness values should also meet the minimum requirements given in Table 4.8. Asphalt pavements are layered systems and typically consist of a wearing course, base course, and a sub-base course. Each of the courses are specifically designed and then considered for the total asphalt pavement structure. The maximum allowable thickness can be computed from the differences between the computed structural numbers. For example, the maximum allowable structural number for the sub-base layer would be equal to the structural number required over the sub-base subtracted from the structural number that is required over the subgrade. From a cost-effective view, if the ratio of pavement costs for layer one to layer two is less than the corresponding ratio of layer coefficients, a_n, times the drainage coefficients, m_n, then the optimal design is one where the

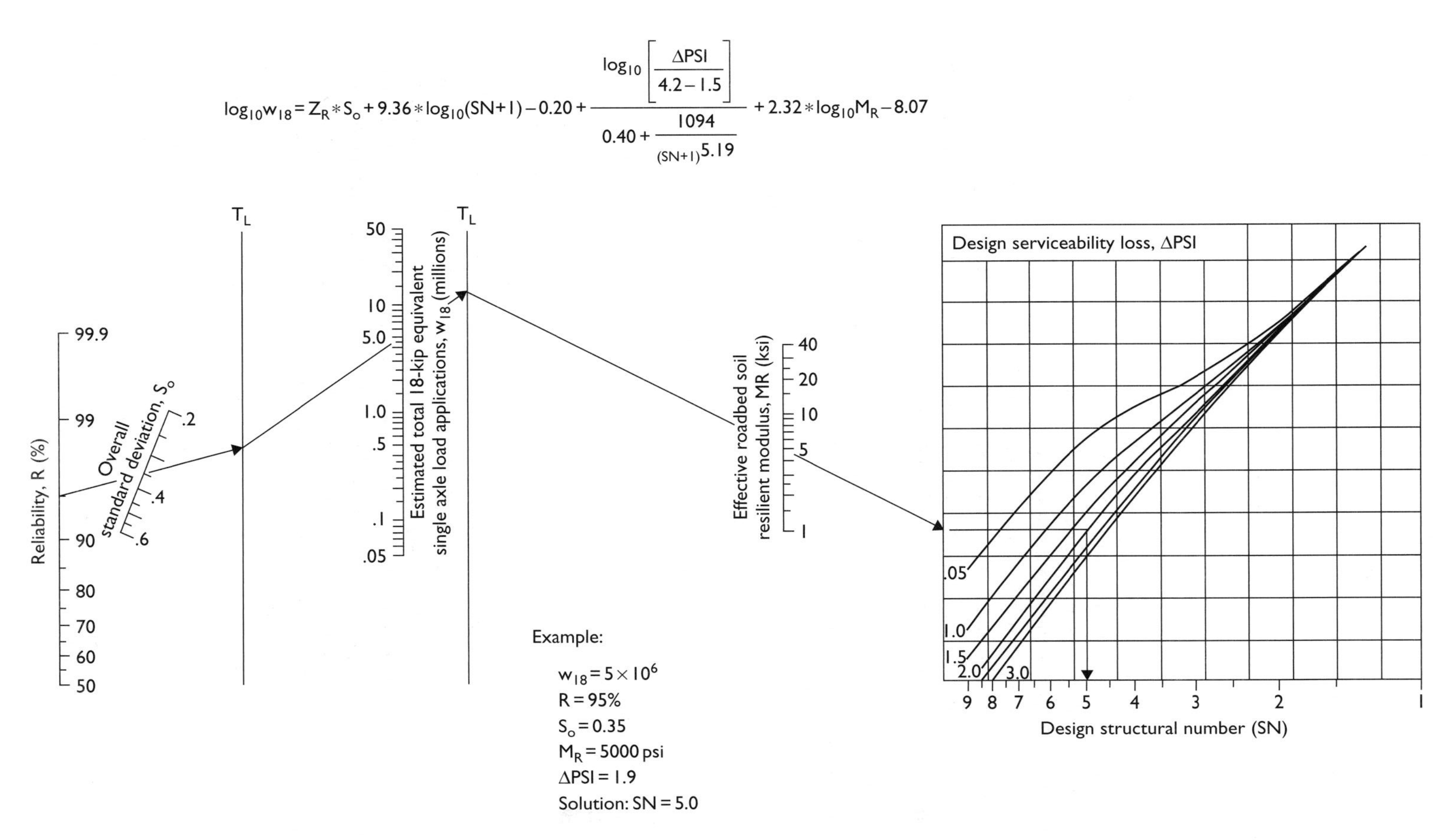

Figure 4.3 AASHTO flexible pavement design nomograph.

Source: From *Guide for the Design of Pavement Structures*; © 1991, by the American Association of State Highway Transportation Officials, Washington, DC. Used by permission.

Table 4.8 Typical minimum layer thickness values

ESALs	*Aggregate base, untreated or treated* (mm)	*Dense graded asphalt mixture* (mm)
< 50,000	25	100
50,000–150,000	50	100
150,000–500,000	65	100
500,000–2,000,000	75	150
2,000,000–7,000,000	90	150
> 7,000,000	100	150

minimum base thickness, D_2, is used (AASHTO 1993). Equation 4.11 is an illustrative version of the described cost comparison. If

$$\frac{\text{Cost of wearing course}}{\text{Cost of base course}} < \frac{a_1}{a_2 m_2} \tag{4.11}$$

then the optimum economical design is the design where the minimum base course thickness is used.

Figure 4.4 and equations 4.8 and 4.12–4.15 are used to determine the thickness of each asphalt pavement layer when using a layered analysis approach.

$$D_1^* \geq \frac{(SN_1)}{a_1} \tag{4.12}$$

$$SN_1^* = a_1 D_1^* \geq SN_1 \tag{4.13}$$

$$D_2^* \geq \frac{SN_2 - SN_1^*}{a_2 m_2} \tag{4.14}$$

$$SN_1^* + SN_2^* \geq SN_1 \tag{4.15}$$

$$D_3^* \geq \frac{SN_3 - (SN_1^* + SN_2^*)}{a_3 m_3} \tag{4.16}$$

where * indicates that the values selected must be equal or greater than the values required, such as a minimum structural number or thickness.

When using the layered design approach, the structural number for the asphalt pavement over the subgrade is first determined. Following the same method, the structural number required over the sub-base layer and the base layer are also to be determined. The differences between the computed structural numbers required over each layer, allow the maximum allowable thickness of any given layer to be determined. The thickness of each layer can be determined by using equations 4.12–4.16. Example 4.1 further illustrates the layered analysis approach.

Example 4.1

Determine the asphalt pavement thickness for a large multi-type vehicle parking facility (such as a freeway rest area) that will have 2,000,000 ESALs of loading during its design

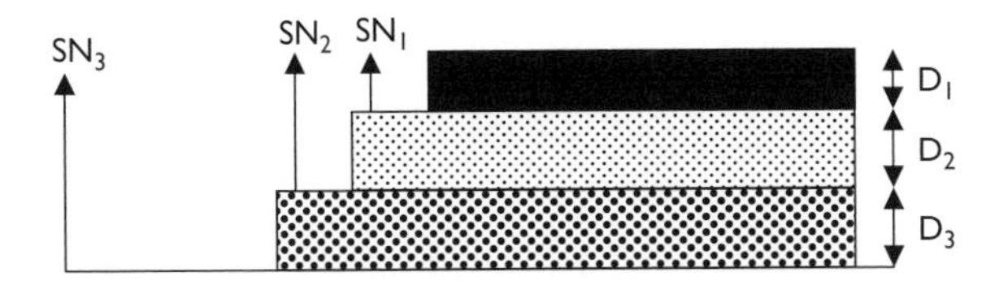

Figure 4.4 Layer thickness diagram.

life. The designer has selected 90 percent reliability with an overall standard deviation of 0.45. The initial serviceability index is 4.5 and the terminal serviceability index is 2.0. The wearing course is a dense graded asphalt mixture. The base course is a crushed stone base to be placed directly over the soil subgrade. The soil subgrade has a CBR value of 3.3 percent. The base course thickness should not be greater than 450 mm. There is no sub-base course. The moisture conditions can be described as good if for 20 percent of the time the pavement structure is exposed to moisture levels approaching saturation.

Solution

The first step is to determine the structural number over the soil subgrade using the AASHTO design nomograph in Figure 4.3. The following information summarizes what is needed to solve for the structural number.

Reliability, R = 90 %
Overall standard deviation, $S_o = 0.45$
$w_{18} = 2{,}000{,}000$ ESALs
Resilient Modulus, $M_R = 1{,}500\ (\text{CBR}) = 1{,}500(3.3) = 5{,}000$ psi (34.5 MPa)
Design serviceability index, $\Delta PSI = \rho_o - \rho_t = 4.5 - 2.0 = 2.5$

Using the given values and the AASHTO design nomograph in Figure 4.3, the structural number has been determined to be 4.0.

From Table 4.5, the layer strength coefficient, a_1, for a dense graded hot mix asphalt wearing course is 0.017. The crushed stone base course a_2 value is 0.0055. The m_2 value has been determined from Tables 4.6 and 4.7 as 1.05. These problems are typically solved through a trial and error process until acceptable thickness values are determined, although specific equations can also be easily derived. Several solutions may also be valid.

For the wearing course, an initial trial value of 75 mm will be used.

$$SN_1 = a_1 D_1 = (0.017)\ (75\,\text{mm}) = 1.28$$

Solving for the crushed stone base course:

$$D_2 = \frac{SN_2 - SN_1}{a_2 m_2} = \frac{4 - 1.28}{(0.0055)(1.05)} = 470\,\text{mm}$$

In an effort to reduce the crushed stone thickness a 100 mm wearing course will be selected.

$$SN_1 = (0.017)(100\,mm) = 1.70$$

$$D_2 = \frac{4 - 1.70}{(0.0055)(1.05)} = 400\,mm$$

In this example the designer has selected a wearing course of 100 mm of dense graded hot mix asphalt and a base course of 400 mm of crushed stone aggregate placed directly over the subgrade.

This example has shown that several thickness values of the various pavement layers can meet the required structural number. Agency policy, economics, construction restraints, among others usually will dictate which combination to select. For example, 100 mm of a dense graded hot mix asphalt mixture has been adopted in recent years by many designers as a minimum thickness requirement for all loading requirements, in an effort to reduce the effects of low temperature and fatigue cracking.

The AASHTO method of flexible pavement design is probably the most widely used method for determining the thickness of an asphalt pavement. It is an empirical method and has been somewhat modified over the years to take into considerations that may vary from the original AASHO road test at Ottawa, Illinois. Various other methods have been developed around the AASHTO method and has proven to be a successful, however conservative design method. Several thickness design tables have also been compiled by using this method.

National Stone Association method

The NSA method of flexible pavement design is an empirical system based on the criteria originally developed by the United States Army Corps of Engineers and refined over the years to reflect the experience of the military services on highway and airfield pavements throughout the world. The original development of these methods can be traced to the AASHO road test and the AASHTO design method for flexible pavements. The NSA method provides designs that incorporate the use of crushed stone bases and sub-bases as the principal load bearing components in asphalt pavements (NSA 1994b). The pavement cross-section in the NSA method typically consists of a dense graded hot mix asphalt wearing course placed on top of a combined base and sub-base consisting of crushed stone or aggregate. This combination is directly on top of a soil subgrade. Figure 4.2 illustrates a typical cross-section of an asphalt pavement designed by the NSA design method. Crushed stone is the prevalent material used for base and sub-base courses, especially in parking lots, residential streets, and other local roads.

The NSA design method is applicable to various asphalt pavement applications. The method does not require equations or nomographs, instead determining the asphalt pavement thickness through a series of tables. The method also takes into account similar pavement performance factors that the AASHTO design method does. These factors consist of four, which are:

- Subgrade soil strength or support
- Pavement loading or traffic intensity
- Material properties, especially strength
- Climatic conditions, mainly consisting of evaluating frost or freeze conditions.

Subgrade soil support

The load-supporting capacity of the soil used for the asphalt pavement subgrade needs to be determined. The capacity should be determined for any moisture and climatic condition except for severe frost or freeze conditions. The CBR or resilient modulus (M_R) tests are used in the NSA method for evaluating the load carrying characteristics of the soil subgrade under no-frost conditions. Other soil strength tests that may be used for the NSA design method are the Texas Triaxial Classification System, ASTM D3397, and the Resistance value (R-value) method, ASTM D2844.

When it is impractical to actually measure the strength of the subgrade soil, the strength values can be estimated on the basis of standard recognized soil classification systems. The systems require only the determination of basic engineering properties of the soil, such as grain size distribution, liquid limit, plastic limit, and plasticity index. The soil classification systems used in the NSA design method are the Unified Soil Classification System and the AASHTO Soil Classification System. Table 4.9 is a correlation of these systems with CBR values and Resistance values. The table describes the soil strength into four categories, excellent, good, fair, and poor. These categories are further used in the thickness design tables.

Approximations can be made from the table when actual CBR values or R-values are not available. Careful consideration should be given when selecting the proper soil strength values, as they are an essential consideration when determining the thickness of an asphalt pavement. It is recommended that a truly poor soil subgrade be upgraded or enhanced to form a sub-base that can firmly support the crushed aggregate base that will be placed upon it. The soil can be strengthened by several methods, including in-place stabilization with Portland cement or lime (NSA 1994b). In some cases, the poor rated soil is removed and

Table 4.9 Approximate soil support categories

Soil descriptions	*Unified soil class*	*AASHTO soil group*	*~CBR value*	*~R-value*
Excellent				
High percentage of granular material	GW; GM; GC; some SM; SP; SC	A-1: A-2; some A-3	>15	>48
Good				
Some granular material mixed with some silt and/or clay	SM; SP; SC; some ML; CL; CH	A-2; A-3; some A-4: A-6; A-7	10–14	42–7
Fair				
Sand clays, sandy silts, light silt-clays, some plasticity	Ml; CL; some MH; CH	A-4 to A-7, low group indices	6–9	32–41
Poor				
Plastic clays, fine silts, very silt-clays, clay with mica	MH; CH; OL; OH; PT unsuitable	A-4 to A-7, high group indices	<6	<32

replaced with a better performing soil or crushed aggregate. It is possible to place an asphalt pavement over a poor rated soil subgrade, but usually the thickness required to "bridge" the soil makes the design uneconomical, unless some remedial measures are performed on the subgrade.

Traffic loading

The NSA flexible pavement design method incorporates the use of ESALs in determining the traffic loading and intensity to be placed on the pavement. Chapter 3 above gives a through explanation on the determination of ESALs. The NSA method groups ESALs into six separate groups, one through six, each designated as a design index (DI). The type of vehicle traffic is classified into three groups; described in Table 3.10. Table 4.10 designates each design index with a traffic description and daily and total ESALs values.

The design life in the NSA method is set at 20 years. The AASHTO method of flexible pavement design measures traffic loading as a total amount of ESALs over the asphalt pavement life. Equating the ESALs values can allow a comparison between the two design methods.

Thickness design

The NSA flexible pavement design method is centered on determining the total thickness of the crushed aggregate used as the combined base and sub-base courses. The asphalt wearing course is set at a minimum thickness, with the value depending on which DI it is

Table 4.10 Traffic design index categories

Design index	*General description*	*Daily ESALs*	*Total ESALs*
DI-1	Light traffic, few vehicles heavier than passenger cars, no regular use by group 2 or 3 vehicles	<5	< 37,000
DI-2	Medium-light traffic, maximum of 1000 ADT, including not over 10% group 2, no regular use by group 3 vehicles	6–20	37,000–146,000
DI-3	Medium traffic, maximum of 3000 ADT, including not over 10% group 2 and 3 with a maximum of 1% group 3 vehicles	21–75	146,000–548,000
DI-4	Medium-heavy traffic, maximum of 6000 ADT, including not over 15% group 2 and 3 with a maximum of 1% group 3 vehicles	76–250	548,000–1,800,000
DI-5	Heavy traffic, maximum of 6000 ADT, may include 25% group 2 and 3 with a maximum of 10% group 3 vehicles	251–900	1,800,000–6,570,000
DI-6	Very heavy traffic, over 6000 ADT, may include over 25% group 2 and 3 vehicles	901–3000	6,570,000–22,000,000

Source: Used with permission: Courtesy of NSSGA (formerly NSA).

Table 4.11 Basic NSA design thickness (mild climates)

Subgrade soil category	*Design thickness (mm) for indicated traffic loading categories*					
	DI-1	*DI-2*	*DI-3*	*DI-4*	*DI-5*	*DI-6*
Excellent	120	150	180	200	230	250
Good	180	200	230	250	280	300
Fair	230	280	300	350	380	430
Poor	340	420	470	520	580	660
Minimum asphalt wearing course thickness	25[1]	50	60	75	90	100

Source: Used with permission: Courtesy of NSSGA (formerly NSA).

Note
1 Usually a surface treatment for DI-1, but can also be dense graded hot mix asphalt.

serving. In summary the NSA method depends on the crushed aggregate base to function with the soil subgrade and perform the structural requirements of the pavement for the traffic loading. Aggregate interlock, particle friction and cohesion provide the strength of the pavement. NSA thickness values are based on information derived from the United States Army Corps of Engineer's design manuals. An initial thickness for mild climates is determined from Table 4.11 and the design is then checked to determine if it can perform adequately in areas of severe moisture or frost.

Severe climate or frost penetration

In some areas of severe climate or frost penetration, the thickness of the asphalt pavement will need to be increased. For the purposes of the NSA design method, severe climate is defined as any of the following three possibilities:

1 Subgrade soils that are considered frost susceptible.
2 Subgrade soils that are likely to hold or retain a considerable amount of moisture.
3 Frost penetration or freeze conditions that are likely to penetrate deeper than the original design thickness.

Frost penetration or ground freeze information is available from any number of sources, such as weather stations, public works departments, agriculture agencies, etc. Thickness increases are made to the original design thickness to overcome the reduced soil subgrade support or CBR values that occurs during critical freeze–thaw conditions. The frost susceptibility of soil can be estimated through systems such as the United States Army Corps of Engineer's Frost Classification system. This system assigns the soil to one of four groups according to its classification in the Unified Soil Classification system and the percentage of particles that are finer than 0.02 mm (NSA 1994b). The Unified Soil Classification system is described in ASTM D2487 and outlined in Appendix B. Table 4.12 is a description of the four frost groups.

Table 4.13 gives the design thickness values for asphalt pavements that will be placed in service in areas where a freeze or frost is expected to penetrate into the soil subgrade to a significant depth and where moisture is readily available and does not drain easily from the pavement.

Table 4.12 Frost group classifications

Frost group	*% Particle size* <0.02 mm	*Unified soil classifications*	*Frost susceptibility*
F-1			
Gravel soils	3–10	GW; GP; GW-GM; GP-GM	Low
F-2			
Gravel soils	10–20	GM; GW-GM; GP-GM	Low to medium
Sands, sand clays	3–15	SW; SP; SM; SW-SM; SP-SM	
F-3			
Gravel soils	>20	GM; GC	High
Sands, coarse to medium	>15	SM; SC	
Clays, PI > 12		CL; CH	
F-4			
Silts, very fine silt-sands clays, PI < 12	>15	ML; MH; SM; CL; CL-ML; CH	Very high

Table 4.13 Design thickness for severe climate-frost penetration areas

Subgrade soil frost group	*Design thickness* (mm) *for indicated traffic loading categories*					
	DI-1	*DI-2*	*DI-3*	*DI-4*	*DI-5*	*DI-6*
F-1	230	250	300	330	380	430
F-2	250	300	360	400	460	510
F-3	380	460	560	640	710	760
F-4	*Subgrade improvement recommended*					
Minimum asphalt wearing course thickness	30	70	80	100	120	130

Source: Used with permission: Courtesy of NSSGA (formerly NSA).

Example 4.2

The same criteria given in Example 4.1 will be used for this example. However, not all the information given is needed for the NSA design method.

ESALs, $w_{18} = 2,000,000$
Subgrade CBR = 3.3%

Good subgrade moisture drainage conditions.

Solution

From the CBR value and Table 4.9, the soil subgrade (estimated at best with the limited soil information) is considered *Poor*. Table 4.10 and the 2,000,000 ESALs indicates that the traffic design index is *DI-5*. Table 4.11 defines that a DI-5 traffic level over a poor subgrade should have 580 mm of pavement thickness of which there should be a minimum of 90 mm of a dense graded hot mix asphalt wearing course.

The asphalt pavement structure should consist of 490 mm of crushed aggregate base/subbase and a 90 mm hot mix asphalt wearing course, not much different than what was determined in Example 4.1. There is not enough information given to determine if the subgrade is penetrated by frost.

The Asphalt Institute design method

The Asphalt Institute design method characterizes an asphalt pavement as a multi-layered elastic system. The method is based on two assumed stress–strain conditions:

1. The vehicle load, W, is transmitted to the pavement surface through the tire as a uniform vertical pressure, P_0. The induced stresses are then spread down through the pavement structure that results in a reduced vertical stress at the subgrade surface. The maximum vertical pressure intensity decreases with the depth or thickness of the pavement structure (Figure 4.5).
2. The vehicle load causes the pavement structure to deflect, which creates both compressive and tensile stresses and strains in the asphalt pavement portion of the pavement structure. Induced horizontal tensile strains, ϵ_t, are calculated at the bottom of the asphalt layer and the vertical compressive strains, ϵ_c are calculated at the interface with the top of the subgrade (Figure 4.6). On a full-depth asphalt pavement, both these strains are calculated at the same location, the top of the subgrade.

The Asphalt Institute design method distinguishes between pavement structural damage, such as fatigue or fatigue cracking, permanent deformation that results from the deformation of the soil subgrade, and plastic deformation that may occur in the asphalt pavement layer. The principal criterion for the fatigue of an asphalt pavement is the horizontal strain at the bottom of the asphalt layer while the principal criterion for permanent deformation due to subgrade deformation is the vertical strain at the top of the subgrade. Cracking of an

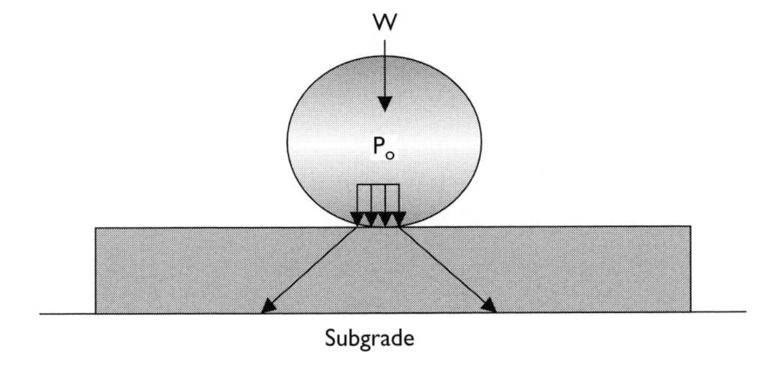

Figure 4.5 Vehicle load pressures on an asphalt pavement.

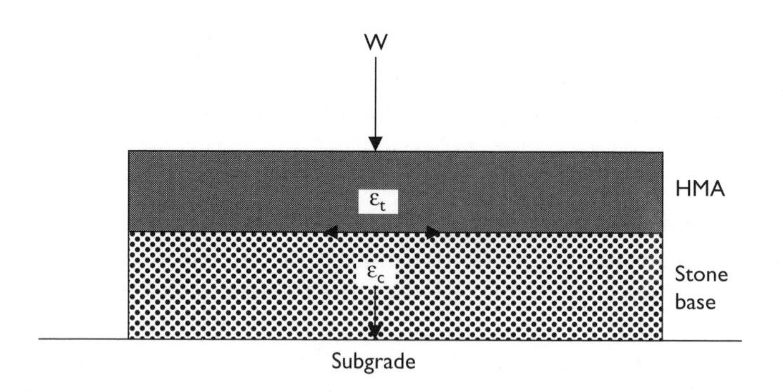

Figure 4.6 Structural layers illustrating compressive and tensile strains.

asphalt pavement layer occurs from repeated tensile strains, of which the maximum occurs at the bottom of the asphalt layer as shown in Figure 4.6. The crack, once initiated, propagates upward causing gradual weakening of the asphalt pavement. The development of the permanent pavement deformation, such as rutting, due to subgrade deformation occurs from the accumulation of permanent strain throughout the entire pavement structure (Whiteoak 1991). The Asphalt Institute has determined that if the vertical strain in the subgrade is below a certain value, excessive subgrade deformation will not occur, thus reducing the chance of subgrade related deformation occurring upward in the pavement structure.

The design method determines the thickness of the asphalt pavement structure so that the critical levels of strain for the selected pavement material will not be exceeded during the pavement's design life. The multi-layered elastic design method requires that the individual layer's materials be characterized by their modulus of elasticity and Poisson's ratio. The materials are assumed to be homogenous and isotropic. Poisson's ratio of asphalt mixtures has been determined to be relatively consistent with a typical value of 0.4. Several computer programs are available that calculate the asphalt pavement's stresses and strains at any position in a multilayer pavement system. The Asphalt Institute's *DAMA* program and the Shell Oil Research Limited BISAR-PC (Bitumen Stress Analysis in Roads–Personal Computer) are two such programs. Two thicknesses, one for each critical strain value, are calculated for various combinations of subgrade and traffic loading conditions. The larger of the two values are used to prepare thickness design graphs. These graphs reduce the need for individual stress and strain calculations for a specific asphalt pavement structure. The Asphalt Institute design method requires the following four criteria to determine the proper thickness of an asphalt pavement:

1 The traffic loading of the pavement in ESALs.
2 Material properties, especially the resilient modulus of the soil subgrade.
3 The mean annual air temperature (MAAT) of the pavement location.
4 The desired base materials: granular or hot mix asphalt.

Traffic loading

Similar to most other design methods, the traffic loading of a pavement is measured in the total number of 80 kN ESALs that will occur during the pavement's design period. The same methods of analysis for total ESAL determination are used as discussed in Chapter 3.

Material properties

The Asphalt Institute design method also characterizes the strength properties of the soil subgrade in terms of its resilient modulus, M_R. The determination of the actual resilient modulus of the soil is preferred rather than converting CBR values. In a standard method of test for the resilient modulus of subgrade soils, a specially prepared and conditioned soil specimen is subjected to repeated applications of axial deviator stress of fixed magnitude, duration, and frequency. During the test, the specimen is subjected to a static all-around stress in a triaxial pressure chamber. The test is intended to simulate the conditions that exist in pavements subjected to moving wheel loads (Wright and Paquette 1987). The resilient modulus of soil depends on the magnitude of the stress under the design load. The approximate confining pressure and deviator stress used in determining the resilient modulus should closely match the anticipated in-place subgrade stresses. If the anticipated in-place stresses are difficult to determine, then a confining pressure of 14 KPa and a deviator stress of 41 KPa are reasonable values for most applications (The Asphalt Institute 1986b). If it is not possible to determine the actual resilient modulus, the modulus can be estimated from the CBR percentage using equation 4.3 or from the R-value using equation 4.4. These conversion equations are not applicable to granular or crushed untreated materials used for base and sub-base courses. The proper determination of the subgrade strength is a key element in the design of asphalt pavements. A sampling and testing plan of the proposed subgrade should be developed as part of the pavement design process. The samples should also be obtained randomly. The following guidelines will assist the designer in developing a sampling and testing plan:

1. Test all subgrade materials that are expected to be within 0.5 m of the planned subgrade elevation. If fill is going to be used, test the soil at the borrow or fill sources. The test should represent as closely as possible the condition of the subgrade that is likely to control the design.
2. If the soil profile indicates that there is a nonsystematic variation in the subgrade soil type along the pavement location or alignment, a random sampling plan should be used within the boundary of each soil type. If the soil profile indicates that there is a systematic variation in soil type, the subgrade alignment can be subdivided by soil type for testing. Separate pavement designs may be needed for each soil type, so adequate samples so be taken. In these situations it is desirable to establish a minimum pavement length or area for which separate designs are considered feasible. If the various soil type areas are not large enough to justify separate pavement designs, a single design should be made based on the worst soil type.
3. Schedule a sufficient number of tests for a statistical selection of the design subgrade value. Six to eight tests are recommended for each soil type. More than nine tests are unnecessary.
4. If a sample from a test location has a value so low that it indicates an extremely weak subgrade area, then additional samples should be taken and tested to determine the boundaries of that area. Such areas may require increases in pavement thickness, or removal and replacement of the weak subgrade with an improved subgrade material, to provide uniform support for the entire length or area of the weak section. Test values that were representing those weak areas should now be omitted (since they were removed) from the design subgrade resilient modulus determination.

(The Asphalt Institute 1999)

Table 4.14 Subgrade resilient modulus design limits

Traffic levels, 80 kN ESALs	*Design subgrade percentile value*
<10,000	60
10,000–1,000,000	75
>1,000,000	87.5

Source: The Asphalt Institute (1999); © The Asphalt Institute; reprinted with permission.

Individual subgrade test values from six to eight tests are used to determine a design subgrade resilient modulus, M_R. The individual results can be arranged in descending order and plotted graphically as a cumulative distribution (Example 4.3). The values may also be entered into a calculator with statistical functions. The design subgrade resilient modulus is defined as a subgrade resilient modulus value that is less than 60, 75, or 87.5 percent of all the test values in the section. The applicable percentile depends on the particular traffic level as shown in Table 4.14.

The Asphalt Institute design method is a conservative approach to the thickness determination of an asphalt pavement. For any given set of test values, the design resilient modulus should be selected as the traffic varies. If a high volume of traffic is anticipated the design resilient modulus is adjusted by the applicable percentile to a lower value than would be used for a lower volume of traffic. This adjustment provides a greater pavement thickness for high volume traffic than is actually required, by artificially lowering the soil subgrade resilient modulus values. To ensure a more conservative design, these lower values increase the asphalt pavement thickness for higher volumes of anticipated traffic.

Example 4.3

The following resilient modulus values were obtained from multiple samples from the subgrade of a proposed subdivision entrance street: 72, 62, 82, 41, 62, 93, 52, and 103 MPa. Determine the design resilient modulus for traffic loading of 8,000 ESALs.

Solution

Put the test resilient modulus values in descending order:

Resilient modulus (MPa)	*Number equal to or greater than*	*Percent equal to or greater than*
103	1	(1/8) = 12.5%
93	2	(2/8) = 25%
82	3	(3/8) = 37.5%
72	4	(4/8) = 50%
62	6	(6/8) = 75%
62	6	(6/8) = 75%
52	7	(7/8) = 87.5%
41	8	(8/8) = 100%

Table 4.15 Untreated aggregate base and sub-base quality requirements

Test	*Test requirement*	
	Base	*Sub-base*
CBR, minimum (%)	80	20
Liquid limit, maximum (%)	25	25
Plasticity index[1], maximum (%)	NP[2]	6
Sand equivalent, minimum (%)	35	25
Maximum passing the 75 μm sieve (%)	7	12

Source: The Asphalt Institute (1999); © The Asphalt Institute; reprinted with permission.

Notes
1 Depending on the material, test the plasticity index or the sand equivalent.
2 Non-Plastic.

From Table 4.14 the design resilient modulus for a traffic level of 8,000 ESALs should be in the 60 percentile. The design resilient modulus can be determined by graphically plotting the results or by interpolation.

From interpolation, the design resilient modulus, M_R, is *68 MPa*.

The Asphalt Institute thickness design charts are based on the assumption that for the asphalt concrete portion of the pavement, the mixture be a high quality dense graded hot mix asphalt. The charts were derived from extensive studies of the dynamic modulus–temperature relationships for dense graded asphalt concrete. It is not necessary to test the strength properties of the hot mix asphalt in order to use the design charts. Assumptions are made that the hot mix asphalt will contain some crushed aggregate with a minimum of 50 percent crushed particles for the wearing course and that the compactive effort on the mixture in the field will give a maximum air void content of 8 percent.

The Asphalt Institute thickness design method also includes some designs using untreated (unbound) aggregate as the base and sub-base courses, with hot mix asphalt just for the wearing course. The untreated aggregate should be crushed and should meet the requirements of Table 4.15.

The untreated aggregate base and sub-base courses should be compacted at the approximate optimum moisture content to achieve a minimum density of 100 percent of the maximum laboratory density. The compaction load and contact pressure should be as high as that which the material being compacted will support without displacing the base or sub-base course or damaging the subgrade below. As stronger layers are placed, the load and contact pressure may be increased to obtain the final compaction density (The Asphalt Institute 1999).

Temperature

The final item to consider before selecting a thickness from the Asphalt Institute design charts is the mean annual air temperature. The dynamic modulus of hot mix asphalt mixtures is highly dependent upon pavement temperatures. The Asphalt Institute developed a temperature versus dynamic modulus of a typical dense graded hot mix asphalt mixture in the development of their thickness design charts. To simulate the effects of temperature as it changes throughout the year, three typical distributions of the mean monthly air temperature,

Table 4.16 Asphalt Institute design air temperatures

Mean annual air temperature (MAAT)	*Freeze/frost effects*
≤7 °C	Yes
15.5 °C	Possible
≥24 °C	No

representing three typical climatic regions were used. The appropriate dynamic modulus values were selected for each temperature. The three mean annual air temperatures are given in Table 4.16. Also taken into consideration in the development of the design charts is the effect of temperature on the resilient modulus of the subgrade and of the granular aggregate base and sub-base materials. For the subgrade, this was accomplished by using an artificially increased subgrade resilient modulus value to represent the freezing period and an artificially reduced subgrade resilient modulus to represent the thawing period. The same process is also used in the development of the design charts for the asphalt pavements containing granular bases and sub-bases (The Asphalt Institute 1999).

Pavement design procedure

The Asphalt Institute design method incorporates two fundamental concepts; that the design method considers the vertical compressive strain at the surface of the subgrade and the horizontal tensile strain on the underside of the lowest pavement layer that incorporates an asphalt binder. The following steps are used in determining the final asphalt pavement thickness:

1. Determine the design traffic in 80 kN ESALs for the pavement design period.
2. Determine the design resilient modulus, M_R, of the proposed subgrade.
3. Determine if the pavement will be a full-depth asphalt pavement or an asphalt pavement incorporating a granular material as the base and sub-base layers. In this design method, untreated granular base and sub-base thickness values are summed together.
4. Determine which of the three MAAT will satisfy the environmental conditions the pavement will be under.
5. If using a granular base, determine if the minimum asphalt concrete thickness values from Table 4.18 are acceptable due to other design and construction influences. If using a full depth asphalt pavement, Table 4.17 gives the minimum wearing course thickness values. If these values are not acceptable, consider an alternative design. The minimum total thickness for full-depth pavements is 100 mm.
6. Determine which Asphalt Institute design chart to use, based on MAAT values, base material selection (untreated granular or asphalt concrete), and granular base thickness values, 150 or 300 mm.
7. Using the design ESAL's value and the design M_R value, pick from the chart curves the minimum asphalt concrete thickness that will meet the conditions determined.
8. Evaluate the thickness value for economical and construction considerations. Possibly redesign using a different granular base thickness value or adjusting the M_R or consider improving the subgrade materials.

Figures 4.7–4.15 are reprints of the design charts that appear in the Asphalt Institute manual on thickness design.

Table 4.17 Minimum thickness of a wearing course for a full-depth asphalt pavement design using the Asphalt Institute design method

Traffic conditions	*Traffic ESALs*	*Minimum thickness of asphalt concrete wearing course* (mm)
Passenger car parking lots, driveways, rural roads	≤ 10,000	25
Medium truck traffic	10,000–1,000,000	40
High truck traffic	≥ 1,000,000	50

Source: The Asphalt Institute (1989); © The Asphalt Institute; reprinted with permission.

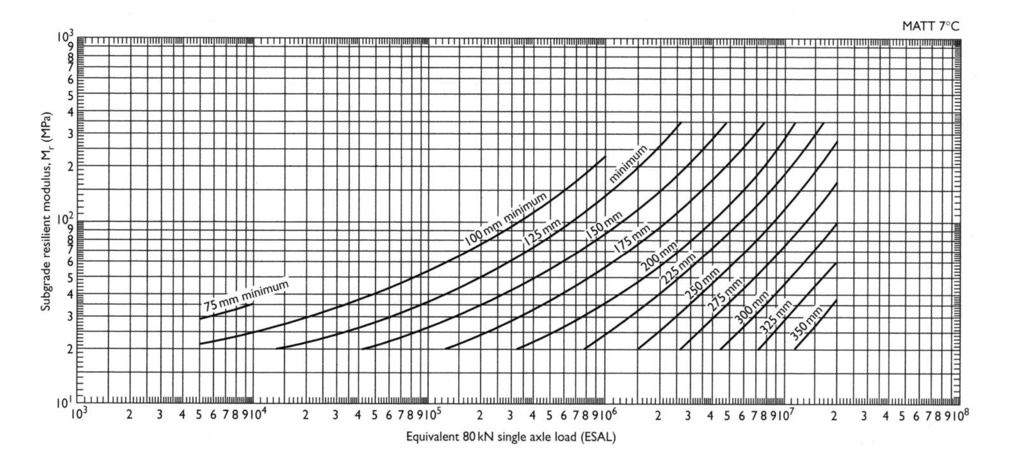

Figure 4.7 Untreated aggregate base, 150 mm thickness.

Source: The Asphalt Institute (1999); © The Asphalt Institute; reprinted with permission.

Example 4.4

The same criteria given in Example 4.1 will also be used for this example. However not all the information given is needed for the Asphalt Institute design method.

ESALs = 2,000,000
Subgrade CBR values are 3.3, 5.0, 6.2, 4.1, 5.4, 3.0
The subgrade has good subgrade drainage conditions
The pavement will be in an environment with a mean annual air temperature of 15 °C.
Determine the pavement thickness for a full-depth design and the thickness for a design when using 300 mm of an untreated aggregate base.

Solution

The pavement design traffic has been stated as *2,000,000 ESALs*.

The next step involves determining the design resilient modulus, M_R for the subgrade.

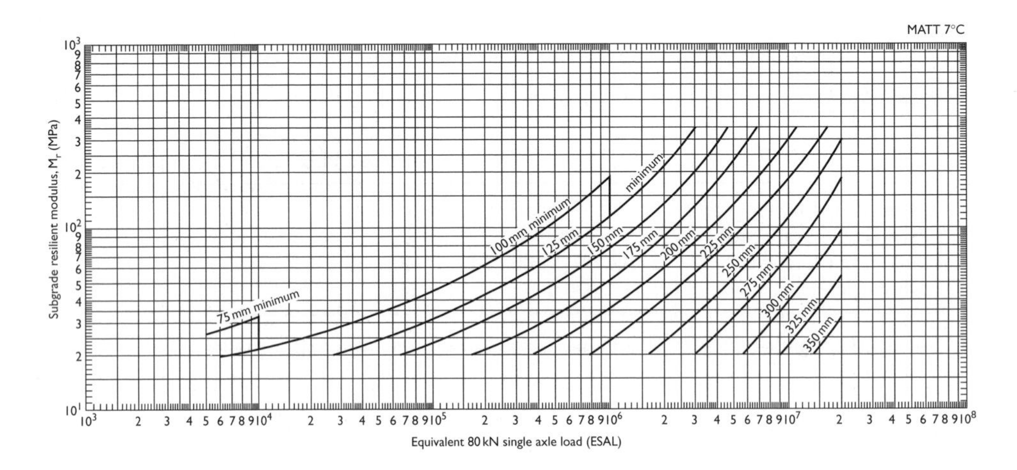

Figure 4.8 Untreated aggregate base, 300 mm thickness.

Source: The Asphalt Institute (1999); © The Asphalt Institute; reprinted with permission.

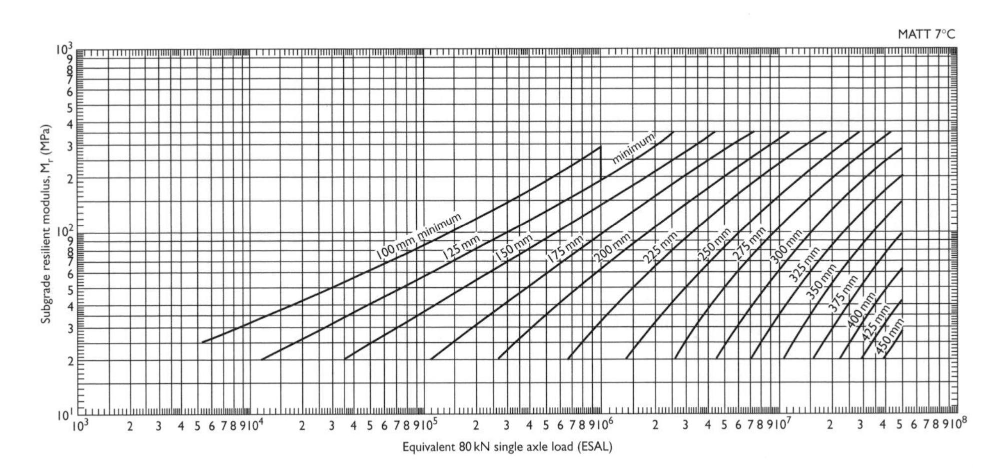

Figure 4.9 Full-depth asphalt concrete.

Source: The Asphalt Institute (1999); © The Asphalt Institute; reprinted with permission.

Tabulate the CBR results, determine the correct value for the traffic percentile and then convert the design CBR value to the resilient modulus using equation 4.3.

CBR (%)	*Number equal to or greater than*	*Percent equal to or greater than*
6.2	1	(1/6) = 16.7%
5.4	2	(2/6) = 33.3%
5.0	3	(3/6) = 50%
4.1	4	(4/6) = 66.7%
3.3	6	(5/6) = 83.3%
3.0	6	(6/6) = 100%

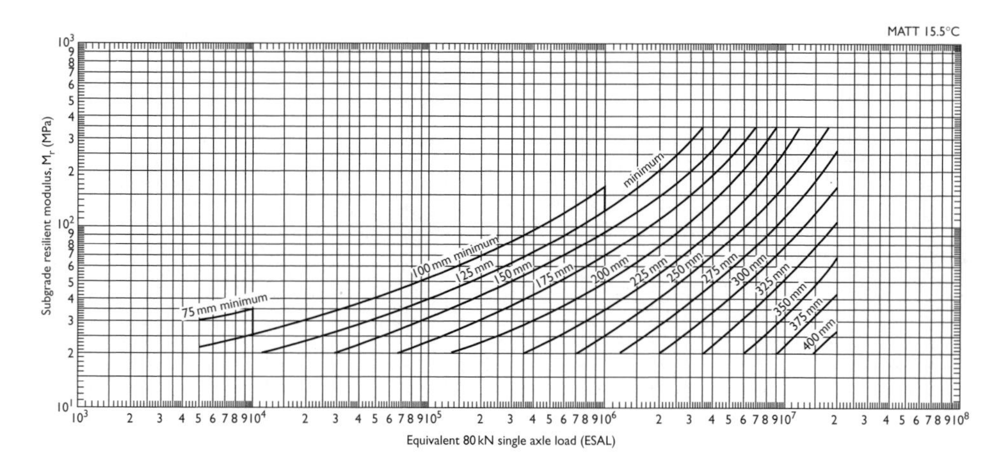

Figure 4.10 Untreated aggregate base, 150 mm thickness.

Source: The Asphalt Institute (1999); © The Asphalt Institute; reprinted with permission.

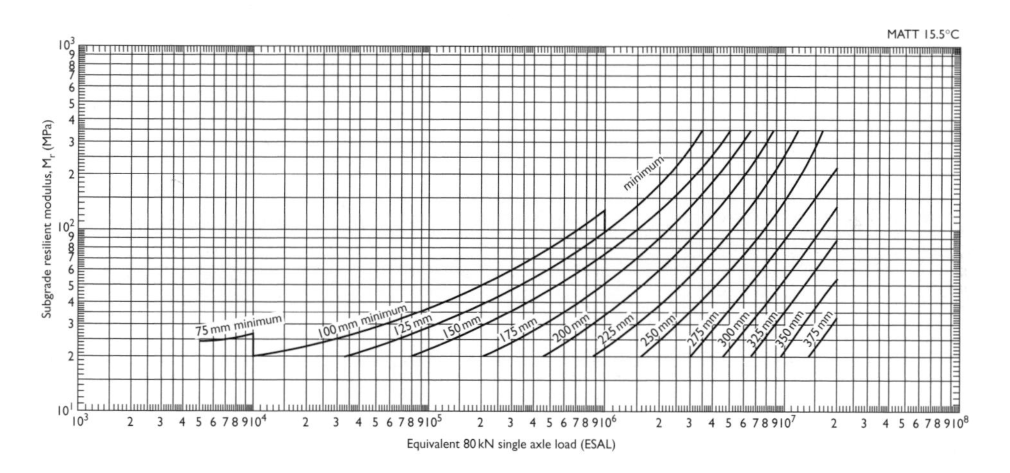

Figure 4.11 Untreated aggregate base, 300 mm thickness.

Source: The Asphalt Institute (1999); © The Asphalt Institute; reprinted with permission.

From Table 4.14, it has been determined that the design resilient modulus based on the traffic is 87.5 percent. From interpolation, the design CBR value is 3.2; thus, the approximate design resilient modulus is *33 MPa*.

Since the mean annual air temperature is approximately 15.5 °C and the design requests a 300 mm untreated aggregate base, Figures 4.11 and 4.12 for the full-depth asphalt pavement will be used.

From Figure 4.11, the design thickness of the asphalt pavement portion is *250 mm* giving a total pavement thickness of *550 mm*.

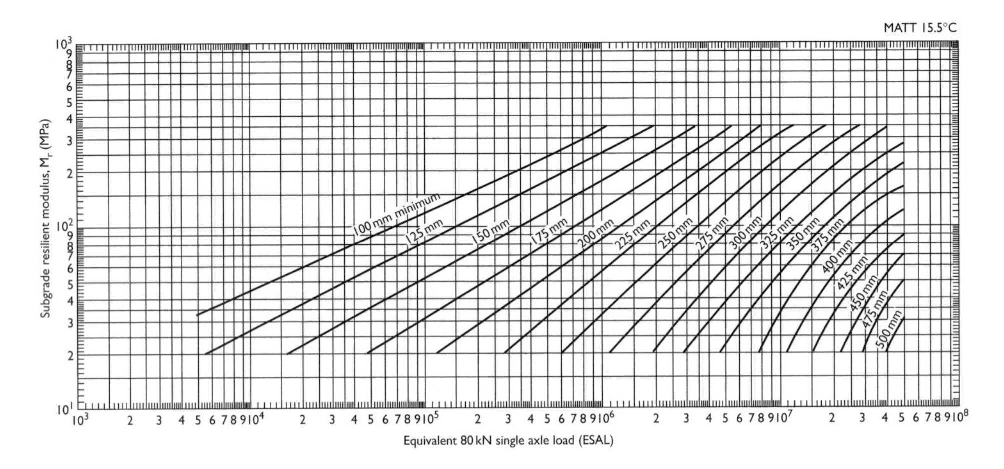

Figure 4.12 Full-depth asphalt concrete.

Source: The Asphalt Institute (1999); © The Asphalt Institute; reprinted with permission.

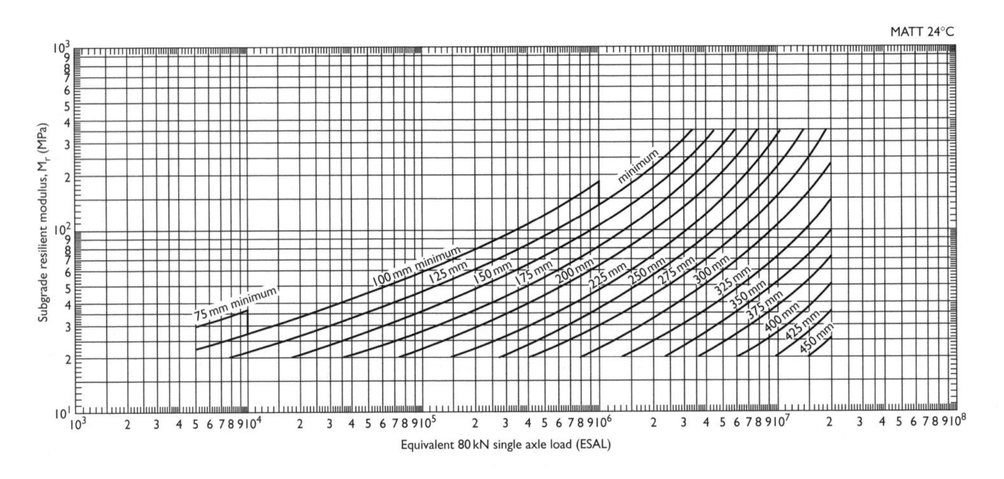

Figure 4.13 Untreated aggregate base, 150 mm thickness.

Source: The Asphalt Institute (1999); © The Asphalt Institute; reprinted with permission.

From Figure 4.12 the design thickness of the full-depth asphalt pavement is *300 mm*. Table 4.18 states that for the given traffic level, the wearing course minimum thickness is 50 mm. One possible design is a pavement that would consist of 50 mm of wearing course asphalt concrete and 250 mm of base course or binder course asphalt concrete.

It should be noted that the design charts do not distinguish between the base and sub-base when using untreated granular aggregate. In this example, it may be more cost effective to break up the 300 mm into a 150 mm sub-base and a 150 mm base, allowing more economical materials that meet the requirements of Table 4.15 to be used as a sub-base. It would also be wise to consider a single 150 mm untreated base course, which would give a

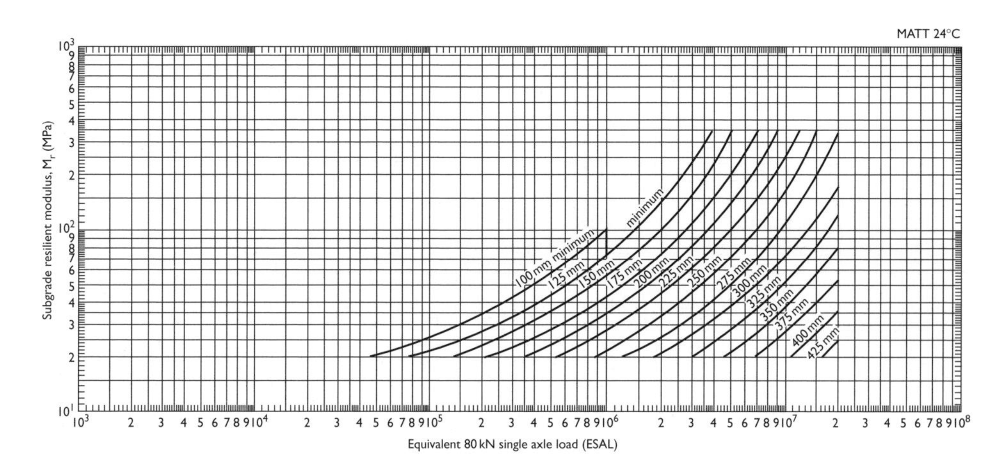

Figure 4.14 Untreated aggregate base, 300 mm thickness.

Source: The Asphalt Institute (1999); © The Asphalt Institute; reprinted with permission.

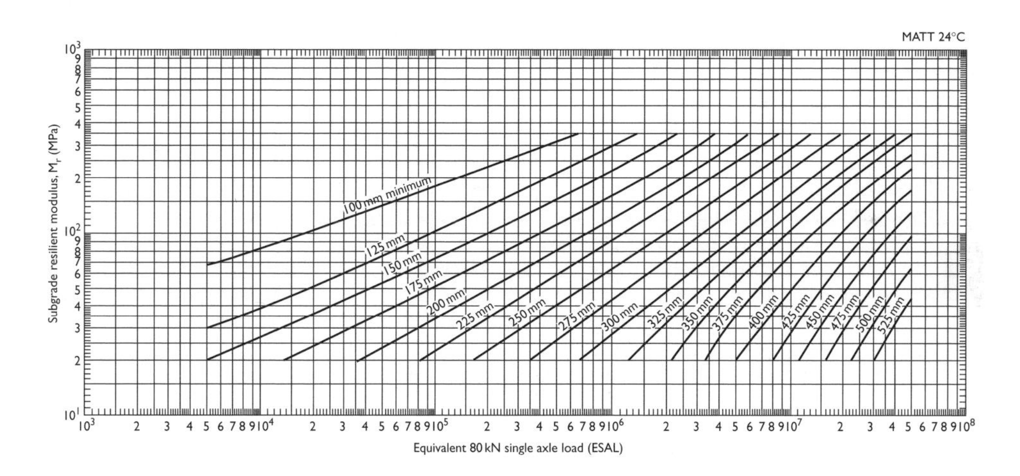

Figure 4.15 Full-depth asphalt concrete.

Source: The Asphalt Institute (1999); © The Asphalt Institute; reprinted with permission.

thicker layer of asphalt concrete. Figure 4.10 gives a value of 275 mm of asphalt concrete when used with 150 mm of untreated base course. This gives a total pavement thickness of *425 mm*.

As was illustrated, three possible pavement thickness designs could have met the traffic and subgrade requirements. An economical analysis, using the cost of materials in-place for the project, along with consideration of construction restraints would determine the most desirable design.

Table 4.18 Minimum thickness of asphalt concrete over granular aggregate bases using the Asphalt Institute design method

Traffic conditions	*Traffic ESALs*	*Minimum thickness of asphalt concrete* (mm)
Passenger car parking lots, driveways, rural roads	≤ 10,000	75
Medium truck traffic	10,000–1,000,000	100
High truck traffic	≥ 1,000,000	125

Source: The Asphalt Institute (1999); © The Asphalt Institute; reprinted with permission.

Table 4.19 Simplified approach to thickness design for parking facilities

Traffic description	*Soil subgrade condition*		
	Excellent/good	*Fair*	*Poor*
Full-depth asphalt pavement thickness (mm)			
Residential driveway	100	100	100
Parking lot < 200 stalls	100	100	100
Parking lot 200–500 stalls	100	100	120
Parking lot with up to 20[1] trucks per day, and entrance and traffic lanes	100	140	190
Truck unloading area	220	260	320
Crushed aggregate base with asphalt wearing course thickness (asphalt/aggregate (mm))			
Residential driveway	50/120	50/150	50/150
Parking lot < 200 stalls	60/150	60/150	60/180
Parking lot 200–500[1] Stalls	75/180	75/230	100/300
Parking lot with up to 20[1] trucks per day, and entrance and traffic lanes	75/150	100/150	120/220
Truck unloading area	120/300	150/350	180/420

Note
1 Pavements with greater than 500 stalls or 20 trucks a day should use a full design method.

In summary, Examples 4.1, 4.2, and 4.4 contained the same traffic and soil conditions. Three different methods were used to determine the required thickness of the asphalt pavement. The results are summarized as follows:

- The AASHTO method determined an asphalt concrete wearing course thickness of 100 mm and a crushed stone base course thickness of 400 mm for a total pavement thickness of 500 mm.
- The NSA method determined an asphalt concrete wearing course thickness of 90 mm and a crushed stone base course thickness of 490 mm for a total pavement thickness of 580 mm.
- The Asphalt Institute method gave three design options. One option was a full-depth asphalt pavement with a total thickness of 300 mm. It also determined a 300 mm crushed

stone base course with 250 mm of asphalt concrete giving a total pavement thickness of 550 mm. If the crushed stone base course thickness was reduced to 150 mm, the thickness of the asphalt concrete becomes 275 mm with a total pavement thickness of 425 mm.

The three design methods described usually give similar thickness requirements along with other thickness design methods. Traffic levels and subgrade soil strengths are the significant variables among all design methods for flexible pavements. Most of the design methods are based on some aspect of the AASHTO method of flexible pavement design. The designer for low-volume roads or parking facilities does necessarily have to use a method that has as many input variables as the AASHTO method. Low-volume type pavements have their thickness governed by the soil subgrade strength with very little impact by traffic. Using thickness design tables that group such types of pavements together is generally adequate for low traffic volumes. Significant truck loading and freeway or other similar pavements should be designed based on their individual requirements using a method such as the AASHTO or Asphalt Institute design methods. In 2002, AASHTO is updating their design method by using a mechanistic or analytical approach.

Table 4.19 is useful as a simplified approach to determining the thickness of an asphalt pavement for parking facilities.

Subgrade modification

All thickness designs should have an economic analysis performed using the cost in-place of the materials for the project. In many instances, the design of an asphalt pavement that is placed over a low strength or poorly rated subgrade is not the most economical. Modification of the subgrade to improve its strength and drainage properties can reduce the thickness of asphalt concrete and/or granular base materials required. The loss of subgrade support is the most frequent cause of premature pavement failures in parking facilities and other low-volume type pavements. When a suitable soil is brought to the correct moisture content, compacted to the required density, shaped to the correct grade and protected from moisture, failure related to the subgrade is significantly reduced. Subgrade modification involves several processes that either physically modifies or replaces the soil used in the subgrade. Modifying the subgrade may be a necessary step due to extremely poor subgrade conditions, such as very poor drainage conditions or CBR values less than 3 percent. It may also be an economic requirement, since improving the subgrade will generally reduce the required thickness of pavement. Improving the soil in the subgrade may be required to

- Add strength to the soil.
- Improve the frost susceptibility where the frost penetration is deeper than the base course.
- Improve expansive, weak, or highly plastic soils.
- Avoid construction delays during rainy conditions by providing a stable working platform.

The poor soil in the subgrade can also be removed and replaced with select higher quality borrow soils, sand, or soil-aggregate blends. Modifying or stabilizing the soil can be done with pozzolonic materials such as flyash, lime, Portland cement, asphalt emulsions, or calcium chloride. All of these stabilizing materials are mixed in-place with the soil. Hydrated lime is especially effective with soils that contain a significant quantity of clay. These types of treatments differ from stabilized bases in that a smaller amount of the lime, flyash, or cement is used. The percent age of the soil passing the 75 μm sieve usually determines which stabilizing material is to be used. Soils that are well graded and granular with less than 35 percent passing the 75 μm sieve and a plasticity index less than 20 can use Portland cement. Portland cement is usually added at a rate of 7–14 percent by volume of the subgrade to a depth of 150 mm. Soils that have less than 20 percent passing the 75 μm sieve and have a plasticity index of less than six can use Portland cement or asphalt emulsions. Asphalt emulsions are added at a rate of 4–7 percent and mixed in to a depth of 50–125 mm. Lime should be used with clay, silt-clay, and similar materials that also have a plasticity index greater than 10. Unified Soil Classification candidates for hydrated lime are CH, CL, ML, MH, CL-MH, SC, SM, GC, and GM. Lime and flyash reduce the soil plasticity, increase the workability, reduce swell, and increase the strength of clay soils. Lime has little beneficial effect on non-plastic soils (NAPA 1991). Quicklime reacts with water to produce hydrated lime, using up water at the rate of 30 percent by weight. This water is drawn out of wet soil and dries it significantly. The lime also provides a source of dry solids to further reduce the moisture content. In the natural state, clay particles have tremendous surface area. This surface area, due to its particularly charged chemical composition, has a weak buffer zone of ions, such as sodium. Instability in these zones allows clay to draw enormous amounts of water around particles. This ability to soak up water can cause destructive swelling and strength loss in the subgrade. The drawing power of the clay mineral can be reduced by the creation of a stable calcium ion buffer in the interlayer regions of the clay layers. A calcium buffer satisfies the demands of the clay mineral surface and equalizes the conditions that permit expansion and strength loss. The water layer the clay minerals can support shrinks due to the calcium ion buffer. As a result, the clay particles come in contact with each other. This contact causes the high plasticity index clay to become low plasticity index or non-plastic material. Cation exchange in expansive clay is driven by the availability of calcium. The more expansive the clay, the more the calcium that is required. Lime provides a significant amount of active calcium. Lime stabilization can be accomplished in soils with no clay content by the addition of flyash. Lime stabilization requires high alkalinity, with a pH of 12.3 and sufficient water to drive the cation exchange. A pH test with the soil and lime is a necessary requirement in the determination of the amount of lime and water that will need to be added to the soil (Chemical Lime Company 2000). Lime is usually added at a rate of 3–6 percent to a depth of 150 mm. Enough water needs to be added to bring the in-place subgrade moisture content to 5 percent above the optimum moisture content of the soil to allow the lime to form quicklime.

Once it has been determined to modify or improve the soil subgrade, the thickness design is evaluated again. Soil modification is usually successful enough to consider it meeting sub-base requirements. The thickness design is now determined by considering the soil as an excellent rating or by actually determining the new resilient modulus of the modified subgrade, through laboratory blending and testing.

Miscellaneous applications

Shoulders

The AASHTO definition of a roadway or highway shoulder is the portion of the roadway that is contiguous with the mainline or travel pavement that exists for the accommodation of stopped vehicles for emergency use and for the lateral support of the mainline base and

Plate 4.3 Roadway shoulder.

sub-base courses (Plate 4.3). The shoulder also assists in draining water away from the wearing course. The 1993 AASHTO method of flexible pavement design does not give any guidelines for the proper thickness determination for paved shoulders. Shoulders will not undergo nearly the number of ESALs that the mainline pavement will, however, heavy truck encroachment onto the shoulder is a major cause of shoulder distress. This loading is typically slow moving or stopped. Economics generally do not permit the shoulder thickness to be of the same thickness as the mainline pavement, nor is it necessary. Construction practices will prefer the thickness to be the same, since it allows the asphalt paver to placing the entire mainline width and the shoulder width all in a single pass. AASHTO has published some guidelines when considering the design of a paved shoulder. The AASHTO Joint Task Force prepared these guidelines on Pavements in 1983:

1 Predict shoulder thickness design upon criteria which will reflect the magnitude and frequency of loads to which the shoulder will be subjected.
2 Integrate shoulder drainage with the overall pavement subdrainage design.
3 Avoid the use of aggregate bases having a significant percentage of 75 μm passing material in order to prevent frost heaving, pumping, clogging of the shoulder drainage system, and base instability.
4 Consider the use of dense graded hot mix asphalt as a paving material for the shoulder.
5 Have a definite program for shoulder maintenance.

(AASHTO 1993)

If the decision has been made to pave the shoulders the same thickness as the mainline pavement, then no other criteria need to be considered. Economic conditions generally dictate the shoulder is placed at a minimal thickness. The shoulder needs to be able to support the infrequent loading of a heavy truck, but generally the strength of the subgrade will determine the proper thickness of the shoulder. For high traffic roadways with a design ESAL of 10 million or more, using an estimate of 2 or 3 percent of the design lane ESALs as the shoulder traffic, will provide some protection from damage by the occasional heavy truck. Using the design indices provided in Table 4.10 and following the thicknesses provided in Tables 4.11 and 4.13 will provide adequate shoulder thickness. If a full-depth asphalt shoulder is desired, Table 4.18 gives some minimal thicknesses, but it also may be desirable from construction practice and economics to have at least the mainline base course and the shoulder base course the same thickness, so that both may be paved at the same time. The reduced construction costs may more than compensate for the additional material costs for the shoulder.

Refuse container pads

The pavement surrounding commercial refuse containers and other similar applications require special consideration. This portion of the pavement can be distressed due to the activities involving the dumping of the container into a usually heavily loaded refuse truck. For example, the thickness design for a fast food restaurant may adequately support the daily movement of passenger vehicles, but may not be able to support a stationary loaded truck dumping the container. The problem is usually overcome by building up the thickness of the pavement and requiring a high traffic type asphalt mixture (Chapter 5) in the area of the refuse container. Some designers may require the use of a Portland cement concrete pad in this area (Plate 4.4), but that is not always necessary if special attention is paid to the proper

Plate 4.4 Portland cement concrete refuse container pad.

mixture type and if the pavement thickness is increased when using asphalt mixtures. Due to relatively limited vehicle movements in these areas, a design ESAL determination will not always give the necessary thickness required.

It is recommended that these areas are constructed of a minimum of 220 mm of full depth asphalt pavement or a minimum combination of a 150 mm high type asphalt wearing course mixture and a 350 mm crushed stone base. The same consideration can also be given to truck trailer landing gear pads where there is a high loading point on the pavement. Portland cement concrete strips are usually used in these applications with the advantage of a contrasting color with the asphalt pavement that allows the driver to easily locate the trailer landing gear on the pad. Construction is also simpler, allowing the pad strip to be poured separately, in a manner similar to that used when constructing an asphalt pavement with Portland cement concrete curb and gutter.

Alleys

Service alleys in commercial areas can receive a significant amount of loading due to the amount of delivery trucks that pass through on a daily basis. The trucks vary in size from panel vehicles to semi-trailer combinations. This area should be designed separately from a mainline parking area, which may consist only of passenger vehicle traffic. The amount of ESALs that may be applied in service areas could generate pavement thicknesses approaching that of freeways. Some designers use an asphalt pavement for the passenger vehicles and a Portland cement pavement for the service alleys (Plate 4.5).

Plate 4.5 Portland cement concrete service alley with an asphalt pavement parking lot.

Plate 4.6 Asphalt pavement used for a service alley.

An asphalt pavement for a service alley (Plate 4.6) will perform adequately if proper attention is given to mixture selection and thickness requirements. These areas are typically constructed separately from the mainline parking areas.

There are several approaches or methods available for the proper structural design of a asphalt pavement. The methods vary from country to country and even vary among the different highway agencies. Many methods are empirical and have developed using local environment and materials and perform adequately under such conditions. Almost all design methods involve around two variables, the subgrade strength and the pavement loading conditions. Most design methods, when inputted with the same variables, will give similar thickness values, with most values on the conservative side. Most empirical methods tend to overdesign the thickness of the asphalt pavement. For the designer, especially one designing asphalt pavements for parking facilities or local roads, any of the methods described earlier will give adequate performing asphalt pavements in regard to thickness, as long as accurate estimates are given for the subgrade strength and traffic loading. In terms of low traffic volume pavements, the subgrade strength will always govern the thickness of the asphalt pavement. For high volume traffic pavements the 1993 AASHTO method of flexible pavement design or the 2002 AASHTO mechanistic method of pavement design will provide a structurally adequate pavement. If the designer is unsure on which design method to select, using the same method the local highway agency uses may provide the best results.

Chapter 5

Mixture design

A vital component in the process of constructing an asphalt pavement is the design of the asphalt mixture that will be used for the pavement. Asphalt mixtures are different from most engineering materials in that the highest strength mixture design is not necessarily the best choice for the particular asphalt pavement application. The ESALs loading, desired surface texture, environmental conditions, and other use factors are all considered in the designing of an asphalt mixture. Several design methods have been established for the proper design of dense graded asphalt mixtures. Dense graded asphalt mixtures or hot mix asphalt (HMA) is the most common material used for asphalt pavements. Dense graded mixtures contain enough fine, small, and medium size aggregate particles to fill the majority of the void space between the largest particles without preventing direct contact between all of the largest particles. This direct contact is also known as "stone on stone" contact. Open graded mixtures or open graded asphalt friction courses (OGFC) are increasingly being specified as high-speed wearing courses due to their ability to drain water quickly and provide skid resistance for vehicles. Open graded mixtures are mixtures that have insufficient fines and sand to fill all of the volume or voids between the aggregate. OGFC are open graded mixtures that also possess special aggregate and texture properties for skid resistance. This texture is not needed nor would it be aesthetically acceptable, nor would it provide the necessary surface smoothness required for a parking lot application. Open graded or OGFC mixtures are also known as porous asphalt or porous mix in many parts of the world. Due to the increase in popularity of OGFC, these are now possible to be designed using a laboratory method. Regardless whether the mixture is a dense graded HMA or OGFC, the mixture design process involves determining the type and proportion of materials. These methods specialize in determining the grading of the aggregate blend and the quantity of asphalt binder to be used. The mixture design is usually determined by the government agency or HMA producer involved with the roadway project. In private work such as parking lots, the mix design is either completed or recommended by an engineering consultant or may be provided by the HMA producer. In many instances involving private work or other small paving projects, the mix design may be adopted from previous projects or a standard "recipe" mix design that the HMA producer uses for most projects.

The stability and durability of the asphalt mixture are the two primary characteristics that are determined at the mix design stage. The workability of the mixture, or ease of placement and compaction, also needs to be balanced with the stability and durability requirements of the particular pavement that the mixture is being designed for. It is important that the asphalt mixture is designed as an economical and practical mixture. The most economical aggregate available that meets all the requirements should be used. Two types of design methodology are used in the designing of asphalt mixtures, recipe and laboratory. The recipe method uses a set proportion of aggregates and asphalt binder that has been determined through a trial and error process until the mixture performs with satisfaction on pavement applications. The "recipe" is then finalized and used for several different materials and applications over the course of several years. A recipe specification defines an asphalt mixture in terms of the aggregate grading, mixture composition, and the method by which the mixture shall be produced, placed, and compacted. Since recipe designs are based on the experience of known mixture compositions, it is not difficult for the designer to specify the mixture that will be suitable. The asphalt mixture supplier has little difficulty in making the mixture to the required composition. Recipe designs have a number of limitations in that the conditions of the traffic, subgrade, and climate to which the mixture will be subjected, will probably not be the same as those existing when the recipe design is evaluated. Variations in the components of the mixture may occur that are outside of the previous experience. If this requires modifications to the recipe, there is no means of assessing what these modifications should be or their effect (Whiteoak 1991).

The laboratory method involves determining in a laboratory the specific combination of aggregate and asphalt binder that will meet specific performance criteria for the materials selected and the specific pavement application. The laboratory methods compensate for material and application variations and are increasingly more popular than the recipe method. Laboratory design methods involves the selection of aggregate type, gradation, proportions, asphalt binder grade, and content which will optimize the engineering properties of the mixture in relation to the desired performance as a pavement. Some recipe designs had their original work as a laboratory design, but should be updated with some frequency in order to consider the variation in aggregate and asphalt binder properties. Most highway agencies responsible for constructing asphalt pavements use the laboratory method of designing asphalt mixtures.

Hot mix asphalt types

"Hot mix asphalt" is a term used generically to describe various types of mixtures of asphalt and aggregate that are produced using heat at a mixing plant. Different mixtures are used to satisfy different performance criteria, such as skid resistance, drainage, strength, and water proofing among others. The grading or gradation specifications determine which type a HMA mixture may meet. HMA can be subdivided into three types of mixtures: dense graded, open graded, and gap graded. These mixtures are mainly distinguished by their aggregate gradation. Dense graded mixtures are further divided into continuously graded or conventional mixtures, large stone mixtures, and sand asphalt. The open graded mixtures are further divided into OGFC or porous asphalt, and asphalt treated permeable bases or drainage layers (TRB 2000). The gap graded mixtures include mixtures such as rolled asphalt and stone mastic asphalt.

The aggregate in dense graded HMA is continuously or uniformly graded. Conventional HMA has a nominal maximum aggregate size of 25 mm or less. These size mixtures are used for most of the wearing and binder or intermediate courses. A large stone mixture is a dense graded mixture that has a nominal maximum aggregate size greater than 25 mm. These mixtures have a large amount of coarse aggregate and are typically used as a base course, where significant strength might be required while being protected from the effects of the environment. Sand or sheet asphalt is a dense graded mixture that has a nominal maximum aggregate size of 5 mm or less. Since these smaller aggregate particles have higher surface area, these mixtures usually have a significant amount of asphalt binder. Thus these mixtures are very durable owing to the high asphalt binder and sand content, but for the same reasons also have low stability. Sand asphalt is usually used as a waterproof membrane on a bridge deck or other similar applications. There is also some use as thin leveling courses, in order to increase the overall smoothness of the asphalt pavement. Open graded mixtures or porous asphalt have mostly coarse aggregates with little or no fine aggregate. The primary purpose of these mixtures is to provide for the rapid removal of water, either from the pavement surface or within the structural components. When used as a wearing course, skid resistance is improved due to the improved water drainage and due to the macro surface texture of an OGFC. These are not used as wearing courses for parking applications or low speed roads. The coarse and open gradation is not acceptable for pedestrian traffic nor is it very aesthetically pleasing. Typically, the nominal maximum aggregate size of an OGFC is 19–25 mm. The asphalt treated permeable base is similar to OGFC or porous asphalt but generally uses a larger nominal maximum size aggregate, usually greater than 25 mm. These permeable bases are used to drain water that may enter the structural section of a pavement, either through the surface or through the subgrade or shoulders. Gap graded mixtures are similar to dense graded mixtures, except that they are missing significant portions of the intermediate sieve (300 μm to 2.38 mm) sizes. A stone mastic mixture is an example of a gap graded mixture that is on the coarse side, while rolled asphalt consists mostly of mortar with some coarse aggregate particles. These mixtures can be used for all parts of the pavement structure and when used as a base course they are a dense, impervious structural component of the asphalt pavement. Rolled asphalt and stone mastic asphalt are mostly used as wearing courses. Rolled asphalt has had little application in the United States and is mostly used in the United Kingdom on lower volume roads. Stone mastic asphalt has had its origin in Germany and is becoming increasingly popular in the United States.

HMA mixture design

Hot mix asphalt is defined as a combination of heated and dried mineral aggregates that are uniformly mixed and coated with a hot asphalt binder. HMA can describe any asphalt mixture that is mixed while hot. In the United States, HMA usually describes a dense graded HMA. The aggregate and asphalt binder is heated prior to mixing in order to dry the aggregate and obtain sufficient asphalt binder fluidity in order to coat the aggregate. The heating, proportioning, and mixing is accomplished in a HMA mixing plant. Asphalt mixtures may be produced from many different aggregate types and combinations. Each mixture has its own characteristics suited to a specific design and construction use. The design of HMA and other mixtures mostly involves selecting and proportioning ingredients to obtain specific construction and pavement performance properties. The goal is to find an economical blend and gradation of aggregates and asphalt binder that give a mixture that has:

- Enough asphalt binder to ensure a durable compacted pavement by thoroughly coating and bonding the aggregate.
- Enough workability to permit mixture placement and compaction without aggregate segregation.

- Enough mixture stability to withstand the repeated loading of traffic without distortion or displacement.
- Sufficient voids or air spaces in the compacted mixture to allow a slight additional amount of added compaction by the repeated loading of traffic. These air voids will prevent asphalt binder bleeding or a loss of mixture stability. The volume of air voids should not be so large to allow excessive oxidation or moisture damage of the mixture.
- The proper selection of aggregates to provide skid resistance in high-speed traffic applications.

(The Asphalt Institute 1990)

Pavement durability is its resistance to weathering, cracking, and traffic abrasion. Weathering includes oxidation or hardening of the asphalt binder, moisture induced damage, and changes in the pavement aggregate soundness caused by freezing and thawing. The pavement's durability under traffic is influenced by the mixture's resistance to vehicle abrasive action and aggregate degrading. Durable asphalt pavements have enough asphalt binder to fully coat and protect the aggregates and to allow adequate compaction of the mixture. Too much asphalt binder in the mixture can lead to the pavement being over compacted and the excess asphalt binder will eventually appear on the surface of the pavement in the form of bleeding or flushing. Too little asphalt binder will produce a very thin film of protection for the aggregate and will have too many connected air voids in the compacted mixture. Design considerations that improve an asphalt pavement's durability are:

- Enough asphalt binder to provide a thick film on the aggregate particles and eliminate interconnected air voids in the asphalt pavement. The asphalt binder also holds or "glues" the aggregate particles together. It prevents penetration of air and water into the asphalt pavement, thus reducing oxidation and moisture damage.
- Selecting aggregate that resists the stripping of the asphalt binder film by water. Adhesion of the asphalt binder to the aggregate in the presence of water can be related to a chemical reaction and/or a mechanical hindrance.
- The use of sound and durable aggregate. Some aggregates tend to degrade under constant traffic and weathering, which eventually leads to a disintegration of the pavement. Mixing plants can also degrade the aggregate and produce excessive dust or minus 75 mm material. Unsound or highly absorptive aggregate absorbs so much of the asphalt binder that the same effect occurs as a mixture designed with not enough asphalt binder.

Mixture stability is its resistance to deformation caused by loading. It must always be adequate for the type and amount of traffic or loading. Stability can be directly measured in the laboratory through tests that determine the strength of the mixture. Stability of the mixture can also be ensured by requiring that it met other physical parameters, such as the amount of crushed aggregate required for the mixture. The minimum stability needed depends on the amount and type of traffic and the conditions that the asphalt pavement is used for (such as residential parking lots, etc.). High stability is required for pavements with a high amount of ESALs or special uses such as truck loading docks or traffic intersections. Asphalt pavements at traffic signals have additional stresses that are induced by accelerating and decelerating vehicles. Parking areas for trucks and heavy equipment also require HMA mixtures that have high stability. To obtain a stable mixture, the aggregate particles in the mixture must resist movement past each other. In other words, asphalt pavement mixtures with a high amount of internal friction, have high stability. Factors that influence pavement stability are, aggregate surface texture, shape and gradation, amount of asphalt binder in the mixture, stiffness of the asphalt binder, and the mixture density. The required resistance to the mixture's internal shear deformation results from a combination of aggregate friction and the cohesive forces of the asphalt binder. Aggregate surface roughness and contact area determine the amount of aggregate friction (The Asphalt Institute 1990).

The asphalt binder provides the cohesive forces that hold the aggregate particles together. The cohesive forces grow with increasing asphalt viscosity, stiffness, or $G^*/\sin\delta$. The proper asphalt binder content influences the mixture stability. Lack of sufficient asphalt binder provides insufficient cohesive forces and excessive asphalt binder overlubricates the mixture and lowers the aggregate interparticle friction. The amount of aggregate particles smaller than 75 μm also influences the stability and compactibility of the asphalt mixture. In the industry these smaller particles can also be referred to as dust, mineral filler, or fines content. A proper amount of mineral filler combined with the asphalt binder is required in a dense graded asphalt mixture to act as a mortar and assist binding the larger aggregate particles together. Mixtures with a low mineral filler content may be difficult to compact. Increasing the mineral filler content will cause the stiffness of the mixture to increase, enabling the mixture to become dense under the compaction roller, rather than just move or "shove" around. The mineral filler also adds in the workability of the asphalt mixture. Excessive mineral filler can fill the mixture's air voids, which will lead to instability and bleeding of the asphalt pavement. Excessive mineral filler mixture can also lead to compaction problems. Asphalt mixtures that are designed to have very high mixture stability can also have poor workability. All the material properties that can increase a mixture's stability can lead to poor workability, if not properly balanced. For example, a mixture that has internal friction that is too high (too harsh or stiff) may be very difficult to place and compact to the proper density. Extremely high stability can also have mixture properties that will not be very durable. Designing an asphalt mixture requires that the materials and their proportions selected will give the proper balance of durability, stability, and workability for the intended pavement application. There is no single asphalt binder content or aggregate gradation that will maximize all of the desirable mixture and pavement properties. The asphalt binder content and the aggregate proportions are selected based on optimizing the properties necessary for the specific pavement application.

Dense graded asphalt mixtures are generally grouped divided into three categories dependent on their end specific use: surface or wearing course mixtures, binder or intermediate course mixtures, and base course mixtures. HMA mixtures are also designed by considering the final thickness that the mixture will be placed. As a general guideline, the maximum size aggregate is the largest in the base course, followed by the intermediate course, and the smallest aggregate size used in the wearing course. This guideline is not a requirement and a wearing course can perform adequately as a base course, if so designed. There is also no universally recognized standard set of dense graded HMA mixture designations. There are similarities with respect to mixture types, but the geographically availability of materials and different climatic design requirements have led to various identifications. In many cases they are identified by the maximum top size aggregate of the mixture. For example, a 12.5 mm mixture would have a nominal maximum top size aggregate of 12.5 mm. Table 2.3 gives some typical dense graded mixture designations. An internationally recognized standard such as ASTM D3515, *Standard Specification for Hot Mixed,*

Hot Laid Bituminous Paving Mixtures, provides some standardized designations for HMA mixtures, both OGFC and dense graded. It is important for the designer to determine what nomenclature preference is used in the geographical area that the pavement will be constructed. For example, an architect for a national hardware store chain based in Chicago, Illinois is specifying that a mixture be designed for an asphalt parking lot for a theme park located in Orlando, Florida. The architect follows the Illinois Department of Transportation specifications and requests a Class I type 3C mixture. The contractor in Orlando, Florida does not recognize that mixture, but would if the architect specified a Florida Department of Transportation FC-3. The SHRP SuperPave® (Superior Performing Asphalt Pavement) mixture designations have alleviated some of this regional confusion by identifying dense graded asphalt mixtures by their maximum top size, similar to ASTM D3515. Most HMA mixtures have a substantial overlap of mixture application. The aggregate top size in an asphalt mixture governs the minimum and maximum pavement layers or lift thicknesses. The minimum thickness for a wearing course usually varies from two to three times the maximum aggregate size. It should be recognized that the actual minimum thickness of any course is that which can be placed in a single lift and compacted to the required density and smoothness. The maximum thickness is usually controlled by the ability of the compactors or rollers to achieve the specified compaction for the particular layer. If a certain layer thickness is specified, the maximum aggregate size and the mixture can sometimes be changed to meet the layer thickness requirement. This change can only be made if the required aggregate sizes are locally available.

Wearing course mixtures must be designed to have sufficient stability and durability to not deform under traffic loads and withstand the detrimental effects of air, water, and temperature changes. The nominal maximum aggregate size of wearing courses typically range from 9.5 to 19 mm. Wearing courses contain a higher asphalt binder content than base or binder courses due to the smaller aggregate sizes. Smaller particles have higher surface areas, thus requiring more asphalt binder to completely coat the particles. Smaller maximum aggregate sizes in a wearing course will produce a tighter, smoother surface finish. This attribute is beneficial to the designer of parking lots, since the tighter surface provides a high degree of environmental protection and is aesthetically pleasing. A coarser wearing course is useful in highway applications, since they tend to provide a higher degree of skid resistance.

Binder mixes are used as an intermediate layer between the wearing course and the underlying asphalt or granular base. Binder or intermediate mixes typically have a larger nominal maximum size aggregate of 19–38 mm with a corresponding lower asphalt binder content. Large stone mixtures are generally mixtures that have a nominal maximum aggregate size greater than 25 mm. Intermediate and base courses are often used interchangeably in pavement design. When an asphalt pavement will undergo heavy wheel loading, an intermediate mixture may be used for a wearing course, if the much coarser texture is not a concern. This is typically used in port facilities using heavy cargo handling vehicles and for truck unloading and industrial areas with heavy forklifts. Larger aggregate mixtures are often more resistant to the scuffing action of these tight radius power turns, than typical wearing course mixtures.

Dense graded asphalt base course mixtures have a maximum aggregate top size up to 75 mm. The asphalt binder content is even lower than intermediate courses and will provide adequate durability since base courses are typically not exposed to the environment. The maximum aggregate top size is usually limited by the maximum aggregate size available locally. Base mixtures can also be designed with an open gradation to allow any water intrusion to drain away from the pavement. Open graded base course mixtures are designed to provide an interconnecting void structure, while using 100 percent crushed aggregate with maximum sizes of 38–75 mm (The Asphalt Institute 1993).

Laboratory mixture design methods are used to determine the amount of asphalt binder required for the mixture and to verify if the proper blending of aggregates have been met. There are several laboratory methods or procedure available to achieve these goals. The most common are the Marshall method, the Hveem method, and the SuperPave method. The most widely used method throughout the world is the Marshall method. Since 1996, the SHRP SuperPave method has been growing to become the most popular method in the United States and is gradually being adopted in the rest of the world. The third and least common method is the Hveem method, once popular in the western United States. The SuperPave method was intended to replace the Marshall and Hveem methods, however due to past experience and the simplicity of the testing equipment, several public agencies have chosen to retain the Marshall method at least for some of their dense graded mixture designations. Many private projects are designed using the Marshall method and reference to the method is still made in many job specifications. All three methods are intended for the design of dense graded asphalt mixtures, namely HMA. However some mixture designers have had success using the methods with asphalt emulsions to design cold dense graded mixtures. This has mostly been done with the Hveem method and with a modified version of the Marshall method. The goal of all three methods is to determine the correct proportions of aggregate and asphalt binder to produce a mixture that when properly constructed will provide a pavement that can perform its intended function and withstand the effects of the environment. Producing a mixture design involves the asphalt binder and the aggregate to be blended together in different proportions in the laboratory. The resulting mixes are evaluated using a standard set of criteria to permit the proper selection of an appropriate asphalt binder content. The type and grading of the aggregate and the stiffness and the amount of the asphalt binder influences the physical properties of the mixture. The optimum asphalt binder content is selected to ensure a balance between the long-term durability of the mixture and its stability or resistance to deformation (NAPA 2001).

Aggregates

The aggregates in a dense graded asphalt mixture have the most significant contribution in the load bearing capacity of an asphalt mixture. The aggregates also determine the surface texture and skid resistance of the pavement. The asphalt binder is used to cement the aggregate particles together. The gradation of the asphalt mixture is the one major variable that the designer can alter to give the properties desired. Aggregate gradation is the distribution of the aggregate particle sizes expressed as a percent of the total weight. The gradation of the aggregate expressed as a percent of the total volume has the most significance, but expressing the gradation as a percent by weight is easier to calculate and is standard practice throughout the world. The gradation is determined by sieve analysis and is expressed as a total percent passing each sieve size in descending order. Total percent passing each sieve is the current recognized method for describing aggregate gradation; however, percent retained and passing retained have been used in the past. The final aggregate gradation or mixture gradation is also known as the "job mix formula" (JMF) or "combined gradation." A dense graded asphalt mixture generally contains a combined gradation of several aggregates, since a single aggregate generally will not provide all the desired properties for a dense graded mixture. Fuller proposed in 1902 that the best gradation for giving the highest strength in Portland cement concrete is one that provides the densest particle packing. The same statement can also be made for dense graded HMA. Fuller developed an aggregate maximum density curve by using his equation (equation 5.1).

$$P = 100(d/D)^n \tag{5.1}$$

where P is the total percent passing the particular sieve; d, the particular sieve size opening diameter; D, the maximum size of the aggregate.

Studies have shown that when n = 0.5, the maximum particle density or maximum packing of smooth spheres will be achieved. Further work done in the 1930s determined that when the smooth spheres are replaced by aggregate, the maximum packing density is achieved when n = 0.45. These were determined to be true whether the aggregate is crushed or uncrushed. The United States Bureau of Public Roads developed the 0.45 power chart in 1962 that is based on equation 5.1, but with n = 0.45. The chart raises each standard sieve size to 0.45 power using equation 5.1. The value of 0.45 for n was chosen in order to allow some minimal room in the aggregate for the asphalt binder than what was determined by Fuller with n = 0.5. The chart allows the designer to easily graphically determine the maximum particle packing or density line and adjust the aggregate grading. The maximum density line is drawn on the chart by starting at the zero origin and drawing a diagonal line to the sieve size corresponding to the maximum aggregate size (Figure 5.1). The maximum aggregate size is one sieve size larger than the nominal maximum size. The nominal maximum aggregate size is one sieve size larger than the first sieve to retain more than the 10 percent of the aggregate by weight. The maximum density line represents the aggregate grading that has the maximum particle packing or density.

The actual gradation of the aggregate of blend of aggregates is plotted on the chart with the maximum density line. The aggregate gradation can then be easily compared to what would be the maximum packing or density of the particular aggregate size. The further the actual aggregate grading line is from the maximum density line, the more room that particular

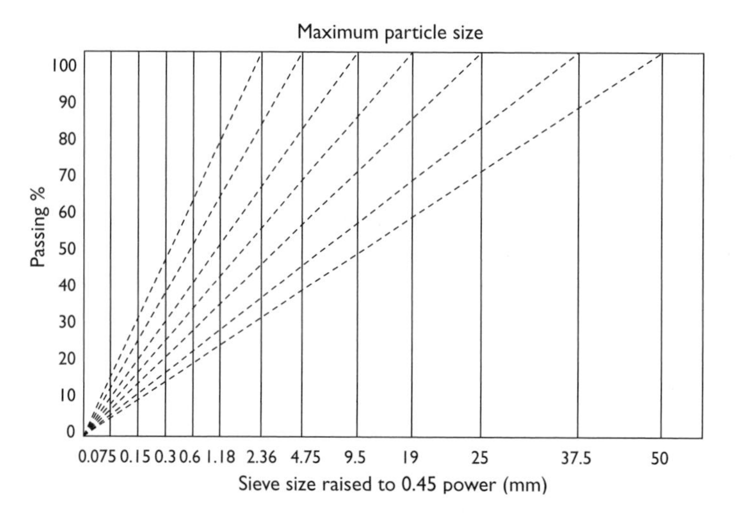

Figure 5.1 0.45 power chart.

grading has for asphalt binder and air voids. This is true whether the grading becomes coarser or finer than the grading of the maximum density line. Since the maximum grading density of the aggregate does not have sufficient room for the asphalt binder and the mixture air voids, it is desirable to have the mixture aggregate gradation either coarser or finer than the grading of the maximum density line. The volume in the aggregate packing that allows for asphalt binder and air voids is known as the voids in the mineral aggregate (VMA) of the asphalt mixture. If possible, the aggregate grading of a dense graded mixture should be parallel to the maximum density line. A grading selected on a line approximately parallel to the maximum density line will produce a uniformly graded mixture that will have little tendency to segregate. When segregation is present in a mixture there is a concentration of coarse aggregate in some areas of the pavement, while other areas contain a concentration of fine aggregate. Segregated mixtures result in pavements that have poor durability and poor structural and textural characteristics. Mixture gradations that tend to cross back and forth across the maximum density line tend to gap-grade the mixture, which also has the potential to segregate. If a fine textured mixture is desired, the mixture gradation should be approximately 2–4 percent above the maximum density line and if a coarse mixture gradation is desired, the mixture gradation should be approximately 2–4 percent below the maximum density line. Many mixtures will start out on the larger sieve sizes above the maximum density line and then cross over at the intermediate sieve sizes and be below the maximum density line for the remainder of the gradation. Mixture gradation lines that also slightly bow up or down tend to be a little more workable (AASHTO 1997). The maximum density line and the 0.45 power chart should only be used as a tool in mixture design and should not be incorporated into any type of specification. It can be used however in the development of specification ranges for dense graded mixtures. The maximum density line is best used as a guide to understand aggregate gradations and how they interact in the asphalt mixture. Dense graded HMA is usually a blend of several aggregates. In most cases no single aggregate gradation will give all the properties that are desired of the asphalt mixture. Economics also requires the designer to use the aggregates that are available locally. Dense graded HMA is a proportion of both coarse and fine aggregates. The 0.45 power chart is a tool to assist the designer in proportioning the blend of aggregates. Blending aggregates allows the designer to adjust the final mixture gradation to provide all the desirable mixture properties. The 0.45 power chart can be used to identify mixture gradations that may be more prone to segregation, have durability problems, or have stability problems. The SuperPave mixture design method added two tools to be used in conjunction with the 0.45 power chart. These added tools for dense graded mixtures are:

1 *Aggregate gradation control points.* The control points use a similar function as specification limits or ranges within which the gradation must pass.
2 *The restricted zone.* The restricted zone is a band that lies along the maximum density gradation line between the intermediate sieve size (either the 4.75 mm or the 2.36 mm, depending on the aggregate's nominal maximum size) and the 300 μm sieve size. Based on SuperPave guidelines, mixture gradations that pass through the restricted zone should be avoided. The restricted zone was established in initial SuperPave guidelines to limit the amount of rounded, natural sand in a dense graded mixture. This sand is thought to contribute to pavement instability and premature rutting and deformation. The restricted zone also reduces the size of the "sand hump" in the gradation line of dense graded mixtures.

The restricted zone was not developed from any testing during the SHRP SuperPave development program, but rather than from the consensus of an expert panel. It should be noted that some mixtures have performed satisfactorily in select applications with high amounts of natural sand or with a gradation that has passed through the restricted zone. These mixtures will generally perform well when used for pavements with low ESAL loading and pavements that require a highly durable mixture, such as parking applications. Rounded particles also increase the workability of the mixture, a desirable property for applications such as curb and gutter, private driveways, or applications that require a lot of handwork. The restricted zone is a mixture design guideline and should not substitute good engineering judgment or successful past experience with local materials.

The aggregate gradation control points are placed on the 0.45 power chart at the following sieve sizes: the nominal maximum sieve size, the 2.36 mm sieve size, and the 75 μm sieve size. The control minimum and maximum passing percentages vary with the SuperPave mixture designation. There are five SuperPave mixture designations, all designated by the nominal maximum aggregate size of their combined gradation. The restricted zone resides along the maximum density gradation line between an intermediate sieve and the 300 μm sieve. The restricted zone's intermediate sieve and the minimum and maximum passing percentages also vary the SuperPave mixture designation. Example 5.1 illustrates a dense graded aggregate blend plotted on a 0.45 power chart using control points and the restricted zone. Tables 5.2–5.6 lists the various aggregate gradation control points and restricted zone points as a function of the SuperPave nominal size mixture designations.

Example 5.1

Determine the nominal maximum aggregate size, draw the maximum density line, locate the control points and the restricted zone, and plot the gradation line of the sieve analysis given in Table 5.1.

Solution

The nominal maximum aggregate size is 19 mm, since it is one sieve size larger than the first descending sieve to retain more than 10 percent of the aggregate particles. That sieve is the 12.5 mm sieve (100 − 77 = 23 percent retained). The maximum aggregate sieve size is one

Table 5.1 Example 5.1 aggregate gradation

Sieve size	*Total passing (%)*
25 mm	100
19 mm	94
12.5 mm	77
9.5 mm	66
4.75 mm	37
2.36 mm	28
1.18 mm	21
600 μm	14
300 μm	8
150 μm	4
75 μm	2.8

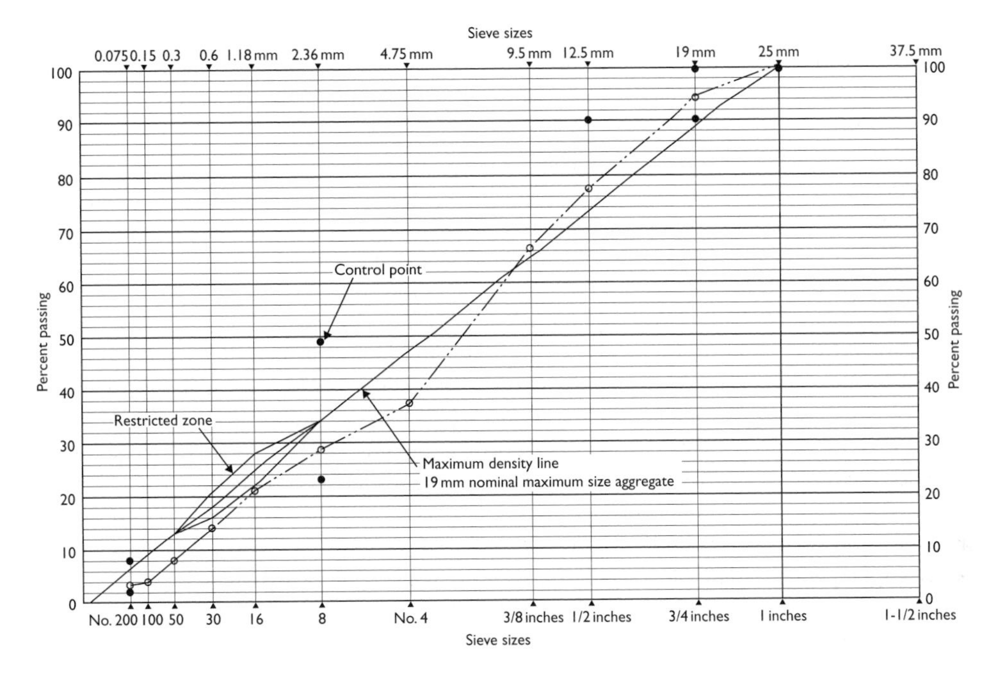

Figure 5.2 Example 5.1 solution, SuperPave gradation.

Table 5.2 37.5 mm SuperPave nominal maximum size control points and restricted zone points, 0.45 power chart

Sieve	*Gradation control points (%)*		*Restricted zone points (%)*	
	Minimum	*Maximum*	*Minimum*	*Maximum*
50 mm	100			
37.5 mm	90	100		
25 mm		90		
19 mm				
12.5 mm				
9.5 mm				
4.75 mm			34.7	34.7
2.36 mm	15	41	23.3	27.3
1.18 mm			15.5	21.5
600 μm			11.7	15.7
300 μm			10	10
75 μm	0	6		

Table 5.3 25 mm SuperPave nominal maximum size control points and restricted zone points, 0.45 power chart

Sieve	*Gradation control points (%)*		*Restricted zone points (%)*	
	Minimum	*Maximum*	*Minimum*	*Maximum*
37.5 mm	100			
25 mm	90	100		
19 mm		90		
12.5 mm				
9.5 mm				
4.75 mm			39.5	39.5
2.36 mm	19	45	26.8	30.8
1.18 mm			18.1	24.1
600 μm			13.6	17.6
300 μm			11.4	11.4
75 μm	1	7		

Table 5.4 19 mm SuperPave nominal maximum size control points and restricted zone points, 0.45 power chart

Sieve	*Gradation control points (%)*		*Restricted zone points (%)*	
	Minimum	*Maximum*	*Minimum*	*Maximum*
25 mm	100	100		
19 mm	90	100		
12.5 mm		90		
9.5 mm				
4.75 mm				
2.36 mm	23	49	34.6	34.6
1.18 mm			22.3	28.3
600 μm			16.7	20.7
300 μm			13.7	13.7
75 μm	2	8		

sieve larger than the nominal maximum aggregate size. The maximum aggregate size is 25 mm. Starting at the origin and drawing a diagonal line up to 100 percent passing at the 25 mm sieve size draws the maximum density line. The gradation is designated by the nominal aggregate size, which is 19 mm. The control points and the restricted zone points are determined for the 19 mm SuperPave gradation, which is obtained from Table 5.3. The actual gradation is plotted on the 0.45 power chart. The solution is illustrated in Figure 5.2.

The SHRP SuperPave mix design method incorporates the use of the control points and the restricted zone with the use of the 0.45 power chart. They can also be used for aggregate blending or gradation analysis with any of the dense graded HMA mixture design methods, including Marshall and Hveem.

The combined gradation of the aggregate or JMF usually must meet a gradation specification requirement. These requirements are in the form of ranges or bands that are given for

Table 5.5 12.5 mm SuperPave nominal maximum size control points and restricted zone points, 0.45 power chart

Sieve	*Gradation control points* (%)		*Restricted zone points* (%)	
	Minimum	*Maximum*	*Minimum*	*Maximum*
19 mm	100			
12.5 mm	90	100		
9.5 mm		90		
4.75 mm				
2.36 mm	28	58	39.1	39.1
1.18 mm			25.6	31.6
600 μm			19.1	23.1
300 μm			15.5	15.5
75 μm	2	10		

Table 5.6 9.5 mm SuperPave nominal maximum size control points and restricted zone points, 0.45 power chart

Sieve	*Gradation control points* (%)		*Restricted zone points* (%)	
	Minimum	*Maximum*	*Minimum*	*Maximum*
12.5 mm	100			
9.5 mm	90	100		
4.75 mm				
2.36 mm	32	67	47.2	47.2
1.18 mm			31.6	37.6
600 μm			23.5	27.5
300 μm			18.7	18.7
75 μm	2	10		

Table 5.7 Combined aggregate gradation specification for dense graded asphalt mixtures

Sieve size	*Dense mixtures*						
	Mix designation and nominal maximum size of aggregate (mm)						
	50	37.5	25	19	12.5	9.5	4.75
	Grading of total mixture aggregate (%) passing by weight						
63 mm	100						
50 mm	90–100	100					
37.5 mm		90–100	100				
25 mm	60–80		90–100	100			
19 mm		56–80		90–100	100		
12.5 mm	35–65		56–80		90–100	100	
9.5 mm				56–80		90–100	100
4.75 mm	17–47	23–53	29–59	35–65	44–74	55–85	80–100
2.36mm	10–36	15–41	19–45	23–49	28–58	32–67	65–100
1.18 mm							40–80
600 μm							25–65
300 μm	3–15	4–16	5–17	5–19	5–21	7–23	7–40
150 μm							3–20
75 μm	0–5	0–6	1–7	2–8	2–10	2–10	2–10

Source: American Society of Testing Materials D3515 (2001b).

each sieve size. Table 5.7 is one example of a combined aggregate gradation specification for dense graded HMA.

The SuperPave control points and the restricted zone are guidelines that are used in conjunction with the 0.45 power chart and were developed under the United States SHRP program. If the specifying agency, architect or designer chooses not to use the SuperPave guidelines, then the following recommendations utilizing the 0.45 chart should be considered:

1. If the fine aggregate is predominately round natural sand, it is preferred for the gradation to stay below the maximum density line.
2. If crushed or very angular fine aggregate is used, the gradation may be above the maximum density line.
3. It is undesirable to have a hump in the gradation in the 600–300 μm size range, especially when natural sands predominate in the fine aggregate. This is important for asphalt mixtures that require high stability such as freeway applications. Low volume parking facilities can tolerate a sand hump, and it will assist in workability of the mixture.

Many combined aggregate gradation specifications have been developed through past experience by trial and error to reflect local conditions. Most specifications for dense graded HMA require aggregate gradations that when plotted on the 0.45 power chart, the middle portion of the curve is approximately parallel to the maximum density line. Dense mixture gradation specifications are also associated with achieving the correct in-place density or air voids (Roberts *et al.* 1991). ASTM D3515 is a popular specification for the gradations of HMA mixtures both dense graded or open graded (porous). These gradations are named or referred to by their nominal maximum aggregate size. The SHRP SuperPave mixtures are also designated by their nominal maximum aggregate size (Table 5.8). This designation is also useful in that it gives guidance in selecting which mixture gradation to use, since the maximum size to use is about half the pavement layer or lift thickness. Since aggregates are natural materials some tolerance is given to allow gradations to slightly deviate from a specified combined gradation or JMF. Table 5.9 gives useful guidelines for production or acceptance tolerances for the JMF when producing a project specification.

It is unlikely that a single aggregate source or gradation will meet a specification band for dense graded asphalt mixtures. Aggregates are natural occurring materials and are processed to meet a single coarse or fine aggregate gradation. Dense graded mixtures are a proportional

Table 5.8 SHRP SuperPave mixture designations

SHRP designation (mm)	*Mixture nominal maximum size* (mm)	*Mixture maximum size* (mm)
37.5	37.5	50.0
25	25.0	37.5
19	19.0	25.0
12.5	12.5	19.0
9.5	9.5	12.5

Table 5.9 ASTM D3515 dense graded sieve tolerances

Sieve size	*Tolerances (±, %)*
≥12.5 mm	8
9.5 mm	7
4.75 mm	7
2.36 mm	6
1.18 mm	6
600 μm	5
300 μm	5
75 μm	3

Source: American Society of Testing Materials D3515 (2001b).

Table 5.10 Example multiple aggregate dense graded combined gradation

Source	*Combined aggregate blending of actual gradations*					
	Individual aggregates				*Job mix formula*	*Mixture specification bands*
	CA11	*CA16*	*FA20*	*FA01*		
Percent of blend	48.0	26.0	16.0	10.0		
Sieve size	*Total percent passing gradations (%)*					
25 mm	100	100	100	100	100	≤100
19 mm	92	100	100	100	96	82–100
12.5 mm	45	100	100	100	74	50–82
9.5 mm	10	97	100	100	56	
4.75 mm	6	30	97	97	36	≤40
2.36 mm	5	7	70	82	24	16–36
1.18 mm	4	4	50	65	17	10–25
600 μm	3	3	30	50	12	
300 μm	3	3	19	16	7	4–12
150 μm	2	2	10	5	4	3–9
75 μm	2.0	2.0	4.0	0.4	2.1	2–6

blend of the various sieve sizes from coarse to fine. This proportional blend is what gives dense graded mixtures its balanced pavement performance properties of durability, stability, and strength. A minimum of two or more aggregates is blended to produce the dense graded mixture. The two aggregates are at the minimum a coarse aggregate fraction and a fine aggregate or sand fraction. Some dense graded mixture can consist of two fine aggregates and two or three coarse aggregates. Table 5.10 is an example of blending of multiple aggregate types for a specific highway agency dense graded mixture. These multiple aggregate blends are usually required for the mixture to meet specification requirements, skid resistance, minimum workability, stability, durability, and economic factors. The cost of the aggregate and the amount of asphalt binder a particular mixture requires are the key economic factors. The amount of asphalt binder in a mixture is proportional to its gradation and the more processing an aggregate undergoes the higher its cost. Skid resistant aggregates for wearing course possess unique properties and also have a higher cost.

Aggregate mixture blending methods are techniques for determining the relative proportion of various aggregates to obtain a desired gradation. Various methods can be used in determining the desired proportions of various aggregates to blend to a specific gradation or specification. All modern gradations and specifications are given in total percent passing by weight. The suitability of the various methods of blending aggregates depends on the types of specification and number of aggregates involves the experience of the designer and the major emphasis of the blending (the closeness to the desired gradation or mixture economics). The basic formula used in the blending of aggregates is given by equation 5.2. Example 5.2 illustrates the application of equation 5.2.

$$P = Aa + Bb + Cc + Dd + \cdots \tag{5.2}$$

where P is the the percent of material passing a given sieve for the combined aggregates, A + B+ C + D, etc.; A, B, C, D, etc. is the percentage of material passing a given sieve for each aggregate, A, B, C, D, etc.; a, b, c, d, etc is the percentage in decimal form of the proportion of aggregates A, B, C, D, etc., to be used in the blend.

Example 5.2

Using equation 5.2, determine the total percent passing for the 2.36 mm sieve:

Aggregate	*Percent in blend*	*Total percent passing sieve*
A	16	18
B	25	42
C	36	65
D	23	80

Solution

$$P = 18(0.16) + 42(0.25) + 65(0.36) + 80(0.23) = 55.2$$

The total percent passing on the 2.36 mm sieve for the combined gradation is 55*percent.* The remainder of the aggregate sizes would be calculated in a similar method.

It is recommended to first plot on a gradation chart all of the individual aggregate gradations along with the gradation specification ranges of the particular mixture (Figure 5.3).

From this graph, a visual determination can be made on the following:

1 Whether a blend can be made using the available aggregates to meet the specification limits.
2 Which are the critical final gradation controlling sieves specific to these aggregates.
3 The approximate trial proportions to be selected.

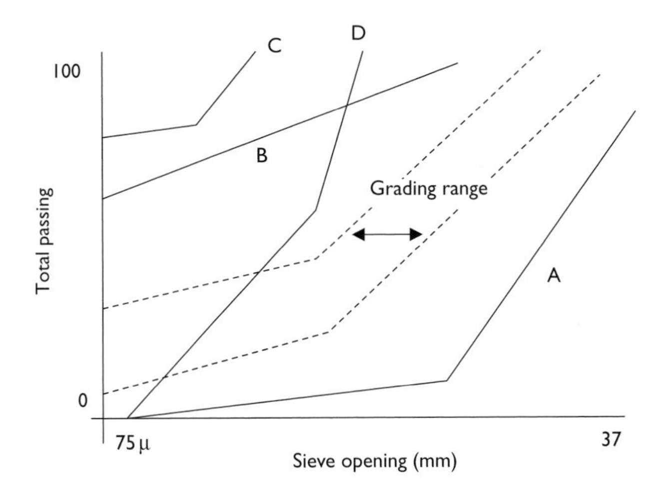

Figure 5.3 Example of aggregate grading with specification range.

Source: *Hot Mix Asphalt Materials, Mixture Design and Construction,* © 1991 by the National Asphalt Pavement Association; used by permission.

The following conclusions can be made from the example in Figure 5.3:

1 The gradation curves for all the possible combinations of aggregates A and B fall between curves A and B. It would be impossible to blend aggregates C and B together and still meet the grading range specification since both have gradations that completely exceed the percent passing values that are included in the grading range.
2 In this example, the curves B and D cross at a point. At this point all possible grading curves for B and D will pass through the point.
3 The curve for a blend that contains more of aggregate A than aggregate B would be closer to the curve of aggregate A than the curve of aggregate B.

(Roberts *et al.* 1991)

Proportioning aggregates usually involves a trial and error selection of various percentages until the desired combined gradation is arrived at. The desired aggregate combination or JMF is usually predetermined by specification ranges. Most designers use the midpoint of the specification ranges when no other guidelines are considered. A trial blend is selected and using equation 5.2, the percent passing each sieve for the selected blend is determined. The grading is then compared with the specifications and can be plotted on a chart to give visual guidance in determining what proportions to change. Adjustments are made for the subsequent trial blends and the calculations repeated until a satisfactory or optimum blend is obtained. The aggregate grading charts can assist the designer in deciding which sieves are critical sieves that will control whether the blend will meet the overall specification ranges. The amount of calculations can be reduced by using only the critical sieves to determine what percentages will give the optimal blend. During the final trial and error blend calculations, the full range of sieve sizes can then be calculated and compared to the specification ranges. The trial and error method is the most commonly used, but relies somewhat on familiarity with the local aggregates and past experience with mixture designs to determine the optimal blend with any expedience. The following summarizes the steps involved in trial and error blending of aggregates:

1 Determine the nominal maximum size of the desired mixture.
2 Determine which aggregate grading will give the nominal maximum size.
3 Determine the gradations of all the aggregates desired for the JMF.
4 Determine which specification ranges are required for the mixture.
5 Select which are the critical sieves for blending the aggregates.
6 Determine an initial set of aggregate proportions or percentages using equation 5.2, which will meet the specification requirements for the critical sieves.
7 Check that the calculated blend using the proportions determined for all the sieves meets the specification ranges.
8 Adjust the individual aggregate proportions as necessary to ensure that the entire combined gradation meets the specification for all the sieves or any other requirements, such as avoiding the SuperPave restricted zone.

Prior mixture design or blending experience with the local aggregates available can greatly reduce the amount of effort in the trial and error process of aggregate blending. When there is no experience with the aggregates, the greatest difficulty involved will be determining which aggregates will govern the nominal maximum size and which sieve sizes are the critical sieve sizes. The SuperPave nominal maximum control limits are an excellent tool in the initial determination of the nominal maximum size and the critical sieves. If these limits meet the specification range requirements, they can be used as stated; otherwise select the same sieve sizes, but use the desired specification range for the control limits. Another difficulty is determining the amount of fine and coarse aggregate that is required for a dense graded mixture. The 2.36 mm sieve is an excellent sieve in determining the initial amount of both coarse and fine aggregate required. The SuperPave limits can be used or use the midpoint from the 2.36 mm sieve specification range. Trial and error blending calculations will determine if just two aggregates or more are required to meet the specification ranges. The following steps can be used when the designer has no other guidelines to help determine the proportion of coarse and fine aggregates required:

1 Determine the nominal size dense graded mixture required.
2 Using the largest two SuperPave control sieve limits required for the particular mixture, determine which coarse aggregate available will meet the required gradation. If the SuperPave control sieve limits are not applicable, use the first two largest sieve sizes in the mixture specification that has any particles retained. Also determine which fine aggregates are available and the grading.
3 Using the midpoint of the ranges for the 2.36 mm sieve size, determine the proportion of the coarse aggregate and fine aggregate that will meet the midpoint value. If no blend can meet the value, consider a blend of two different gradations of fine aggregates. An initial starting point can be a 50/50 blend of the two fine aggregates. Typically a crushed fine aggregate and a natural fine aggregate will be used together.
4 Confirm that the blend meets the 75 μm sieve size specification ranges. An alternation of the fine aggregate proportions may be required. If the values are still too low to meet the requirement, a reduction in the total amount of fine will be required. The additional amount would be compensated for by coarse aggregate. Remember that the proportions

are required to sum up to 100 percent. If the values are too low to meet the requirements with the existing aggregates, the addition of mineral filler may be required. An initial starting proportion value for mineral filler would be 1 percent.

5 Recalculate the total percent passing for the remainder of the control sieves and check for conformance with the specification ranges. An additional control sieve can be selected, if there is concern over a tender mixture or avoiding the SuperPave restricted zone. This control sieve would be the 600 μm sieve.
6 If the final proportions are acceptable from the control sieve requirements, calculate the total percent passing for the remainder of the sieves and check for conformance with the specification ranges. If still in conformance, plot the combined gradation on a 0.45 power chart and "fine tune" the gradation, if necessary.

Blending of several aggregates to meet a specified target is usually done by hand through the trial and error process. The blending can also be solved algebraically through the use of multiple equations and matrices. Several computer programs are available to solve the blending equations for the designer. One such program is *CAMA (Computer **A**ided **M**ixture **D**esign)*, which was developed by The Asphalt Institute, Lexington, Kentucky, USA. The proportions of each of the aggregates can be optimized to meet the midpoint of a gradation specification range, the most cost effective blend or similar criteria. These programs can generate mathematically correct proportions, but good engineering judgment should also be applied. For example, the matrices or computer program may give proportions of a mixture that is too harsh to place, or proportions that are so precise that they are beyond the capabilities of a HMA mixing plant.

Example 5.3

Determine a blend of a coarse aggregate and fine aggregate with gradations given in Table 5.11. The blend should be for a dense graded mixture meeting the specifications also given in Table 5.11.

Solution

Table 5.11 Example 5.3 gradations and specification ranges

Sieve size	*Aggregate A*	*Aggregate B*	*50% A*	*50% B*	*Total blend*	*Specification range*
Percent used	100	100	50	50		
19 mm	100	100	50	50	100	100
12.5 mm	90	100	45	50	95	80–100
9.5 mm	59	100	29.5	50	79.5	70–90
4.75 mm	16	96	8	48	56	50–70
2.36 mm	3	82	1.5	41	42.5	35–50
1.18 mm	0	51	0	25.5	25.5	18–29
600 μm	0	36	0	18	18	13–23
300 μm	0	27	0	13.5	13.5	—
150 μm	0	16	0	8	8	8–16
75 μm	0	9	0	4.5	4.5	4–10

The gradations given indicate that it is possible to find a blend that will meet the specification ranges. A 50/50 blend should be considered as an initial starting point. The initial trial blend can be determined through the use of some critical sieves. By evaluating the gradations, it is given that all material retained on the 9.5 mm sieve must come from aggregate A and all material finer than the 1.18 mm sieve has to come from aggregate B. The control sieves should be the 9.5 mm sieve and the 1.18 mm sieve. The midpoint of the specification range for the 9.5 mm sieve is 80 percent and the midpoint of the specification range for the 1.18 mm sieve is 24 percent. Fifty percent of aggregate A and 50 percent of aggregate B gives a value of 79.5 percent passing for the 9.5 mm sieve. Likewise 50 percent of aggregate A and 50 percent of aggregate B gives a value of 25.5 percent passing for the 1.18 mm sieve. Both of these values meet the specification ranges. Calculating the remaining sieves illustrates that they also meet the specification ranges. The values are given in Table 5.11.

Example 5.3 illustrates that many possible solutions may be tried to determine the proportion of the aggregates. The solution used a 50/50 blend, however, a 51/49, 49/51, 53/47, etc. would have met the specification ranges and been correct. The experience of the designer assists in "fine tuning" the proportions. For example, two different proportions may meet the gradation specification ranges, but one blend may be significantly more workable than the other. Three or more aggregate blends are very common in the production of dense graded mixtures. Example 5.4 illustrates the blending of a three aggregate mixture.

Example 5.4

Determine the blend of the three aggregates (one coarse aggregate, one fine aggregate, and mineral filler) using the information given in Table 5.12.

Solution

The critical sieves in this example are the 4.75 mm, the 600 μm, and the 75 μm. The 75 μm sieve is critical since the aggregate used as mineral filler has 88 percent passing the 75 μm sieve and the specifications only allow up to 11 percent. The desired 52 percent (100−48) of the material larger than the 4.75 mm sieve must come from aggregate A. The percentage of aggregate A that should be used for the initial trial blend is:

$$\text{percent of A} = \frac{52}{81} = 64\%$$

Table 5.12 Example 5.4 gradation and specification

Sieve size	*Aggregate gradation*			*Specification range*	*Specification midpoint*
	A	*B*	*Mineral filler*		
25 mm	100	100	100	94–100	97
12.5 mm	63	100	100	70–85	78
4.75 mm	19	100	100	40–55	48
2.36 mm	8	93	100	30–42	36
600 μm	5	55	100	20–30	25
150 μm	3	36	97	12–22	17
75 μm	2	3	88	5–11	8

The information on the 600 μm sieve indicates that 75 percent of the material larger than the 600 μm must come from aggregates A and B. Since 75 percent is the total amount of material desired to be retained on the 600 μm sieve and 64 percent of the 95 percent retained (100-5) is that portion that is provided by aggregate A,

$$\text{percent of B} = 75 - (0.64 \times 95)$$
$$\text{percent of B} = 75 - 61 = 14$$

Based on these initial calculations, it appears that the best first blend estimate would be 64 percent of aggregate A, 14 percent of aggregate B, and the remainder 22 percent being the mineral filler. However, 22 percent mineral filler would produce too much material passing the 75 μm sieve. The percent of mineral filler should probably be no more than 8 percent, since the mineral filler alone will contribute 7 percent (8 percent of 88 percent passing) passing the 75 μm sieve. The first initial blend that meets the specification range would be 71 percent of aggregate A, 21 percent of aggregate B, and 8 percent mineral filler. Table 5.13 gives these gradation results.

This blend can be optimized or altered to produce a blend that is closer to the midpoint of the specification range. Reducing the mineral filler by 2 percent and aggregate A by 5 percent and increasing aggregate B by 7 percent should put the blend on target. These results are also given in Table 5.13. These blend proportions were estimated by using the following "rule of thumb": For each 1 percent that two aggregate proportions are changed, the resulting blend will change by the difference between the percent passing for any given sieve for the two aggregates times 1 percent. For example, increasing aggregate B by 5 percent and decreasing aggregate A by 5 percent will result in a change of 5 percent of 81 percent (100−19) retained on the 4.75 mm sieve. Thus the 4.75 mm sieve had a net increase of 4 percent passing. The final blend becomes (Roberts *et al.* 1991):

Aggregate A: 66%
Aggregate B: 28%
Mineral filler: 6%

Table 5.13 Example 5.4 blend trials

Aggregate	*Blend (%)*	*Sieve size*						
		25 mm	12.5 mm	4.75 mm	2.36 mm	600 mm	150 mm	75 mm
First trial								
A	71	71	44.7	13.5	5.7	3.6	2.1	1.4
B	21	21	21	21	19.5	11.6	7.6	0.6
Mineral filler	8	8	8	8	8	8	7.8	7.0
Blend		100	74	42	33	23	18	9.0
Final trial								
A	66	66	41.6	12.5	5.3	3.2	2.0	1.3
B	28	28	28	28	26.0	15.4	10.1	0.08
Mineral filler	6	6	6	6	6	6	5.8	5.3
Blend		100	76	46	37	25	18	7.4
Target		97	78	48	36	25	17	8.0

Table 5.14 Initial blend percentages for ASTM standard gradations

Dense graded mix designation (mm)	*Coarse aggregate size number, ASTM D448*					*Sand, ASTM D1073, number 3*	*Mineral filler, ASTM D242*
	4	*5*	*68*	*8*	*89*		
	Initial trial aggregate amount (%)						
37.5	34			38		28	
25.0		28		36		36	
19.0			60			36	4
12.5				52		50	3
9.5					46	50	4

The blends can also be optimized to produce the most economical combination based on the cost of each of the aggregates. If blending aggregates are to meet ASTM gradation specifications, Table 5.14 can be useful in selecting the initial trial blend of aggregates. The Marshall and SuperPave design procedures involve mixture analysis that will also provide information that may further require the initial aggregate proportions to be adjusted.

Asphalt mixture volumetric properties

The volumetric properties of a compacted asphalt mixture provide indications of the potential performance of the mixture as a pavement. The volumetric properties of the compacted asphalt mixture are determined at the laboratory design stage and can also be incorporated into specifications. Volumetric properties are used in the two most common laboratory design procedures, Marshall and SuperPave. These volumetric properties can also be correlated to the in-service properties of the asphalt pavement. The fundamental volumetric properties of a compacted asphalt mixture are:

- air voids, P_a
- voids in the mineral aggregate, VMA
- voids filled with asphalt, VFA
- effective asphalt content, P_{be}.

These properties can be determined on any asphalt mixture, whether it is in-place such as the asphalt pavement cores or samples, or on laboratory compacted mixtures. Determining volumetric properties involves weight–volume relationships. The basic physical property of any material in weight–volume relationships is the specific gravity. The specific gravity is used to convert from weight to volume or volume to weight. Specific gravity is a ratio of the mass of a material of a given volume to the weight of an equal volume of water, both at same temperature. There are five different types of specific gravity measurements used in the volumetric analysis of asphalt mixtures:

- Apparent specific gravity, G_{sa}: the ratio of the mass in air of a unit volume of an *impermeable* aggregate or stone at a stated temperature.
- Bulk specific gravity, G_{sb}: the ratio of the mass in air of a unit volume of a *permeable* (*including both permeable and impermeable voids*) aggregate at a stated temperature.

- Effective specific gravity, G_{se}: the ratio of the mass in air of a unit volume of *permeable (excluding voids permeable to the asphalt binder)* aggregate at a stated temperature.
- Bulk specific gravity of the compacted asphalt mixture, G_{mb}: the ratio of the mass in air of a unit volume of a compacted specimen of an asphalt mixture at a stated temperature.
- Theoretical maximum specific gravity of an asphalt mixture, G_{mm}: the ratio of the mass in air of a unit volume of an uncompacted or loose asphalt mixture at a stated temperature. It is also known as the Rice specific gravity, named after James Rice, the developer of the test procedure to measure the maximum specific gravity.

All specific gravity values are ratios that are at a given temperature to the mass in air of equal density of an equal volume of gas-free distilled water. For simplicity and practical use in asphalt mixture design, the specific gravity of water is assumed to be 1.00 (grams/cubic centimeter). Using that assumption then, the specific gravity becomes simply the mass divided by its volume (when using metric units). At this point, some mixture design nomenclature has been introduced involving letters and subscripts. The capital letter describes the property being measured, the first subscript letter refers to the material being measured and the second subscript, when present, refers to the type of specific gravity being measured. For example, G_{se} refers to the effective specific gravity of the combined stone. G_{mb} refers to the bulk specific gravity of the mixture. Table 5.15 gives a complete listing of this nomenclature used in the volumetric analysis of asphalt mixtures.

The specific gravity of a material is used to bridge the gap between weight and volume relationships in asphalt mixture design. A component diagram is a visual tool used to assist the designer in understanding the volumetric relationships in mixture design. The component diagram is also used in understanding volumetric relationships in soil mechanics. Figure 5.4 is a component diagram used in asphalt mixture design.

Volumetric equations

The specific gravities of all the components in an asphalt mixture are required to be known prior to the volumetric analysis of the mixture. There are established test procedures for

Table 5.15 Asphalt mixture volumetric terminology

Nomenclature	*Description*
G	Bulk specific gravity of an aggregate
G_b	Specific gravity of the asphalt binder
G_{sb}	Bulk specific gravity of the *combined* aggregate
G_{se}	Effective specific gravity of the *combined* aggregate
G_{sa}	Apparent specific gravity of the *combined* aggregate
G_{mb}	Bulk specific gravity of the compacted asphalt mixture
G_{mm}	Maximum theoretical specific gravity of the asphalt mixture
P	Percent by weight of aggregate
P_b	Asphalt percent by weight of the total mixture
P_s	Aggregate percent by weight of the total mixture
P_{mm}	Uncompacted mixture percent by weight of the total mixture (usually 100%)
P_{be}	Effective asphalt binder percent by weight of total mixture
P_a	Air void percent of compacted mixture
P_{ba}	Percent of absorbed asphalt binder, by total weight of aggregate

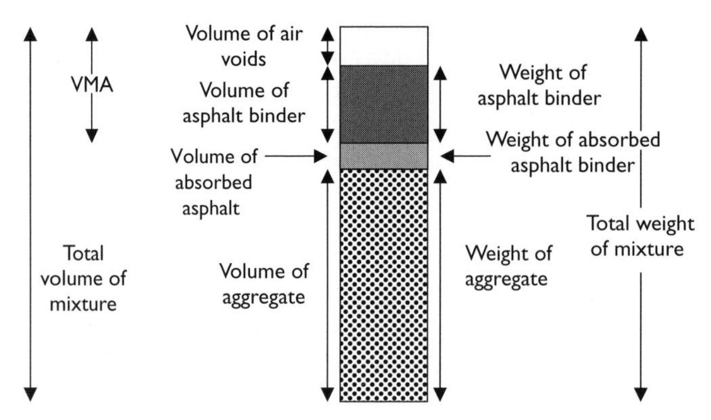

Figure 5.4 Component diagram used in asphalt mixtures.

measuring the specific gravity that may be obtained from the local highway agency or by following an international recognized procedure. The following describes the measurements and calculations required for the volumetric analysis:

1 Measure the bulk specific gravities of the coarse aggregate (ASTM C127) and of the fine aggregate (ASTM C128).
2 Measure the specific gravity of the asphalt binder (ASTM D70) and of the mineral filler (ASTM D854).
3 Calculate the bulk specific gravity (G_{sb}) of the total combined aggregate in the mixture.
4 Measure the maximum specific gravity (G_{mm}) of the loose or uncompacted asphalt mixture (ASTM D2041).
5 Measure the bulk specific gravity (G_{mb}) of the compacted asphalt mixture (ASTM D1188/D2726).
6 Calculate the effective specific gravity (G_{se}) of the aggregate.
7 Calculate the asphalt binder absorption of the aggregate (P_{ba}).
8 Calculate the effective asphalt content (P_{be}) of the asphalt mixture.
9 Calculate the percent VMA of the compacted asphalt mixture.
10 Calculate the percent air voids (P_a) in the compacted asphalt mixture.
11 Calculate the percent voids filled with asphalt (VFA) in the compacted asphalt mixture.

(The Asphalt Institute 1993)

Bulk specific gravity of aggregate: When the combined or total aggregate portion of the asphalt mixture consists of separate fractions of coarse aggregate, fine aggregate and mineral filler, all having different bulk specific gravities, the combined bulk specific gravity for the total aggregate is calculated using equation 5.3.

$$G_{sb} = \frac{P_1 + P_2 + P_n}{P_1/G_1 + P_2/G_2 + P_n/G_n} \tag{5.3}$$

where P_1, P_2, P_n is the individual percentages by weight of each aggregate; G_1, G_2, G_n, the individual bulk specific gravities of each aggregate.

Effective specific gravity of aggregate: The effective specific gravity of the aggregate includes all the void spaces in the aggregate particles except for those that absorb asphalt binder. The effective specific gravity of the combined aggregate, G_{se}, is given in equation 5.4.

$$G_{se} = \frac{P_{mm} - P_b}{P_{mm}/G_{mm} - P_b/G_b} \tag{5.4}$$

where G_{mm} is the measured maximum specific gravity of the uncompacted asphalt mixture; P_{mm}, the percent by weight of total uncompacted mixture = 100 percent; P_b, the percent weight of asphalt binder of the mixture; G_b, the specific gravity of the asphalt binder.

Asphalt binder absorption: The absorption of the asphalt binder by the aggregate is expressed as a percentage by weight of aggregate. Asphalt binder absorption, P_{ba}, is determined using equation 5.5.

$$P_{ba} = (100)\frac{(G_{se} - G_{sb})}{(G_{sb})(G_{se})}(G_b) \tag{5.5}$$

where G_{se} is the effective specific gravity of the combined aggregate; G_{sb}, the bulk specific gravity of the combined aggregate; G_b, the specific gravity of the asphalt binder.

Effective asphalt content of an asphalt mixture: The effective asphalt content of an asphalt mixture is the total asphalt binder content minus the amount of asphalt binder that is absorbed into the aggregate particles. It is the portion of the asphalt binder that remains as the coating or film surrounding the outside of each of the aggregate particles. The absorbed asphalt binder provides no performance characteristics for the asphalt pavement. The effective asphalt content, P_{be} is given by equation 5.6.

$$P_{be} = P_b - \frac{P_{ba}}{100}(P_s) \tag{5.6}$$

where P_b is the asphalt binder content percent by weight of the total mixture; P_{ba}, the absorbed asphalt binder content percent by weight of aggregate; P_s, the percent by weight of the total amount of aggregate in the mixture.

Percentage of voids in the mineral aggregate, VMA, in a compacted asphalt mixture: The VMA are defined as the intergranular void space between the aggregate particles in a compacted asphalt mixture that includes the air voids and the effective asphalt content, expressed as a percent of the total volume. The VMA is calculated on the basis of the bulk specific gravity of the combined aggregate and is expressed as a percentage of the bulk volume of the compacted asphalt mixture. VMA is calculated by subtracting the volume of the aggregate determined by its bulk specific gravity from the volume of the compacted asphalt mixture. VMA is calculated by equation 5.7.

$$\text{VMA} = 100 - \frac{G_{mb}(P_s)}{G_{sb}} \tag{5.7}$$

where G_{sb} is the bulk specific gravity of the combined aggregate; G_{mb}, the measured bulk specific gravity of the compacted asphalt mixture; P_s, the percent by weight of the total amount of aggregate in the mixture.

Percentage of air voids in the compacted asphalt mixture: The air voids in the total compacted asphalt mixture consists of small air spaces between the asphalt binder coated aggregate particles. The volume percentage of air voids, P_a, in a compacted asphalt mixture is determined by equation 5.8.

$$P_a = \frac{G_{mm} - G_{mb}}{G_{mm}}(100) \tag{5.8}$$

where G_{mm} is the measured maximum specific gravity of the uncompacted asphalt mixture; G_{mb}, the measured bulk specific gravity of the compacted asphalt mixture.

Percentage of voids filled with asphalt in the compacted mixture: The percentage of the VMA, that are filled with asphalt binder is the VFA. VFA is determined by equation 5.9.

$$\text{VFA} = \frac{\text{VMA} - P_a}{\text{VMA}}(100) \tag{5.9}$$

where VMA is the VMA; P_a, the percent air voids in the compacted asphalt mixture.

Maximum specific gravity of mixtures with different asphalt binder contents: When designing a paving mixture with a given aggregate blend, the maximum specific gravity, G_{mm}, at each asphalt binder content is needed to determine the percentage of air voids, P_a, for each asphalt binder content. The maximum specific gravity or Rice gravity can be determined for each asphalt binder content by ASTM D2041; the precision of the test is best when the mixture is close to the optimum asphalt binder content. After calculating the effective specific gravity, G_{se}, of the aggregate from each measured maximum specific gravity and averaging the G_{se} results, the maximum specific gravity can be calculated for any other asphalt binder content using equation 5.10. The equation assumes that the effective specific gravity of aggregate is constant, since asphalt absorption does not vary appreciably with changes in asphalt binder content.

$$G_{mm} = \frac{P_{mm} - P_b}{P_s/G_{se} - P_b/G_b} \tag{5.10}$$

where P_{mm} is the percent by mass of total mixture = 100; P_s, the aggregate content, percent by total mass of mixture; P_b, the asphalt content, percent by total mass of mixture; G_{se}, the effective specific gravity of the aggregate; G_b, the specific gravity of the asphalt binder (The Asphalt Institute 2001).

Example 5.5

Using the information provided in Table 5.16, determine the asphalt absorption, effective asphalt binder content, VMA, air voids, and VFA of the given asphalt mixture.

Table 5.16 Example 5.5 volumetric calculation information

Material	*Measured specific gravity*	*Mixture composition*	
		Weight of total mixture (%)	*Weight of total aggregate (%)*
Coarse aggregate	2.315	61.8	65
Crushed sand	2.432	16.2	17
Natural sand	2.521	16.2	17
Mineral filler	2.789*	0.9	1
Asphalt binder	0.993	4.9	
Compacted mixture, bulk	2.138		
Uncompacted mixture, measured maximum	2.221		

Note
* The apparent specific gravity, but due to the small particle size of the mineral filler and being impermeable, its bulk and apparent specific gravity are close to equal.

Solution

First determine the bulk specific gravity of the combined aggregates:

$$G_{sb} = \frac{65 + 17 + 17 + 1}{\dfrac{65}{2.315} + \dfrac{17}{2.432} + \dfrac{17}{2.521} + \dfrac{1}{2.789}} = 2.371$$

Determine the effective specific gravity of the combined aggregates:

$$G_{se} = \frac{100 - 4.9}{\dfrac{100}{2.229} - \dfrac{4.9}{0.993}} = 2.382$$

Now, the percentage of absorbed asphalt binder can be determined:

$$P_{ba} = \frac{2.382 - 2.371}{(2.382)(2.371)}(0.993)(100) = \mathit{0.19\%}$$

The effective asphalt binder content is:

$$P_{be} = 4.9 - \frac{0.19}{100}(95.1) = \mathit{4.72\%}$$

The VMA is:

$$\text{VMA} = 100 - \frac{(2.138)(95.1)}{2.371} = \mathit{14.2\%}$$

The percentage of air voids in the mixture is:

$$P_a = (100)\frac{(2.229 - 2.138)}{2.229} = \mathit{4.1\%}$$

Table 5.17 Example 5.5 volumetric results summary

Test parameter	*Results (%)*
Absorbed asphalt binder, P_{ba}	0.19
Effective asphalt binder content, P_{be}	4.72
VMA	14.2
Mixture air voids, P_a	4.1
VFA	71.1

The voids filled with asphalt binder is:

$$\mathrm{VFA} = (100)\frac{(14.2 - 4.1)}{14.2} = 71.1\%$$

The results are summarized in Table 5.17.

Marshall method of mixture design

The Marshall method of mixture design is the most commonly used procedure in the world for the design of dense graded HMA mixtures. It is gradually being replaced, however, by the SHRP SuperPave method of mixture design. The Marshall method was developed by Bruce Marshall, an employee of the state of Mississippi Highway Department in 1939. The United States Army Corps of Engineers adopted and refined the procedure in 1942 as a mixture design procedure for bomber runways during the Second World War. The increase in tire pressures and aircraft wheel loads during the Second World War required an improved and portable mixture design procedure. The Corps of Engineers began experimenting with Bruce Marshall's procedure both in the laboratory and field experiments. One key feature of the method is that asphalt mixtures are compacted in a 4 inch (102 mm) diameter mold using a 4.5 kg (10 lb) weight that drops a distance of 457 mm (18 inches). The experiments consisted of varying the number of times the weight or "hammer" (Plate 5.1) dropped on the specimen. This number of "drops" is also referred to as "blows." These impacts cause the asphalt mixture in the specimen mold to densify or become compacted. The goal of the laboratory compaction experiment was to determine a laboratory preparation procedure that would involve minimum effort and time, but would provide a basis for selecting the optimum asphalt binder content for the mixture. Bruce Marshall's compaction hammer met the portability requirements. A compactive effort of 50 blows per side of the cylindrical specimens became the standard. This was the original basis of the standard 50 blow Marshall mixture design method. Further research related different compaction levels to the different aircraft tire pressures. For airfields handling aircraft with tire pressures in the range of 1,400 KPa, 75 blows of the compaction hammer per side are used to compact the specimen. It was determined that different ranges of blows of the compaction hammer produce asphalt mixture densities representative of those resulting from the initial construction compaction by rollers and followed by repeated traffic loads. The Marshall procedure was further refined by the Asphalt Institute for the design of asphalt mixtures for highway and other vehicular pavements and is used by many highway agencies. The Marshall method is the procedure that has also been used for the mixture design for rolled asphalt. The procedure is outlined in ASTM D1559 and is known as *Resistance to Plastic Flow of Bituminous Mixtures using Marshall Apparatus*.

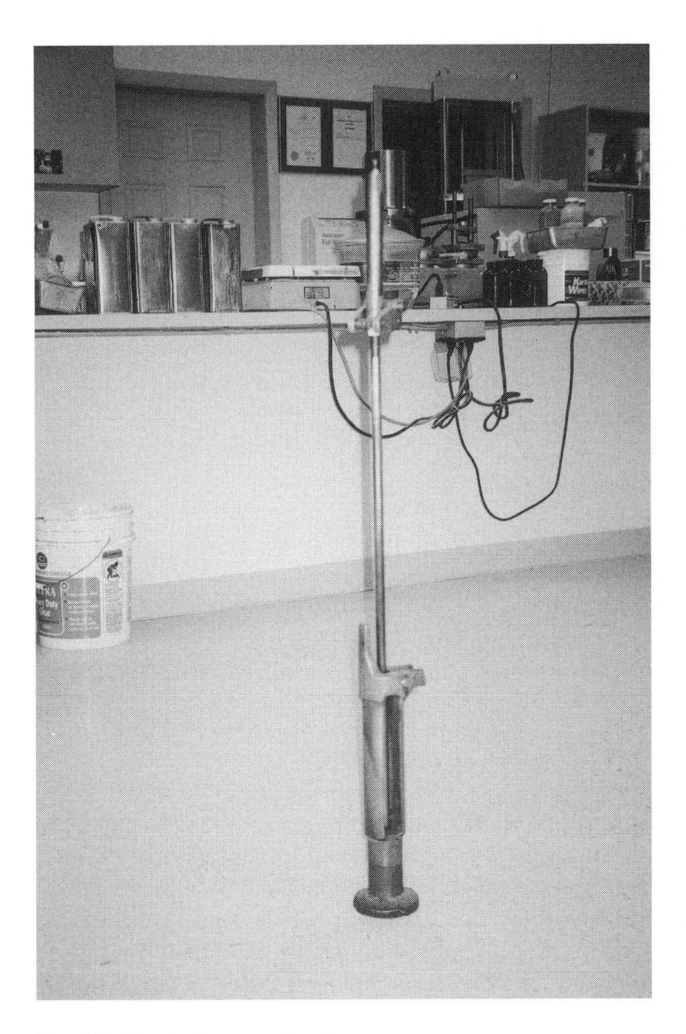

Plate 5.1 Marshall compaction hammer.

At the time of this publication, ASTM had not renewed the D1559 procedure for updated manuals. If the designer or engineer wishes to use a current published test method, the AASHTO test method, T-245, also titled, *Resistance to Plastic Flow of Bituminous Mixtures using Marshall Apparatus* is a straight substitution for ASTM D1559. Many highway agencies have adopted these methods into their own procedures. The Marshall method is applicable only to HMA that uses penetration; viscosity, aged residue, and PG graded asphalt binders. There is some laboratory procedures that significantly modified the Marshall procedure for use on designing dense graded mixtures containing asphalt emulsions. The Marshall procedure was developed using standard specimens that are 64 mm (2.5 inches) high and 102 mm (4 inches) in diameter. These specimens are produced in cylindrical molds (Plate 5.2) that also have a diameter of 102 mm. This standard size restricts the original Marshall method to mixtures that have a nominal maximum size of 25 mm or less, since the

Plate 5.2 Marshall specimen molds.

mold diameter to the particle size ratio should be four or greater. The procedure has been modified to handle larger stone mixtures by using a standard specimen size of 95 mm (3.75 inches) height and 152 mm (6 inches) diameter. The standard compactive effort was increased by increasing the hammer weight to 10.21 kg (22.5 lb) and the number of blows to 75. This procedure also became universally adopted and can be followed using ASTM D5581, *Resistance to Plastic Flow of Bituminous Mixtures using Marshall Apparatus (6-inch specimen)*. The purpose of the Marshall method is to determine the optimum asphalt binder content for a particular blend of aggregate. This is accomplished by preparing specimens at five different asphalt binder contents and then selecting or interpolating the asphalt binder content that provides the desirable properties. The method also provides additional test information about the compacted asphalt mixture specimens. The two principal features of the Marshall method are the volumetric analysis of the compacted specimens and their stability and flow properties.

Marshall method procedure

Detailed procedures for laboratory personnel can be found in publications such as ASTM D1559, AASHTO T245, The Asphalt Institute's *Mix Design Methods for Asphalt Concrete and Other Hot-Mix Types, MS-2*, BS 598 or from some highway agencies. The procedure goal is to produce a compacted asphalt mixture specimen that will have its volumetric properties determined and then have its Marshall Stability and Flow determined. Marshall stability is defined as the maximum load carried by a compacted specimen tested at 60 °C

Plate 5.3 Mechanical Marshall compaction hammer.

at a loading rate of 51 mm (2 inches) per minute. Marshall stability is a measure of the mass viscosity of the asphalt binder and aggregate mixture and is affected by the angle of internal friction of the aggregate and by the viscosity of the asphalt binder at 60 °C. Marshall flow is measured on the specimen at the same time as the Marshall stability. The flow is defined as the vertical deformation of the specimen while being loaded. It is measured from the start of loading to the point at which the specimen's stability begins to decrease. Its unit of measure is 0.25 mm (0.01 inches). The unique features of the Marshall method of mixture design are the Marshall apparatus and the stability-flow test. The original procedure developed during the Second World War had the hammer dropped on the apparatus by hand, which has now generally been replaced by a mechanical hammer (Plate 5.3).

Plate 5.4 Marshall testing machine and Marshall testing head.

The stability and flow test is completed using a Marshall testing machine (Plate 5.4), which is a compression testing device that applies loads to the specimen through cylindrical testing heads at a constant rate of strain of 51 mm per minute. A universal testing machine equipped with load and deformation measuring instruments and the Marshall testing head may be used in place of the Marshall testing machine.

The following is a summary of the procedure used to design HMA mixtures using the Marshall method:

Material acceptance and evaluation

1 Determine the traffic ESALs that the project will undergo. Determine the nominal maximum aggregate size of the mixture and the type of mixture (wearing, base, etc.)

required for the project. Determine what aggregates are available locally for the project that will meet the required nominal maximum size, coarse aggregate angularity, fine aggregate angularity, flat and elongated particles content, clay content, toughness, soundness, deleterious content, and any other specification that may be required for the aggregate.

2 Determine the following physical properties for each aggregate submitted for the mixture design: bulk specific gravity, absorption, and washed gradation.
3 Determine the proportion or blends of the various aggregates as described previously in the chapter. The gradation should be plotted on a 0.45 power chart and also have the final gradation compared with the specification ranges or targets for the particular mixture being designed.
4 Determine the appropriate asphalt binder grade for the mixture type, ESAL loading, climatic conditions and what grades are available locally for the project. Verify specification compliance for the asphalt binder.
5 Determine the specific gravity of the asphalt binder. Determine the asphalt binder viscosity in centistokes at 60 and 135 °C and plot on log–log centistokes graph. This graph is used to determine the mixing and compaction temperatures of the specific asphalt binder (Figure 5.5).
6 The specific gravity and the viscosity graph can usually be obtained from the supplier of the asphalt binder.
7 Determine the proper mixing and compaction temperatures from the graph using the following two assumptions:

   (a) The mixing temperature is the temperature of the asphalt binder that provides a kinematic viscosity of 170 ± 20 centistokes.
   (b) The compaction temperature is the temperature of the asphalt binder that provides a kinematic viscosity of 280 ± 20 centistokes.

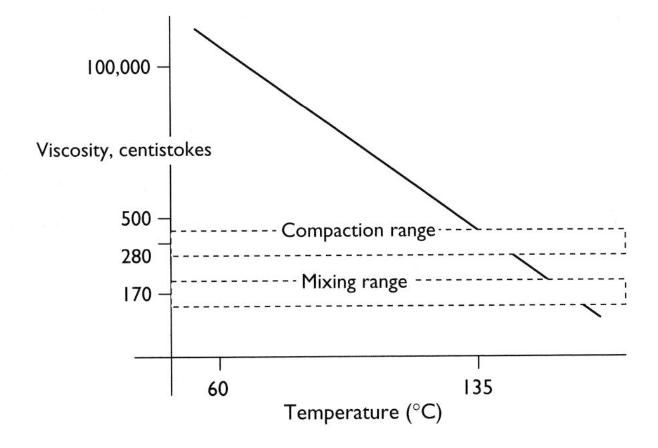

Figure 5.5 Temperature-viscosity log–log graph example.

Specimen preparation

1 Obtain enough materials to make 18 specimens that weigh 1,100–1,200 grams each. Dry and sieve the various aggregates into individual sieve sizes. To reduce the amount of sieving, the individual sieve sizes can also be grouped into the following size fractions:

 25–19 mm
 9–9.5 mm
 9.5–4.75 mm
 4.75–2.36 mm

 Total passing the 2.36 mm sieve

 Some laboratories eliminate the sieving all together because this can be a time consuming step. However, it is recommended to use the entire individual sieves or the grouped sieve sizes in order to have a representative and accurate blend of all the aggregates. This step also reduces the chance for aggregate segregation during the specimen blending process.

2 Using the aggregate blend proportions, determine the amount of material required for each sieve from each aggregate to make a specimen that weighs approximately 1,250 grams.

3 Weigh out the aggregate for 18 specimens and place in individual containers. Heat the individual aggregate specimens to the selected mixing temperature. Some laboratories weigh out enough material for a total of 18 specimens and combine it all together. This bulk material would be heated and then mixed with the asphalt binder as one large sample. This sample would then be split into individual specimens prior to compaction. Individual specimen preparation reduces the occurrence of segregation, especially when designing binder or base mixtures.

4 Determine the number of blows required for the type mixture. This can come from the specifications, type mixture being produced, estimated ESAL loading on the pavement or from customary practice. When no information is available use 50 blows for mixtures with a nominal maximum aggregate size less than 25 mm and use 75 blows for sizes greater than 25 mm.

5 It is recommended that a trial specimen be mixed and compacted prior to preparing all the aggregate mixtures. Use the midpoint of the asphalt binder content ranges or the expected content for the amount of asphalt binder to mix into the aggregate for this trial specimen. Measure the height of the trial specimen and compare to the requirement of 63.5 ± 1.27 mm for the 102 mm size specimens. If the height of the trial specimen is outside of these limits, adjust the amount of aggregate in the mixture by using equation 5.11.

$$\text{Adjusted mass of aggregate} = \frac{(63.5)\ \text{actual mass of aggregate used}}{\text{trial specimen height obtained (mm)}} \tag{5.11}$$

6 Heat sufficient asphalt binder at the mixing temperature to prepare a total of 18 specimens. Three compacted specimens should be prepared at five different asphalt binder contents. Asphalt binder contents should be selected at 0.5 percent increments with at least two asphalt binder contents above the estimated optimum content and at least two below the optimum content. This estimated optimum content is based on

previous experience with the aggregates, midpoints of specification ranges, producer recommendations, or any other available information. When blending the aggregate and asphalt binder together, the asphalt content may be expressed either, as the percentage by weight of total mixture or as a percentage by weight of total aggregate. The most common method is expressing the asphalt binder content as percentage by weight of total mixture. The volumetric equations also use the asphalt binder content as percentage by weight of total mixture. Example 5.6 shows the calculations both ways.

Example 5.6

Determine the amount of aggregate and asphalt binder required for a 1,250 gram mixture specimen that has asphalt binder content of 5.6 percent, both by total weight of mixture and total weight of aggregate.

Solution

By weight of total mixture: 5.6% of 1,250 grams = *70 grams of asphalt binder.*
1,250 grams − 70 grams = *1180 grams of aggregate.*

By weight of total aggregate: $\frac{1{,}250 \text{ grams}}{(1 + 5.6\%)} = \frac{1{,}250}{1.056} = 1{,}183.7$ grams of aggregate.
1,250 grams − 1,183.7 grams = *66.3 grams of asphalt.*

7 Three loose mixtures (Plate 5.5) should also be made near the optimum asphalt binder content to measure the theoretical maximum or Rice specific gravity. It is preferable to measure the Rice specific gravity in triplicate and then averaged. Some agencies require

Plate 5.5 Loose mixture for Rice specific gravity.

the theoretical or Rice specific gravity measured at all five asphalt binder contents. The precision of the test is best however when the mixture is close to the optimum asphalt binder content. Obtain the average effective specific gravity of the aggregate from the theoretical or Rice gravity measured at the estimated asphalt binder content. Using equation 5.10, the G_{mm} can now be accurately estimated at the four other asphalt binder contents (Roberts *et al.* 1991).

8 Verify that the aggregate temperature is at the required mixing temperature, then remove it from the oven, place it on a scale, and add the proper amount of asphalt binder for the selected asphalt binder content.
9 Mix the aggregate and asphalt until it is a well coated and a homogenous mixture. Individual specimens can be mixed by hand or through a mechanical mixer (Plate 5.6). Large bulk mixtures will require a large mechanical mixer (Plate 5.7).
10 Proceed onto the compaction phase and repeat until three samples are completed for each of the five asphalt binder contents. If the mixture is mixed in bulk, split out to the proper size samples and then repeat for the other asphalt binder contents.

Specimen compaction

1 The compaction temperature for the asphalt mixture is always below the mixing temperature. The specific Marshall procedure as outlined in ASTM D1559 or similar test procedures requires the mixture to be compacted immediately after mixing. Check the temperature of the mixture and allow it to cool to the compaction temperature. If its temperature is below the compaction temperature, discard the material and prepare another specimen. Some highway agencies have modified the Marshall procedure to

Plate 5.6 Mechanical mixer.

Plate 5.7 Large bulk mixer.

allow the just mixed material to cool down to ambient temperature and then be reheated to the compaction temperature. Other agencies have added a step of short-term aging of the mixture immediately after the mixing stage. The mixture is stored in an oven at the compaction temperature for 2–4 h prior to compaction. The thinking is that this time allows the mixture to absorb some asphalt binder and gives a realistic simulation of short term aging that occurs during construction, such as hauling of the mixture to the project site.

2 Place the individually prepared mixture (around 1,250 grams or adjusted as necessary) into the preheated Marshall mold that has been assembled on the Marshall hammer base plate (Plate 5.8). Paper or wax discs are used to keep the mixture from sticking to the hammer or the base plate. The mixture is spaded in the mold with a small trowel 15 times around the perimeter and 10 times around the center of the mixture. Place the

Plate 5.8 Specimen mold prior to compaction.

hammer into the mold on top of the mixture and apply the specified number of blows. Flip the mold over and apply the same number of the blows to the opposite side.

3 Remove the specimen and mold from the base plate and allow it to cool to the point where there is no deformation when firmly touched by hand. Extrude the specimen from the mold using a jack. Allow the specimens to cool further overnight and mark each specimen for identification.

Volumetric measurements

1 Determine the bulk specific gravity, G_{mb}, for each specimen by the following:

   I Determine the air dry weight of the specimen.
   II Submerge the specimen in water and obtain its submerged weight.
   III Remove the specimen from the water, towel dry it to a saturated surface dry (SSD) condition and then weigh it, obtaining the SSD weight.
   IV The volume of the sample is the SSD weight minus the submerged weight.
   V The bulk specific gravity is the air dry weight of the specimen divided by its volume.

   This procedure can be specifically followed by using ASTM D1188 or D2726. Average the bulk specific gravity for the three specimens for each asphalt binder content.

2 Measure the theoretical maximum, G_{mm}, or Rice specific gravity on the loose asphalt mixtures for at least two of the asphalt contents, preferably near the optimum asphalt binder content. ASTM D2041 or AASHTO T209 may be used as a laboratory procedure

to determine the Rice specific gravity. Tabulate the average of the bulk specific gravity of the specimen for each of the asphalt binder contents. Also, report the corresponding maximum theoretical gravity for each asphalt binder content, whether the G_{mm} is measured or calculated using equation 5.10.

3 Measure each specimen's height and record.
4 Determine the volumetric properties for each asphalt binder content using the average bulk specific gravity, G_{mb} and the average maximum theoretical gravity, G_{mm}. The volumetric properties that need to be calculated include: percent air voids, P_a, percent asphalt binder adsorption, P_{ba}, percent voids filled with asphalt, VFA, and VMA.

Marshall stability and flow

1 Heat the water bath (Plate 5.9) to 60 °C and place the specimens in the bath for 30–40 min.
2 After submerging the specimen for the required amount of time in the hot water bath, remove it and lightly dry it with a towel to remove excess water. Immediately place it into the Marshall testing head (Plate 5.10).
3 Apply the testing load at a constant rate of 51 mm/min until failure occurs. The point of failure is defined by the maximum load reading obtained. The total number of Newtons (lbs) required to produce failure of the specimen shall be recorded as its Marshall stability value.

Plate 5.9 Marshall water bath.

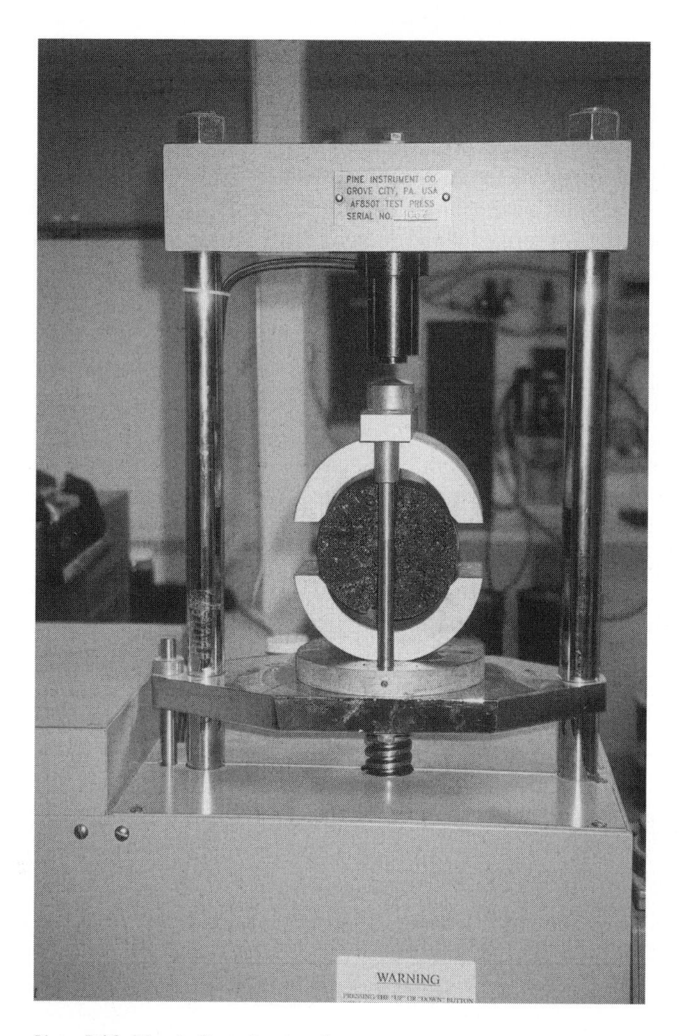

Plate 5.10 Marshall testing head.

4 While the stability test is in progress, if not using an automatic recording device, hold the flow meter firmly in position over the guide rod and remove as the load begins to decrease. Take the reading and record. This reading is the flow value for the specimen, expressed in units of 0.25 mm (0.01 inches). For example, if the specimen deformed 2.8 mm (the reading on the meter), the flow value is 11. The automatic recording device uses graph paper (Figure 5.6) allowing the flow value to be taken directly off the graph with no conversions.
5 The entire procedure for the stability and flow measurements, starting with the removal of the specimen from the water bath, shall be completed within a period of 30 s.
6 Repeat the steps until all the compacted specimens have been tested.

(The Asphalt Institute 1993)

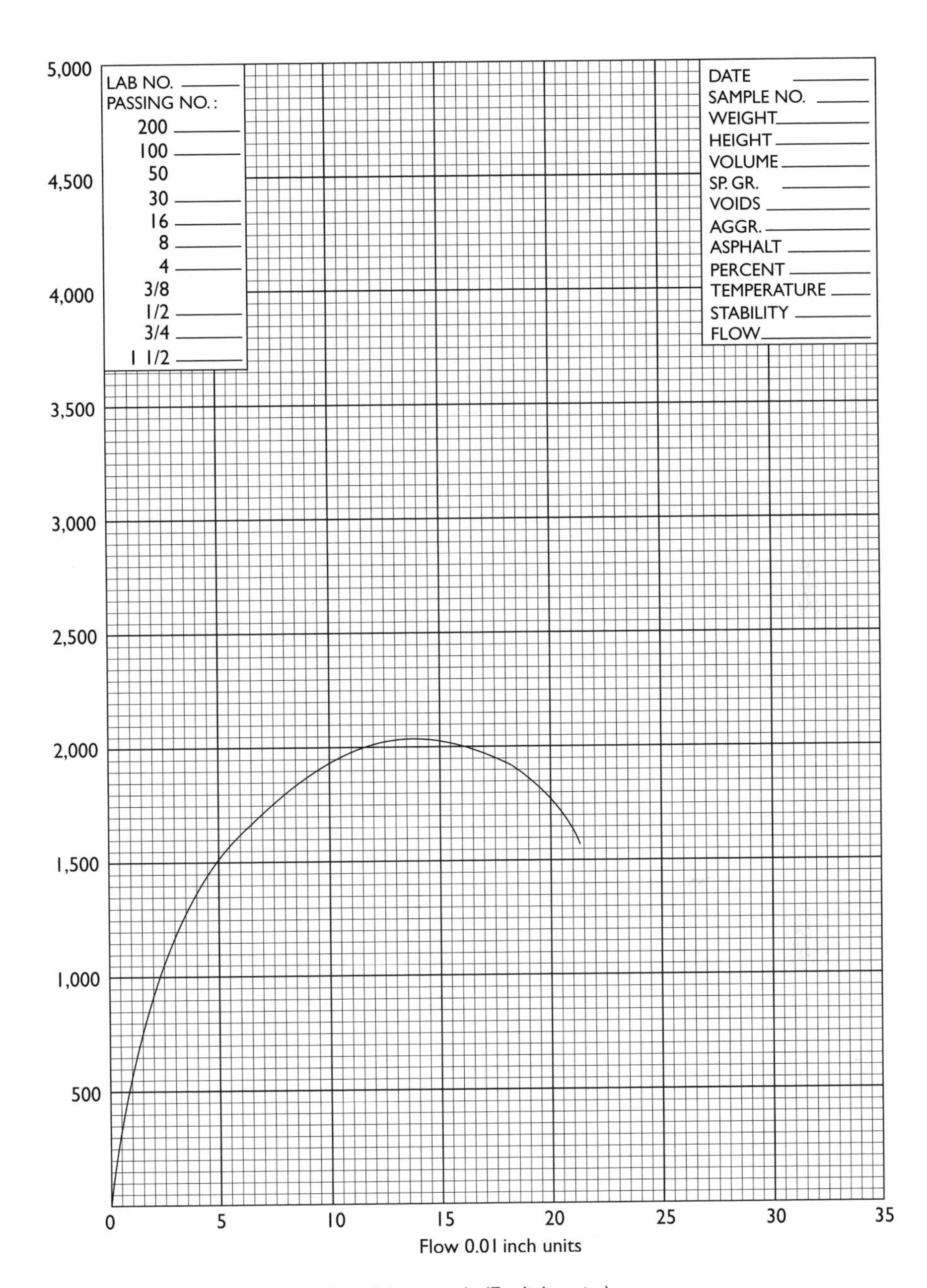

Figure 5.6 Example of Marshall stability graph (English units).

Test information

The volumetric properties and the stability and flow values of the specimens need to be for each asphalt binder content. The Marshall stability values that are determined make an assumption that the specimen height or thickness is 63.5 mm. Specimens that depart from this value will need to have the Marshall stability value converted to the standard 63.5 mm through the use of a conversion factor. The Marshall stability value is multiplied by a conversion factor provided in Table 5.18. The conversion factor corresponds to the specimen's actual thickness. The flow value does not need to be modified, since it is a measurement of the specimen's deformation through its diameter. The flow value is generally reported with no units, even though it is actually a distance measurement.

The stability value is adjusted by the thickness conversion value for each specimen and these values are then averaged for each asphalt binder content. The individual specimen flow values are also averaged for each asphalt binder content. The bulk specific gravity, G_{mb}, is determined for each specimen. The average theoretical or Rice specific gravity, G_{mm}, is determined for each asphalt binder content. Using the average specific gravity values for each asphalt binder content, the following volumetric properties are determined: percent air voids, P_a, VMA, and VFA. In the determination of VMA and VFA, the bulk specific gravity,

Table 5.18 Marshall stability conversion factors

Specimen thickness (mm)	*Conversion factor, 102 mm diameter specimen*
25.4	5.56
27	5.00
28.6	4.55
30.2	4.17
31.8	3.85
33.3	3.57
34.9	3.33
36.5	3.03
38.1	2.78
39.7	2.50
41.3	2.27
42.9	2.08
44.4	1.92
46.0	1.79
47.6	1.67
49.2	1.56
50.8	1.47
52.4	1.39
54.0	1.32
55.6	1.25
57.2	1.19
58.7	1.14
60.3	1.09
61.9	1.04
63.5	1.00
65.1	0.96
66.7	0.93
68.3	0.89
69.8	0.86
71.4	0.83
73.0	0.81
74.6	0.78
76.2	0.76

G_{sb}, of the combined aggregate blend will also need to be calculated. Table 5.19 is an example of the test information compiled for each asphalt binder content. If using equation 5.10 to determine the maximum theoretical specific gravity at all the asphalt binder contents, the effective specific gravity, G_{se}, of the combined aggregate will also need to be determined. G_{se} is also required to determine the asphalt binder absorption and the effective asphalt binder content, if that information is required. The information that is required to determine the optimum asphalt binder content is

- Marshall stability
- Marshall flow
- Percent air voids in the total mixture, P_a
- VMA
- VFA.

During the development of the Marshall method of mixture design, maximum unit weight of the mixture that was obtainable was also a desirable property and used in the selection of the optimum asphalt binder content. Most designers today consider the volumetric properties and the stability and flow values to a lesser extent, the key to the proper selection of the optimum asphalt binder content and have eliminated the unit weight analysis from the mixture design process.

Table 5.20 is an example of a summary of the test results and volumetric properties.

Using the information in Table 5.20, graphs are made of asphalt binder content on the x-axis versus the following on the y-axis:

- Marshall stability
- Marshall flow
- Percent air voids in the total mixture, P_a
- VMA
- VFA.

Table 5.19 Example of specimen information compiled by asphalt binder content

Asphalt binder content (%)	*Specimen ID*	G_{mb}	*Height* (mm)	*Corrected stability* (kN)	*Flow,* 0.25 mm	*Ave.* stability (kN)	*Ave. flow,* 0.25 mm	G_{mm}
4.0	1	2.371	63.5	7.9	10.6	7.9	10.4	2.494
	2	2.371	63.5	8.0	10.4			
	3	2.372	61.9	7.9	10.2			
4.5	4	2.395	63.5	8.2	9.8	8.2	9.7	2.473
	5	2.395	63.5	8.3	9.8			
	6	2.390	63.5	8.2	9.6			
5.0	7	2.401	65.1	8.7	10.1	8.7	9.8	2.464
	8	2.400	60.3	8.6	9.7			
	9	2.402	63.5	8.7	9.6			
5.5	10	2.422	63.5	7.4	11.2	7.4	11.6	2.441
	11	2.423	61.9	7.4	11.8			
	12	2.424	61.9	7.5	11.7			
6.0	13	2.421	65.1	6.1	12.1	6.2	12.2	2.423
	14	2.420	65.1	6.3	12.5			
	15	2.420	63.5	6.1	12.0			

Table 5.20 Test result summary

Asphalt binder content (%)	G_{mb}	G_{mm}	*Stability* (kN)	*Flow,* 0.25 mm	P_a	*VMA (%)*	*VFA (%)*
4.0	2.371	2.494	7.9	10.4	4.9	13.1	62.6
4.5	2.393	2.473	8.2	9.7	3.2	12.8	75.0
5.0	2.401	2.464	8.7	9.8	2.6	12.9	79.8
5.5	2.423	2.441	7.4	11.6	0.7	12.6	94.4
6.0	2.420	2.423	6.2	12.2	0.1	13.2	99.2

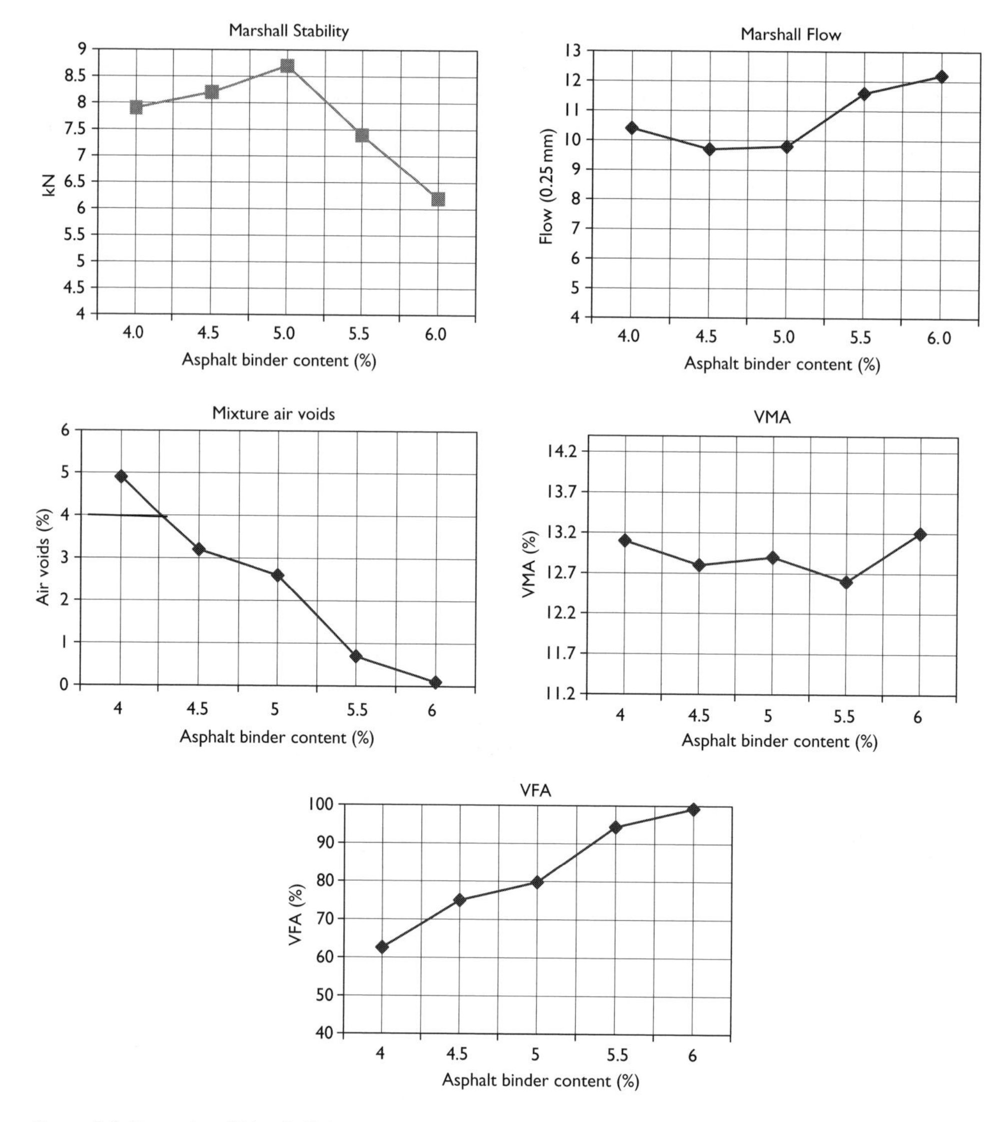

Figure 5.7 Example of Marshall design curves.

Figure 5.7 shows graphs of the information in Table 5.20 versus the corresponding asphalt binder content. These graphs are also known as test property curves or Marshall design curves.

Interpretation of the test information

By examining the Marshall design curves in Figure 5.7, information can be obtained as to how sensitive the aggregate and the mixture gradation are to the asphalt binder content. Curves that have a high slope value, or are steeply curved are general indicators that the mixture is sensitive to asphalt binder content changes. Information that can be initially determined from both Table 5.20 and Figure 5.7 is that, one more asphalt binder content should have been done at 3.5 percent and the 6 percent content could have been eliminated. The lower 3.5 percent content test information especially would have been valuable for the mixture air void analysis. Experience generally dictates what asphalt binder contents to select for a particular aggregate combination. In this particular case, the additional lower asphalt binder content or "point" could be completed later for supplemental information and to verify that the optimum asphalt binder content was correctly chosen. As many laboratories become more experienced in mixture design, they may reduce the number of asphalt binder content levels to four or even less. The idea is to "bracket" the optimum asphalt binder content with one or two lower and higher values. In Figure 5.7, initial indications showed that the mixture air void design curve could have been bracketed with one lower asphalt binder value.

It has been shown that Marshall design curves generally follow a reasonably consistent pattern for dense graded asphalt mixtures. However, there will be exceptions to the following trends:

- The Marshall stability value increases with increasing asphalt binder content up to a maximum after which the stability value decreases. Some design curves also have shown stability values decreasing with increasing asphalt binder content and do not have a peak value.
- The Marshall flow value increases with increasing asphalt binder content. In other words, the higher the asphalt binder content, the greater the specimen deforms under load.
- The mixture air voids, P_a, decreases with increasing asphalt binder content, until it reaches a minimum air void content.
- By definition, the VMA, should not be influenced or changed by changing asphalt binder content. However, the asphalt binder acts as a lubricant while the mixture is being compacted, thus leading that mixtures with a higher asphalt binder content should have a lower VMA percentage. What is usually observed is that the VMA generally decreases to a minimum value, then increases with increasing asphalt binder content.
- The percent voids filled with asphalt, VFA, increases with increasing asphalt binder content. This should be expected since the VMA is being filled with the asphalt binder.

(The Asphalt Institute 1993)

Optimum asphalt binder content determination

The Marshall design tools are the easiest tools for the designer to use in the selection of the optimum asphalt binder content. The concept of selecting the optimum content is to use the design curves and using select criteria, determine which asphalt binder content meets the criteria. The criteria consist of:

- Minimum Marshall stability value
- Range of Marshall flow values
- Range of mixture air void, P_a, values
- Minimum VMA value
- Range of VFA values.

The designer will generally select one set of criteria, such as mixture air voids, determine the optimum asphalt binder content for that criteria, and then verify whether binder content meets the other criteria. For example, the designer determines that an optimum asphalt binder content of 4.3 percent gives a mixture with 4 percent air voids. The designer then verifies that the 4.3 percent binder content will meet the Marshall stability, Marshall flow, VMA, and VFA criteria. Figure 5.8 illustrates that the designer first determined what asphalt binder content gives a mixture air voids of 4 percent and then determined the other properties of the mixture with an asphalt binder content of 4.3 percent.

The other properties of the example mixture determined from the Marshall design curves are given in Table 5.21. The information is visually determined from the design curves, but can also be interpolated using the specific test data.

The most common method in determining the optimum asphalt binder content is to first determine the asphalt binder content that has the desired amount of mixture air voids. The desired amount is usually the midpoint of the air void specification range, which is typically 3–5 percent. The amount of air voids in the mixture is directly related to the amount of asphalt binder in the mixture. Air void content is also one of the key performance indicators of the mixture. For these two reasons, the mixture air voids are usually the determining factor in selecting the optimum asphalt binder content. Using the initial optimum asphalt binder content, select the following values from the Marshall design curves:

- Marshall stability
- Marshall flow
- VMA
- VFA.

Compare each of these values against the specification requirements and if all are in specification, then the initial optimum asphalt binder content is satisfactory. If any value does not meet the specification requirements, determine from the Marshall design curves if slight shifts in the optimum asphalt binder content will make all the values meet the specification requirements. If this is not possible, consider redesigning the mixture, with emphasis on changing the combined aggregate gradation by adjusting the aggregate blends.

Another method for selecting the optimum asphalt binder content is to select each individual asphalt binder content from the Marshall design curves based on the following criteria:

- The asphalt binder content at the midpoint of the specification range for the mixture air voids.
- The asphalt binder content at the maximum stability value.
- The asphalt binder content at the midpoint of the specification range for VFA.

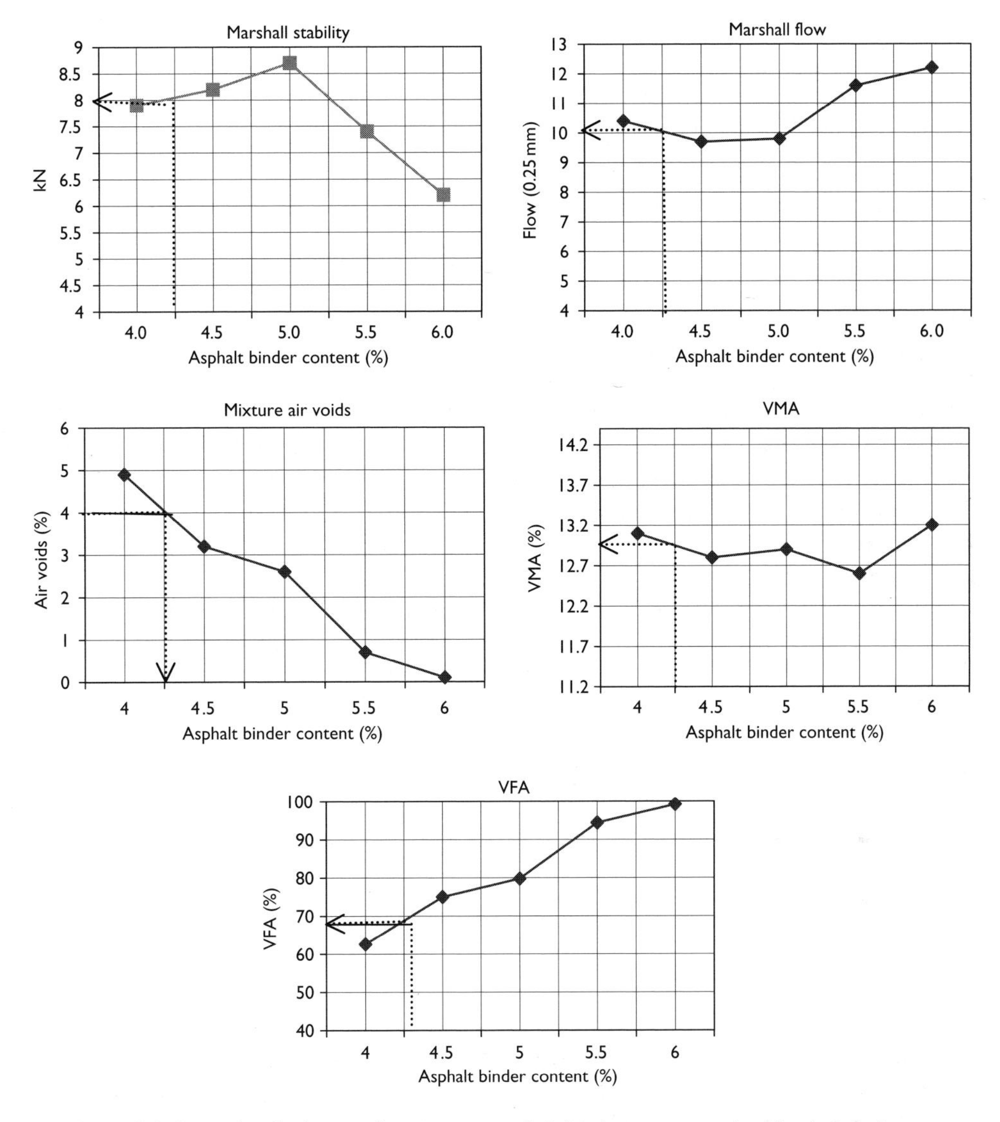

Figure 5.8 Example of selecting the optimum asphalt binder content using Marshall design curves.

Table 5.21 Information derived from the example Marshall design curves

Parameter	*Asphalt binder content (%)*	*Value*
Mixture air voids, P_a	4.3	4.0%
VMA	4.3	13.0%
VFA	4.3	69.0%
Marshall stability	4.3	8.1kN
Marshall flow	4.3	10.0%

The three possible asphalt binder content values are then averaged to give the optimum asphalt binder content. The Marshall flow and the VMA are checked to see if they meet specification at the optimum content. If all values are still in specification, obtain from the Marshall design curves the mixture air voids, Marshall stability, and VFA for the averaged optimum asphalt binder content value.

Mixture design specifications

The United States Army Corps of Engineers during the Second World War originally developed Marshall mixture design specifications for the construction of airfields used by heavy aircraft such as bombers. The Asphalt Institute and highway agencies further tailored these specifications to meet the needs of all types of roadways. The mixture specifications are grouped by traffic levels or ESALs. Since these specifications consider pavement loading, they can also be used for parking facilities and other low volume applications. It is important to consider that these specifications were developed and refined over the course of several years of experience in the design, construction, and performance of the asphalt mixtures. Designers should not deviate from the given specifications unless local experience or demand requires it. Some highway agencies have modified the specifications to fit their own conditions and experience. Designers for small projects, such as parking facilities or local roads, should use the specifications as written or use the local highway agency specifications if they are more applicable. Table 5.22 gives an example of Marshall mixture specification that may be used during the mixture design process.

The aggregate gradation and particle shape mostly determines the VMA. The asphalt binder content plays a minor role, mainly as a lubricant during the compaction process. The specifications for VMA are based on the nominal maximum size of the aggregate in the mixture. Table 5.23 is an example of a typical VMA specification. The VMA specifications given in Table 5.23 are based on the VMA calculation using the bulk specific gravity of the combined aggregate. Some highway agencies use the effective specific gravity of the combined aggregate. VMA values calculated with the effective specific gravity are higher than the VMA values calculated with the bulk specific gravity of the combined aggregate.

Table 5.22 Marshall design laboratory mixture specifications

Marshall mixture criteria	*Design traffic level, ESALs*		
	$<10{,}000$	10,000–1,000,000	$>1{,}000{,}000$
Number of hammer compaction blows per side of specimen	35	50	75
Mixture air voids, P_a (%)	3–5	3–5	3–5
Stability, kN	≥ 3.3	≥ 5.3	≥ 8.0
Flow, 0.25 mm	8–18	8–16	8–14
VMA (%)	Based on nominal maximum aggregate size		
VFA (%)	70–80	65–78	65–75

Table 5.23 Minimum Marshall VMA requirements

Mixture nominal maximum particle size (mm)	*Minimum VMA* (%)
63.0	10.0
50.0	10.5
37.5	11.0
25.0	12.0
19.0	13.0
12.5	14.0
9.5	15.0
4.75	17.0
2.36	20.0

Table 5.24 Example VMA specification by mixture type

Mixture type	*Minimum VMA* (%)
2.36 mm sand asphalt	20.0
4.75 mm leveling or scratch course	17.0
9.5 mm wearing course	15.0
19 mm intermediate course	13.0
25 mm base course	12.0

The VMA calculated with the effective specific gravity considers that the absorbed asphalt binder contributes to the performance of the asphalt mixture and be included in any specification. However, it is a general consensus among the asphalt industry and in the SHRP SuperPave method of mixture design, that VMA be calculated using the bulk specific gravity of the combined aggregate. If there is any possibility of confusion, the VMA specification should mention the method of calculation used.

Many Marshall mixture design specifications, especially highway agency or architect specifications, do not use as many VMA specifications as shown in Table 5.23. Since VMA is specified by the mixture's nominal maximum aggregate size, and mixture types such as wearing, binder, or base course mixtures are also specified by the nominal maximum aggregate size, it is possible to combine the two. For example, if wearing courses are specified as having a nominal maximum aggregate size of 9.5 mm, then wearing course VMA requirements would be a minimum of 15 percent. Thus, a specification may be written for VMA by just describing what mixture type is being designed–wearing course, base course, etc. Table 5.24 is an example of a VMA specification by mixture type.

Evaluation of Marshall criteria

The finished Marshall mixture design is usually the most economical one that will satisfactorily meet all of the required criteria. The mixture should not be designed to optimize one particular property. For example, a mixture that is designed only to give the highest Marshall stability values may tend to give an asphalt pavement that is less durable than desired and may crack prematurely under heavy volumes of traffic. The design asphalt binder content should be a compromise selected to balance all of the desirable mixture properties. Mixture design criteria will usually produce a narrow range of acceptable asphalt binder contents that will pass all of the guidelines. This property of the mixture applies to most design methods, including Marshall, Hveem, and SuperPave. The asphalt binder content selection can be adjusted within this narrow range to achieve a mixture property that will satisfy a requirement of a specific specification or application. The size of the adjustment range varies with the particular asphalt mixture and is dependent on how sensitive the mixture is to the asphalt binder content. Different properties are more critical for different circumstances, depending on traffic, structure, climate, construction equipment, and other factors. The balancing process of the various desirable properties of the asphalt mixture is not the same for every pavement and every mixture design (The Asphalt Institute 1993). The influence of each of the mixture properties on the performance of the asphalt pavement needs to be determined in order to balance all the properties for the specific asphalt mixture application.

Marshall stability

Marshall stability is defined as the maximum compressive load carried by a compacted specimen tested at 60 °C at a loading rate of 51 mm/min. The Marshall stability is reported in Newtons (lbs). The Marshall stability is a measure of the mass viscosity of the asphalt mixture and is controlled by the angle of internal friction of the aggregate and the viscosity at 60 °C of the asphalt binder. The Marshall stability values can be increased by using a "stiffer" or more viscous asphalt binder grade in the mixture. For example, changing the mixture's asphalt binder grade from an AC-10 to an AC-20 will increase the Marshall stability values. Using aggregates that are more angular or a blend of all crushed aggregates will give higher Marshall stability values than rounded gravel type aggregates. The designer can select various asphalt binder grades and aggregate properties to give an asphalt mixture that can have a high or low Marshall stability. Any material than can stiffen an asphalt binder will also increase the Marshall stability. For example, a small increase in the mineral filler (dust) content of the asphalt mixture will stiffen the asphalt binder and stiffen the total asphalt mixture, leading to higher Marshall stability values. However, if too much mineral filler is used, the filler can act as an asphalt binder extender, affecting the mixture as if it had a higher asphalt binder content. This net effect could actually cause the mixture to have a decrease in the Marshall stability values (Roberts *et al.* 1991) This phenomena among others is the reason some agencies require or designers use as a guideline a maximum dust (passing the 75 μm sieve) asphalt binder content ratio. Most specifications use a maximum value of 1.0 or 1.2 at the laboratory design stage. For example, if the optimum asphalt binder content is 5.2 percent, the gradation of the blended aggregates in the asphalt mixture can not contain more than 5.2 percent passing the 75 μm sieve. The maximum value of 1.0 or 1.2 usually applies only to dense graded asphalt mixtures.

The Marshall stability and the stability of the asphalt mixture during construction or long-term pavement performance are not necessarily related. The stability of the mixture in place is affected by the ambient temperature, types of loading, rate of loading, tire contact pressures, subgrade strength, and various other factors. Marshall stability uses only one temperature and one loading rate. The primary use of the Marshall stability is in evaluating the change in the stability values with the change in the asphalt binder content of the mixture and to set some minimum strength values for the various applications of the asphalt mixture.

Marshall flow

The Marshall flow is the vertical deformation of the compacted specimen from the start of the Marshall stability loading until the stability values begin to decrease. Flow is reported in increments of 0.25 mm. High flow values indicate an asphalt mixture that has plastic behavior and has the potential for permanent deformation, such as rutting or shoving, under loading. Low flow values indicate a mixture that may have insufficient asphalt binder, which may lead to durability problems with the pavement. Low flow values may also indicate a mixture with a binder so stiff, that the pavement experiences low temperature or fatigue cracking. Marshall flow is a function of the asphalt binder stiffness and the asphalt binder content of the mixture. If the flow values need to be increased, increase the asphalt binder content or choose a "softer" asphalt binder grade. If the flow value needs to decrease, reduce the asphalt binder content or use a "stiffer" asphalt binder grade.

Marshall Quotient

Some highway agencies, namely in Europe use a ratio of the Marshall stability to the Marshall flow. This ratio is known as the "Marshall Quotient" and represents the ratio of load to deformation. The Marshall Quotient can be used to give an indication of the mixture's stiffness. The higher the ratio, the stiffer the mixture is. One or more of the following can increase the Marshall Quotient (Whiteoak 1991):

- The addition of coarse aggregate
- The use of a coarser sand
- The use of crushed sand
- A stiffer asphalt binder.

Mixture air voids

The mixture air voids, P_a, are inversely proportional to the asphalt binder content. The higher the asphalt binder content is in the mixture, the lower the air void content is, the lower the asphalt binder content, the higher the mixture air void content is. Most specifications for dense graded mixture air voids are in a range of 3–5 percent. The designer generally targets the midpoint of the range in selecting the optimum asphalt binder content. The mixture air voids determined in the laboratory during the design process consider the compactive effort on the mixture from *both* the rollers and from several years of traffic. This combination of compactive efforts is the premise for the different number of blows used by the Marshall hammer for the different traffic levels. The initial, after construction, air voids in the pavement should always be higher than the specification range given for the laboratory design. Mixture compaction is required to interlock the aggregate particles, creating internal friction in the mixture, leading to shear resistance for the asphalt mixture. This shear resistance reduces the deformation of the mixture caused by loading. Compaction can also be used to create an asphalt mixture that is relatively impervious to both air and water. A 5 or 6 percent maximum air void content in the laboratory mixture design process will give a mixture that should not have any interconnected air voids, but are rather individual air voids scattered throughout the mixture. Interconnected air voids will allow air and water to penetrate throughout the pavement, leading to long-term durability problems. A minimum air void content is desirable to allow the pavement to expand and contract under loading and temperature changes. Pavements that have very low air void contents tend to have asphalt binder bleeding or flushing on the surface and tend to deform under traffic. The mixture air void range is used to ensure enough asphalt binder to provide a durable pavement, but not so much that the pavement will deform under loading or have excess asphalt binder flushing or bleeding on the surface. The asphalt binder in conjunction with mineral filler or dust (minus 75 μm) can be used to fill air voids or to create air voids. The designer will reduce the asphalt binder content or the dust content to increase the mixture air voids. Increasing the asphalt binder content and/or the dust content can decrease the air mixture air voids. The gradation specifications, VMA criteria, and the dust/asphalt binder ratio limit changes in the dust content.

Voids in the mineral aggregate

The voids in the mineral aggregate, or VMA, is the total volume of voids within the mass of the compacted aggregate. It is the volume of the mixture air voids plus the volume of the effective asphalt binder in the mixture. VMA is important in that it allows room for enough asphalt binder to make a durable mixture plus enough room for mixture air voids to ensure a stable mixture. If there is not enough VMA in the compacted aggregate, there is not enough room for required amount of asphalt binder to provide a durable mixture. A certain amount of asphalt binder is required to provide an adequate film thickness around each aggregate particle. A durable asphalt mixture requires an adequate film thickness. VMA is an indirect way to specify the amount of the asphalt binder film thickness in a mixture. If the VMA amount is too large, an uneconomical amount of asphalt binder will be required to reduce the mixture air voids to an acceptable level. This amount of asphalt binder in the mixture may also lead to mixture stability problems. As the nominal maximum particle size of the aggregate increases, the minimum VMA required decreases. This occurs because the total void space between large aggregate particles is smaller than the void space between small aggregate particles (Roberts 1991). Since VMA is actually a property of the aggregates in the mixture, Changes in the aggregate gradation or shape provide the significant changes in VMA. Generally as a gradation moves further away from the 0.45 power maximum density line, VMA will increase. This occurs whether the gradation becomes finer or coarser, as long as it gets further from the maximum density line. The definition of the 0.45 power maximum density line is the gradation of a particular aggregate that provides the most packing or the minimum amount of VMA. Gradations that are further from this gradation are not "packed" as well or have a larger amount of VMA in them. Changing the amount of minus 75 μm in the asphalt mixture has a significant affect on VMA. A 1 percent decrease in the amount of 75 μm in the mixture will increase VMA around 1 percent (Murphy and Bentsen 2001). Aggregate shape or texture that makes a mixture harder to compact or "pack" will also increase the VMA amount. A cubical shaped, rough textured aggregate should provide a mixture with a higher VMA than a round or flat smooth textured aggregate. The designer can combine both the effects aggregate gradation and aggregate shape to alter the VMA in the mixture. VMA is also somewhat controlled by the asphalt binder content, but only to the extent that the asphalt binder acts as a lubricant and has an effect on how well the aggregate particles are compacted together. If the aggregate particles pack together easily under compaction, such as when increasing the asphalt binder content, the VMA will decrease. If the aggregates do not pack very easily under compaction, such as when using a stiffer asphalt binder, the VMA will increase. Each mixture has unique VMA characteristics and over time as a designer becomes familiar with the particular materials trial and error will provide what amount of change in the mixture gradation that will be required to influence changes in VMA.

Voids filled with asphalt

The VFA are determined by amount of VMA and air voids in the mixture. It is a percentage of the VMA that is filled with asphalt binder. Since it is a percentage of the VMA, VFA can limit the amount of VMA in a mixture. If the VMA is limited in a mixture and the mixture air voids are kept constant, then the amount of asphalt binder in a mixture becomes limited. VFA also restricts the allowable air void content for mixtures that are near the minimum VMA criteria. Mixtures that are designed for lower traffic volumes, such as parking facilities, will not pass a VFA criteria with a relatively high percentage of mixture air voids (such as 5 percent) even though the air void specification criteria has been met. The purpose is to avoid less durable mixtures in light traffic applications. Mixtures designed for heavy traffic, such as freeways, will not pass the VFA criteria with relatively low percent of mixture air voids, even though the amount of air voids is within a specification range. Because low air void contents can be very critical in terms of mixture deformation, the VFA criteria helps to avoid those mixtures that would be susceptible to rutting or shoving in heavy traffic applications (The Asphalt Institute 2001). To increase the VFA, increase the amount of asphalt binder in the mixture. To decrease the VFA, decrease the amount of asphalt binder in the mixture. It should be apparent now, that change in the asphalt binder content will also directly change the mixture air void content, *unless* the VMA is also changed. That concept is how VFA works to limit the amount of VMA in a mixture.

The SHRP SuperPave method of mixture design

The Strategic Highway Research Program (SHRP) was established by the United States Congress in 1987 as a five-year $150 million research program to improve the performance and durability of roads and to make those roads safer for both motorists and highway workers. $50 million of the SHRP research funds were used for the development of performance based asphalt binder and mixture specifications to directly correlate laboratory methods with pavement performance. The SHRP asphalt mixture research program developed the laboratory mixture design method known as SuperPave, an acronym from the phrase, **Su**perior **Per**forming Asphalt **Pave**ments. The SuperPave method incorporates performance based asphalt mixture materials characterization with the design environmental conditions, in order to improve pavement performance by controlling rutting, low temperature cracking, and fatigue cracking. SuperPave consists of three major components: the asphalt binder specification, the mixture design and analysis system, and a computer software system that integrates the binder and mixture components. The SuperPave mixture design and analysis system uses three progressively rigorous degrees of testing and analysis to provide a well performing mixture for a given pavement project and climate (The Asphalt Institute 2001). At present, only the first degree or level one, SuperPave method is being practiced routinely by highway agencies and mixture designers. The SuperPave level one mixture design procedure incorporates careful material selection and volumetric proportioning to produce a mixture that will perform satisfactorily. The level two and three methods incorporate all of the level one procedures and also include two additional pieces of laboratory equipment to predict the various pavement distresses that may or may not occur involving the designed mixture. The additional equipment is known as the "SuperPave Shear Tester" (SST) and the Indirect tensile Tester (IDT). This equipment is used to ensure that SuperPave designed mixtures exhibit acceptable amounts of the following pavement distress types:

- low temperature cracking
- fatigue cracking
- permanent deformation.

Once the mixture was designed using the level one method, specimens were prepared and tested by the SST and the IDT. The amount of traffic predicted for the pavement application dictated whether the specimens received the level two or intermediate performance predictions or the complete or level three performance predictions. The significant differences between level two and level three is the amount and type of testing on the specimens using both the SST and the IDT.

The SST can perform the following six tests on the mixture:

- volumetric test
- uniaxial test
- repeated shear test at constant stress ratio
- repeated shear test at a constant height
- simple shear test at a constant height
- frequency sweep test at a constant height.

Whether the mixture is being designed to level two or to level three determines which of the SST tests will be used. The IDT is used to measure the creep compliance and the strength of the mixture using indirect tensile loading at intermediate to low temperatures. Additional test temperatures are used to differentiate the level three methods from the level two methods. Due to the current cost and complexity of the SST, level two and level three mixture designs are not routinely performed by mixture designers or highway agencies. Level one of the SuperPave design method must be completed first before proceeding to level two and three. The original SuperPave mixture design protocol required a level one design for pavements with a 20-year design ESAL of less than 1,000,000; a level two for 1,000,000–10,000,000 ESALs and a level three for pavements greater than 10,000,000 ESALs. For the purposes of this text, only the SuperPave level one method will be discussed. The level one method is currently in practice by many agencies and designers and has the ability to be performed in the field. The SuperPave mixture design method and the Marshall mixture design method share some common properties. Both require a compactive effort on an asphalt mixture specimen and use its volumetric properties to determine the optimum asphalt binder content. Both are used to design dense graded mixtures, but can be used to design other mixtures such as gap graded and open graded. However, the SuperPave method does not perform any stability or flow testing on the specimen or consider unit weight or density in determining the optimum binder content. The significant difference between the Marshall design method and the SuperPave method is the method of compaction. The Marshall method used an impact type of compaction using a Marshall hammer and the SuperPave method uses a combination of a static load and a gyrating mold to "knead" the specimen to

Plate 5.11 SuperPave gyratory compactor.

a given density. The SuperPave accomplishes this type of compaction through the use of a SuperPave gyratory compactor (Plate 5.11).

The SuperPave gyratory compactor (SGC) uses a specially designed mold (Plate 5.12) to hold the loose asphalt mixture. This mold is rotated through a 1.25° angle by a jig (Plate 5.13). While the mold is being rotated at 30 gyrations/min, a 600 KPa static load is placed on the specimen through the use of a ram.

The impact type of compaction provided by the Marshall hammer does not realistically duplicate the type of compaction that an asphalt mixture undergoes by both construction rollers and traffic. The SHRP program determined that a gyrating or "kneading" type of compaction provides a more realistic duplication of mixture compaction in the field than an impact type. By duplicating the compactive effort in the field, the specimens produced should represent an actual compacted asphalt pavement, and therefore the optimum asphalt binder content selected using those specimens should be more accurate. The SGC is also capable of handling large stone mixtures by using a specimen mold with a diameter of 150 mm. The development of the SGC was based on the Texas gyratory compactor combining the compaction characteristics of the French gyratory compactor. The SGC is an electrohydraulic machine consisting of the following components:

- reaction frame, rotating base, and motor
- loading ram and pressure gage
- specimen height measuring and recording system
- mold base and plate.

Plate 5.12 SGC mold.

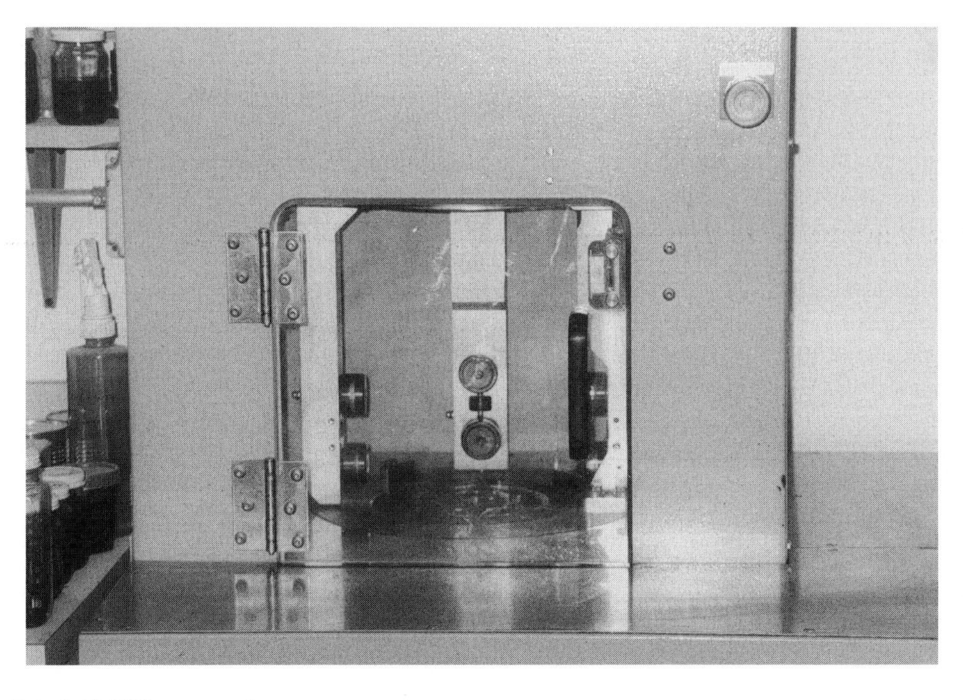

Plate 5.13 SGC gyratory jig.

The SGC compacts a specimen through the following sequence of events:

1 The loading mechanism presses against the reaction frame and applies a load to the loading ram to produce a 600 kPa compaction pressure on the specimen.
2 An electronic pressure gage measures the ram loading to maintain a constant 600 kPa pressure during the compaction process.
3 The compaction mold has a base plate in its bottom to provide specimen confinement during compaction.
4 The SGC base rotates at 30 revolutions per minute during compaction with the mold positioned at a compaction angle of 1.25°.

The SGC is also capable of compacting smaller specimens using a 100 mm mold (Plate 5.14). However the current SuperPave design method specifies a 150 mm mold. The measurement of the specimen height during the compaction process is a key feature of the SGC. As the specimen is compacted, or as the number of gyrations are completed, the specimen height will get smaller. The specimen density can be estimated at any point during the compaction process by knowing the mass of the mixture placed in the mold, the inside diameter of the mold and the specimen height. The inside diameter is fixed at 150 mm and the loose mixture is weighed before it is placed in the mold. The specimen height is measured by the SGC by recording the position of the loading ram throughout the compaction process (The Asphalt Institute 2001). Using the fixed plane area of the specimen and its varying

Plate 5.14 100 mm and 150 mm SGC molds.

height, h_x, the specimen volume, V_{mx}, can be determined at any point during the compaction process through equation 5.12.

$$V_{mx}\,(cm^3) = \frac{\pi\,(d)^2(h_x)}{4000} \quad (5.12)$$

where d is the specimen or mold inside diameter (mm); h_x, the specimen height at x number of gyrations.

This calculation is completed automatically by the SGC software. The specimen volume determined in this method is actually an estimated volume, since the volume calculation considers the specimen to be a smooth walled cylinder. The actual specimen has an irregular surface caused by the aggregate. Since the mass of the specimen and its volume are now known the specific gravity and the density of the specimen can now be determined. The SGC calculated specific gravity or density is corrected by physically completing a bulk specific gravity test on the extruded specimen after the desired number of gyrations has been completed. This correction factor is a ratio given by equation 5.13.

$$\text{Correction factor, } C = \frac{G_{mb}\,(\text{measured})}{G_{mb}\,(\text{estimated})} \quad (5.13)$$

This information is used to determine the compactive effort that is placed on the specimen and the specimen's estimated air voids, P_a, at any point during the compaction process. As the number of gyrations are completed, the specimen's density is increased and its air voids are decreased.

Similar to the Marshall and other asphalt mixture design methods, the asphalt mixtures are designed at a specific level of compactive effort, based on the traffic or loading to be expected by the pavement. In the SuperPave design method, the compactive effort is a function of the number of gyrations completed by the SGC. The traffic or loading specific number of gyrations is known as the design number of gyrations, or N_{des}. The traffic or loading is represented by the design ESALs of the pavement. For the purposes of the SuperPave design method, the design ESALs are the anticipated amount of ESALs in the design lane over a 20-year period. If considering a shorter or longer period as the design life of the pavement, extrapolate or interpolate the ESALs to a 20-year period. The number of gyrations for each traffic level is given in Table 5.25 (AASHTO 2000a).

Table 5.25 introduces two other gyrations parameters using the SGC, the initial number of gyrations, $N_{initial}$, and the maximum number of gyrations, $N_{maximum}$ or N_{max}. $N_{initial}$ and

Table 5.25 SuperPave gyratory compactive effort

Design ESALs 20 years	*SGC compactive effort (number of gyrations)*		
	$N_{initial}$	N_{design}	N_{max}
$<$300,000	6	50	75
300,000 to $<$3,000,000	7	75	115
3,000,000 to $<$30,000,000	8	100	160
$\geq$30,000,000	9	125	205

N_{max} are related to the design number of gyrations, N_{des}. These relationships are given by equations 5.14 and 5.15.

$$\text{Log } N_{initial} = 0.45 \text{ Log } N_{des} \quad (5.14)$$

$$\text{Log } N_{max} = 1.1 \text{ Log } N_{des} \quad (5.15)$$

The specimen is compacted by the SGC to N_{des}. The initial SuperPave design method required the specimen to be compacted to N_{max}, but has been changed and published in AASHTO specifications (AASHTO 2000a) to N_{des}. In the past, the gyratory compaction procedure required that specimens are compacted to N_{max} and the densities and volumetric properties are back calculated at N_{des}. This caused an error in the calculated volumetric properties at N_{des}. Since the mixture is designed upon its volumetric properties at N_{des}, compacting the specimens to N_{des}, and not N_{max}, now completes SuperPave volumetric designs (NCAT 1998). The initial number of gyrations, $N_{initial}$, is used to give an estimate of the asphalt mixture's ability to be compacted by rollers during placement of the mixture in the field. The mixture is compacted by the SGC to $N_{initial}$, and an estimated bulk specific gravity is determined by the SGC. The design ESALs determines the maximum density amount at $N_{initial}$. The mixture is then further compacted to N_{des}, where another estimated bulk specific gravity is determined by the SGC. The specimen is then extruded and its actual or measured bulk specific gravity is determined. The specific gravity of the specimen at the maximum number of gyrations, N_{max} is extrapolated from the information provided at $N_{initial}$ and N_{des}. The density at N_{max} is relevant in that the specific gravity or density of the specimen at N_{max} should not be greater than 98 percent of the theoretical maximum specific gravity, G_{mm}. A high density at N_{max} is undesirable, since N_{max} represents a traffic level much higher than that for which the project is designed. By limiting the density at N_{max}, it is expected that the mixture will not densify to extremely low air voids with unexpectedly high traffic or ESALs (TRB 2000). Table 5.26 summarizes the various density requirements at the different gyration levels.

The significant differences between a modern Marshall design method and the SuperPave design method (level one) is the method of compaction. Both methods, with the exception of Marshall Stability and Flow, use the volumetric properties of the specimens to determine the optimum asphalt binder content. The SuperPave design method does complete stability or flow testing on the specimens. Volumetric properties of a mixture are physical properties that are measured in the same manner regardless of how the specimen was mixed or compacted. For example, VMA will be measured the same, whether the asphalt mixture was compacted with a Marshall hammer, a SuperPave gyratory compactor, a vibratory roller or with

Table 5.26 Density requirements

Design ESALs (20 years)	*Required relative Density (% of G_{mm})*		
	$N_{initial}$	N_{des}	N_{max}
$<$300,000	$\leq$91.5	96.0	$\leq$98.0
300,000 to $<$3,000,000	$\leq$90.5	96.0	$\leq$98.0
3,000,000 to $<$30,000,000	$\leq$89.0	96.0	$\leq$98.0
$\geq$30,000,000	$\leq$89.0	96.0	$\leq$98.0

the tire of a semi-trailer truck. Like the Marshall design method and most other methods, the SuperPave design method is a two stage design process, first the mixture gradation is determined and then the optimum asphalt binder content is determined. The SuperPave level one design method consists of four major steps:

1. Selection of the aggregate and the asphalt binder
2. Selection of the mixture design gradation or design aggregate structure
3. Selection of the optimum or design asphalt binder content
4. Evaluation of the moisture sensitivity of the designed mixture.

Selection of the aggregate and the asphalt binder consists of determining the traffic or loading and the environmental factors for the asphalt pavement. The PG of the asphalt binder is then selected. The aggregate requirements are determined based on traffic and the materials selected are based on their ability to meet or exceed established criteria. It should be noted that the SuperPave design method can be physically completed with any asphalt binder or aggregate selected, not necessarily the ones that meet the SHRP SuperPave criteria. For example, a mixture design using the SGC and SuperPave procedures can be completed on a asphalt mixture containing a 60–70 penetration asphalt binder instead of the SHRP PG 64-22 and provide an adequate performing optimum asphalt binder content. A complete SuperPave mixture procedure requires the use of the SHRP consensus aggregate properties and PG asphalt binders.

Selection of the mixture gradation or design aggregate structure consists of determining the aggregate stockpile gradations and the corresponding combined aggregate gradation. The combined aggregate gradation and the optimum asphalt binder content should yield acceptable volumetric and compaction properties when compared to the mixture criteria. This also applies to the Marshall design method.

Selection of the optimum asphalt binder content consists of varying the amount of asphalt binder with the combined aggregate to obtain acceptable volumetric and compaction properties when compared to the mixture criteria that is based on loading and environmental conditions. This step also verifies if an adequate or correct combined aggregate gradation has been determined. This step also allows the determination of changes in the mixture's volumetric and compaction properties with changes in the amount of asphalt binder in the mixture. This is also similarly used in the Marshall design method. The final step required by the SuperPave design method is the evaluation of the moisture sensitivity of the mixture by using AASHTO T-283 (AASHTO 2000b) to determine if the mixture will be susceptible to moisture damage (NATC 1994).

The SHRP SuperPave consensus aggregate properties are select aggregate characteristics that are needed to provide a high performing asphalt pavement. They are known as consensus properties because there is wide agreement among researchers and users in their use and specified values. The consensus properties are:

- coarse aggregate angularity
- fine aggregate angularity
- flat and elongated particles
- aggregate clay content.

The criteria for these aggregate properties are based on ESALs and their location within the pavement structure. Aggregates that are near the pavement surface, which are subjected to high traffic levels, require more stringent properties than aggregates used for low volume roads or base courses. The consensus aggregate property criteria are intended to be applied to the combined aggregate blend rather than to individual components. However, it is simpler to require the individual aggregates to meet their respected properties, so that undesirable aggregates may be eliminated from consideration at the beginning of the mixture design process (The Asphalt Institute 2001).

Coarse aggregate angularity

A certain amount of angularity of the coarse aggregate will ensure a high degree of aggregate internal friction, which helps provide rutting or permanent deformation resistance. Coarse aggregate angularity is defined as the percent by weight of aggregates that are larger than 4.75 mm and have at least one fractured face. Coarse aggregate angularity can be measured by following the ASTM procedure D4821.

Fine aggregate angularity

Fine aggregate angularity ensures a high degree of fine aggregate internal friction, which also provides permanent deformation resistance. Fine aggregate angularity is defined as the percent air voids present in loosely compacted aggregates that are smaller than 2.36 mm. Higher air void contents in the fine aggregate indicate more fractured faces. Fine aggregate angularity can be measured by following the AASHTO procedure T-304. Fine aggregate angularity is measured indirectly by pouring a sample of washed fine aggregate into a cylinder of known volume through a standard sized funnel. By measuring the mass of the fine aggregate in the cylinder, the void content can be calculated as the difference between the cylinder volume and the fine aggregate volume collected in the cylinder. The fine aggregate bulk specific gravity, G_{sb} is used to determine the fine aggregate volume.

Flat and elongated particles

Flat and elongated particles are characteristic of the coarse aggregate. It is the percent by mass of the coarse aggregates that have a maximum to minimum dimension ratio greater than five. Too many flat and elongated aggregate particles in a mixture are undesirable since they have a tendency to break during construction or under traffic. The aggregate is measured by comparing the particle length to the particle thickness. Aggregate greater in size than 4.75 mm are measured using a proportional caliper device. The procedure is physically measuring a sampling of the coarse aggregate using the caliper device. The procedure can be completed by following ASTM D4791, *Standard Test Method for Flat and Elongated Particles in Coarse Aggregate*.

Clay content

Clay content is the percent of clay material contained in the aggregate fraction that is finer than 4.75 mm. It is measured by following ASTM D2419, *Sand Equivalent Value of Soils and Fine Aggregate*, or AASHTO T-176. This test is also known as the sand equivalent test. The terms sand equivalent or clay content is used interchangeably when referring to the amount of clay in an aggregate source. The allowable clay content for the fine aggregate is expressed as a minimum percent of the sand equivalent test.

Table 5.27 SuperPave aggregate consensus requirements, particle angularity and shape

Traffic, million ESALs (20 years)	*Mixture depth from pavement surface*				*Flat and elongated particles criteria ASTM D4791* maximum *(%)*
	Coarse aggregate angularity criteria[1] *ASTM D5821* (mm)		*Fine aggregate angularity criteria*[2] *AASHTO T-304* (mm)		
	≤100	>100	≤100	>100	
<0.3	55/–	–/–	—	—	—
0.3 to <3	75/–	50/–	40	40	10
3 to <10	85/80	60/–	45	40	10
10 to <30	95/90	80/75	45	40	10
≥30	100/100	100/100	45	45	10

Notes

1 95/90 denotes that 95 percent of the coarse aggregate has one or more fractured faces and 90 percent has two or more.

2 Percent air voids in loosely compacted fine aggregate.

Table 5.28 SuperPave sand equivalent (clay content) criteria

Traffic, million ESALs	*Sand equivalent, minimum* (%)
<0.3	40
0.3 to <3	40
3 to <10	45
10 to <30	45
≥30	50

Source: The Asphalt Institute (2000); © The Asphalt Institute; reprinted with permission.

Table 5.27 illustrates the current aggregate consensus property requirements for the SHRP SuperPave design method. Table 5.28 is the SuperPave sand equivalent requirements for the fine aggregate.

The SuperPave design method also considers other aggregate properties that are also known as source aggregate properties. These source aggregate properties are the physical properties of the aggregate that are usually considered as acceptance criteria when purchasing aggregate. Many highway agencies are already using acceptance or source criteria when specifying aggregate. The SuperPave design method considers the following three source properties of importance to the performance of an asphalt pavement:

- Toughness
- Soundness
- Deleterious materials.

Since most highway agencies already have specifications in place for their local materials, the SuperPave design method does not specify any specific criteria for these properties. Chapter 2 describes toughness, soundness, and deleterious materials in more detail.

SuperPave design method procedure

The purpose of the mixture design procedure is to determine the optimum asphalt binder content for the asphalt mixture and to verify if the selected aggregate blend will perform adequately for the intended pavement function. There are several detailed laboratory procedures available that outline the steps needed to perform a level one SuperPave design.

The Asphalt Institute's *SuperPave Mix Design, SP-2, third edition*, the AASHTO test procedure, *T-312, Preparing and Determining the Density of Hot-Mix Asphalt (HMA) Specimens by Means of the SuperPave Gyratory Compactor*, and the AASHTO practice, *PP-28-2000, Standard Practice for SuperPave Volumetric Design for Hot-Mix Asphalt (HMA)* are three available procedures. At the time of this publication, ASTM has not published a procedure, but is in the process of doing so. It will be similar to the AASHTO procedure. The aggregate proportioning or combined aggregate gradation is determined in the same fashion as was previously described for the Marshall procedure, but including the use of the gradation control points and the restricted zone (Tables 5.2–5.6). The preparation of the aggregates and the mixing of the asphalt mixture are similar to what is completed for the Marshall method. Also similar to the Marshall method is that four trial asphalt binder content batches are produced at 0.5 percent intervals. The estimated optimum asphalt binder content should be at the middle of this range. The following is a summary of the current SuperPave design procedure and is not intended to replace a detailed laboratory procedure, such as AASHTO T-312.

Specimen preparation

1. Prepare the aggregates and the asphalt binder in the same method as was done for the Marshall method.
2. Determine the traffic level and the desired $N_{initial}$, N_{des}, and N_{max}. Determine that the aggregate meets the required consensus properties for the traffic level. Verify that the selected asphalt binder grade is appropriate for the climate and traffic or application. Determine the mixing and compaction temperatures for selected asphalt binder grade as was previously discussed for the Marshall design method. If using a modified asphalt binder such as a polymer modified asphalt binder, obtain the mixing and compaction temperatures from the binder supplier.
3. The SuperPave design method uses a larger mold than the typical Marshall design. Enough aggregate is required to produce a compacted specimen that is 150 mm in diameter by 155 mm in height. This generally requires about 4,700 g of aggregate for each specimen. Two specimens will be required for each asphalt binder content. A maximum theoretical specific gravity (average of two tests) will also need to be determined for each asphalt binder content. After all the specimens are compacted for the four asphalt binder contents, the optimum content is selected and two additional specimens are then compacted to N_{max} at the optimum asphalt binder content. Six to nine additional mixture specimens are also needed for moisture sensitivity testing. To perform one SuperPave level one design, 100–120 kg of aggregate will be required.
4. Determine the mixing and compaction temperatures that are required for the specific asphalt binder.
5. Mix each specimen in the same fashion as that was described previously for the Marshall method. Each specimen should be mixed individually.
6. Immediately after mixing, place each individual mixture in a flat pan in an oven for 2 h of short-term aging at a temperature equal to the mixture compaction temperature.

7 At the end of the short-term aging, weigh each loose specimen mixture and record the mass.
8 Proceed to specimen compaction.

Specimen compaction

1 Preheat the specimen molds and the base plates at the compaction temperature.
2 Once the short-term aged mixture reaches compaction temperature, place it in the preheated mold, level the mixture, and place a paper disk on top of the mixture.
3 Place the loaded mold into the SGC. Center the mold under the loading ram and start the SGC so that the ram extends down into the mold cylinder and contacts the specimen. The ram will stop when the pressure reaches 600 kPa. Apply the 1.25° gyration angle and start the gyratory compaction.
4 Compaction will proceed until N_{des} has been completed. During compaction, the ram will maintain a constant pressure of 600 kPa. The specimen height will be continually measured and recorded on the SGC printer. The height is measured after each revolution.
5 After N_{des} has been reached by the SGC, the gyration angle will be released and the ram will be raised. Remove the mold from the SGC and extrude the compacted specimen from the mold. A five minute cooling period prior to extrusion will facilitate specimen removal without any distortion.
6 Identify the compacted specimen by marking it with a grease pencil or similar marker with a unique number.
7 Repeat the compaction procedure until all the required mixtures (eight) are compacted.
8 Once the optimum asphalt binder content is determined, mix and age two mixtures at the optimum binder content. Compact these two specimens to N_{max}.

SuperPave test information and analysis

SuperPave compaction data is analyzed by computing the estimated bulk specific gravity, G_{mb}, corrected G_{mb} and the corrected percentage of the maximum theoretical specific gravity, G_{mm}, or the corrected density at each desired gyration. During compaction after each gyration, the height is measured and recorded. The weight of the loose mixture prior to compaction is also known. An estimate of G_{mb} can be determined at any desired gyration, x. This estimate (equation 5.16) is determined by dividing the mass of the mixture, W_m, placed in the mold by the volume, V_{mx}, of the specimen in the mold. The mass of the mixture remains constant throughout the compaction process, while the volume becomes smaller as the number of gyrations increase.

$$G_{mb}\,(\text{estimated}) = \frac{W_m/V_{mx}}{\gamma_w} \tag{5.16}$$

where W_m is the mass of loose mixture or specimen, g; V_{mx}, the volume of compacted specimen at x gyrations, cm^3 as determined by equation 5.12; γ_w, the density of water, 1 g/cm^3.

After the specimen has been compacted, the actual bulk specific gravity is measured (G_{mb}, measured) using a procedure such as ASTM D1188 or ASTM D2726. The similar AASHTO procedures are T-166 or T-275. The difference between the two procedures among each organization is that ASTM D1188 and AASHTO T-275 is that the specimen is coated with paraffin and in ASTM D2726 and AASHTO T-166, the specimen is not. The paraffin-coated specimens are for compacted mixtures that have open or interconnected voids and/or absorb more than 2 percent water when immersed.

The estimated G_{mb} assumes that the compacted specimen has a completely smooth surface, which is generally not completely true. V_{mx} is determined by the equation used to determine the volume of a smooth walled cylinder. The estimated G_{mb} needs to be corrected using information that is provided through the measured G_{mb}. Estimated G_{mb} is determined at $N_{initial}$, N_{des}, and N_{max}. The measured G_{mb} is tested at N_{des}. During the mixture verification process, a measured G_{mb} is tested at N_{max} on a mixture with the optimum asphalt binder content. The measured G_{mb} at N_{max} is used to verify that the maximum allowable density is less than 98 percent of G_{mm} at N_{max}. The estimated G_{mb} is corrected by a ratio (equation 5.13) of the measured to estimated bulk specific gravity. The estimated G_{mb} at any gyration level can be determined by equation 5.17.

$$G_{mb}\text{ (corrected)} = C \times G_{mb}\text{ (estimated)} \tag{5.17}$$

where G_{mb} (corrected) is the corrected bulk specific gravity at any gyration; C, the correction factor or ratio (equation 5.13); G_{mb} (estimated), the estimated bulk specific gravity at any gyration.

The percent G_{mm} at any gyration level is then calculated as the ratio of G_{mb} (corrected) to G_{mm} (measured), and the average percent G_{mm} value for that asphalt binder content compacted. Using the $N_{initial}$, N_{des}, and N_{max} gyration levels based on the traffic level for the project, the volumetric mixture design criteria are established based on a 4 percent air void, P_a, content at N_{des}. The volumetric criteria consist of VMA, VFA, and mixture density at $N_{initial}$, N_{des}, and N_{max} for each asphalt binder content (The Asphalt Institute 2001). An effective asphalt binder content to dust (minus 75 μm) ratio is also determined. The SuperPave level one mixture design method does not require the use of stability or flow measurements or any other mixture strength criteria. The following steps should be used in developing the volumetric data required for the gradation verification and the determination of the optimum asphalt binder content:

1 Calculate the percent air voids, P_a, for each specimen at each trial asphalt binder content using equation 5.8. Use the measured G_{mb} at N_{des}. Average the specimen's air voids together for each trial asphalt binder content.
2 Calculate the VMA, for each specimen at each trial asphalt binder content using equation 5.7. Use the measured G_{mb} at N_{des}. Average the specimen's VMA together for each trial asphalt binder content.
3 Calculate the voids filled with asphalt, VFA, for each specimen at each trial asphalt binder content using equation 5.9. Use the VMA and air voids that were determined for each trial asphalt binder content. Average the specimen's VFA together for each trial asphalt binder content.
4 Determine which trial asphalt binder content gives an air void content of 4 percent. Since it is unlikely any of the trial binder contents will give an exact air void content of 4 percent, the use of interpolation or graphing, similar to the Marshall design method, will need to be used to determine the binder content that gives 4 percent air voids.
5 Determine the VMA, either through interpolation or graphing, at the asphalt binder content that provides 4 percent air voids.

6 Determine the VFA, either through interpolation or graphing, at the asphalt binder content that provides 4 percent air voids.
7 Determine the effective asphalt binder content, P_{be}, at each trial asphalt binder content using equation 5.6. P_{be} of the mixture at 4 percent air voids should also be calculated.
8 Determine the dust to effective asphalt ratio for each trial asphalt binder content using equation 5.18.

$$\text{Dust/asphalt binder ratio} = \frac{P_{0.075}}{P_{be}} \tag{5.18}$$

where $P_{0.075}$ is the percent total aggregate passing the 75 μm sieve.

9 Using the corrected G_{mb} determined at $N_{initial}$ for each trial asphalt binder content, determine the density or percent G_{mm} using equation 5.19.

$$\text{Density (\%)}\ G_{mm} = \frac{G_{mb} \times 100}{G_{mm}} \tag{5.19}$$

10 Compare the volumetric properties determined at each trial asphalt binder content and the properties that are estimated at 4 percent air voids with the criteria in Table 5.29 or other provided specification.

The mixture's compaction or densification properties are also determined at each trial asphalt binder content. The compaction properties of the mixture are the estimated (and then corrected) G_{mb} value and its corresponding percent of G_{mm} that is determined at each revolution

Table 5.29 SuperPave VMA and VFA criteria

Traffic, million ESALs (20 years)	*VFA (%)*	*VMA (%) Nominal maximum aggregate size (mm)*				
		37.5	*25.0*	*19.0*	*12.5*	*9.5*
<0.3	70–80	11.0	12.0	13.0	14.0	15.0
0.3 to <3	65–78					
3 to <10	65–75					
10 to <30	65–75					
≥30	65–75					

Table 5.30 Example of data obtained at the trial asphalt binder contents

Trial asphalt binder content	*% of* G_{mm} *at* $N_{initial}$	P_a *at* N_{des} (%)	*VMA, % at* N_{des}	*VFA, % at* N_{des}	P_{be} (%)	*Dust/binder ratio*
4.0	86.5	4.7	12.9	63.6	3.8	1.32
4.5	88.1	3.0	12.5	76.0	4.3	1.16
5.0	90.2	2.5	12.9	80.6	4.8	1.04
5.5	92.3	0.7	12.5	94.4	5.3	0.94

of the gyratory compactor. This information can be graphed to give a visual aid of how easy or difficult the mixture is to compact. This information is also used to determine if the mixture, at each trial asphalt binder content, meets the maximum percent of G_{mm} at $N_{initial}$ for the given traffic level. These requirements are provided in Table 5.26. Table 5.30 provides an example of data obtained for each trial asphalt binder content.

Optimum or design binder content

The optimum or design asphalt binder content is the amount that provides 4 percent air voids, the minimum VMA requirement and meets the VFA range. The dust to effective asphalt binder, P_{be}, should also be determined for the optimum asphalt binder content. The effective asphalt binder content is the amount of asphalt binder that is not absorbed into the aggregate. The amount that is left is what acts as an adhesive. The dust to effective asphalt binder should typically fall in the range of 0.6–1.2. Some agencies require that at the mixture design stage, the ratio does not exceed 1.0 and then uses a 0.8–1.6 range as a specification during mixture production. These dust to asphalt binder ratio limits usually only apply to dense graded mixtures. If the optimum asphalt binder content provides a mixture that meets the required VMA, VFA, and the dust to binder ratio, proceed to mixture verification. If any criteria do not meet the guidelines, the mixture's aggregate structure or gradation will need to be adjusted. The adjustments for VMA and VFA can follow the same methods as previously described for adjusting these properties using the Marshall design method. To adjust the dust to binder ratio, either the effective asphalt binder or the minus 75 μm (dust) can be increased or decreased. In most conditions, the ratio is too high, and the best results obtained are by decreasing the amount of minus 75 μm particles in the mixture. Table 5.31 is an example of data obtained for the optimum or design asphalt binder content.

Table 5.31 illustrates that the optimum asphalt binder content selected for 4 percent air voids gives a dust to binder ratio that is slightly above the SuperPave recommendations. This is easily adjusted by reducing the amount of minus 75 μm material in the aggregate gradation. The current amount of minus 75 μm in this example is 5 percent. The net result of this change is that the gradation becomes slightly coarser, which will also increase the VMA, the other variable in this example that needed to be adjusted. After making these gradation changes, the mixture is verified to confirm that these changes resulted in the desired properties.

Mixture verification

The optimum asphalt binder content and the selected aggregate structure are verified through the preparation of two replicate specimens mixed at the optimum asphalt binder content. The mixture is short-term aged as previously described and then compacted at the compaction temperature. In this case, the mixture is compacted to N_{max} for the desired traffic level. The percent of G_{mm} at $N_{initial}$ and the percent of G_{mm} at N_{max} is determined and

Table 5.31 Example of data for optimum asphalt binder content

P_a at N_{des} (%)	*Optimum asphalt binder content (%)*	*VMA (%)*	*VFA (%)*	*Dust/binder ratio*	*% of G_{mm} at $N_{initial}$*	*% of G_{mm} at N_{max}*
4.0	4.2	12.7	68.6	1.24	87.1	97.7

compared to the density requirements given in Table 5.26. For all traffic levels, the percent of G_{mm} at N_{max} should be equal to or less than 98 percent. If the percent of G_{mm} at N_{max} is greater than 98 percent, the mixture is too sensitive to asphalt binder content and will need to be redesigned. Mixtures with values that are greater than 98 percent may lead to a deformation problem when placed into service as an asphalt pavement. Increasing the VMA in the mixture should help alleviate these problems.

Moisture sensitivity

The final step of the SuperPave level one mixture design method is to determine if the final mixture selected is sensitive to damage by water. Mixtures that are sensitive to water damage can undergo the stripping of the asphalt binder film from the aggregate, which leads to durability problems for the pavement. There are several procedures available to determine moisture sensitivity, from purely empirical methods, such as a boiling water test to strength determining procedures involving compacted specimens. Two widely performed procedures are ASTM D4867 and AASHTO T-283. Both of these procedures are very similar. Many highway agencies have developed their own specific procedure that was modeled on one of these two procedures. AASHTO T-283–2001, *Resistance of a Compacted Bituminous Mixture to Moisture Induced Damage*, is the procedure required by the SuperPave mixture design method. The procedure includes an optional freeze cycle. Six mixture specimens are mixed and short-term aged at the optimum or design asphalt binder content. The specimens are compacted to 7 ± 1.0 percent air voids, which is 93 ± 1.0 percent of G_{mm}. The number of gyrations required to produce this can be determined from the compaction or densification graph made by the SGC during the mixture verification phase of the design method. The following is an outline of the steps needed for evaluating the moisture sensitivity of the mixture:

1. Six specimens of asphalt binder and aggregate are mechanically mixed at the optimum asphalt binder content and allowed to cool at room temperature for 2 h.
2. The mixture is then cured in a 60 °C oven for 16 h.
3. After curing, the mixture is placed in an oven at the compaction temperature for 2 h prior to compaction.
4. The specimens are compacted with the SGC to an air void level of 7.0 ± 1.0 percent. The specimens are compacted using a ram pressure of 600 kPa and a compaction angle of 1.25 degrees. The number of gyrations are determined to give an air void level of 7.0 ± 0.5 percent.
5. After extraction, the specimens are stored at room temperature for 24 h.
6. The specimens are sorted into two subsets of three specimens each, so that the average air voids of the two subsets are approximately equal.
7. One subset is conditioned by vacuum saturating with distilled water to 70–80 percent of the air void volume. These specimens are placed in a freezer at −18 °C for a minimum of 16 h (optional).
8. After removal from the freezer, they are placed in a 60 °C water bath for 24 h.
9. The specimens are then placed in a 25 °C bath for 2 h. Also, at this time, the unconditioned specimens are placed in the 25 °C bath.
10. After these 2 h of temperature stabilization, the indirect tensile strength is determined for all the specimens through the use of a special loading head (Plate 5.15).

A Marshall testing machine (Plate 5.16) can load this indirect tensile breaking head. The indirect tensile strength is calculated by equation 5.20.

$$S_t = (2P)/\pi\ t\ D \tag{5.20}$$

where S_t is the tensile strength, Pa; P, the maximum load, Newton; t, the specimen thickness, mm; D, the specimen diameter, mm.

The tensile strength (TSR) ratio is calculated and a visual stripping rating given to the specimens. The TSR is calculated by equation 5.21.

$$\text{Tensile strength ratio (TSR)} = S_2/S_1 \tag{5.21}$$

where S_1 is the average tensile strength of dry subset; S_2, the average tensile strength of conditioned subset.

A range of zero to five is used for the visual rating, with zero representing no stripping and five being considered as 100 percent stripped (Lavin 1999).

The SuperPave design method requires that the TSR be equal to or greater than 0.80. Some highway agencies have modified this requirement by being slightly higher (0.85) and some have been lower (0.70–0.75). If the TSR of the designed mixture is less than 0.80, anti-stripping additives may be required to increase the ratio. The mixture should then be reevaluated or tested for moisture sensitivity after the addition of the additive to verify the improvement. If the additives improve the TSR, their addition would then become one of the ingredients required by the design.

Plate 5.15 Indirect tensile strength breaking head.

Plate 5.16 Indirect tensile strength being determined by Marshall testing machine.

SuperPave design summary

The moisture sensitivity testing completes the final phase of the level one SuperPave mixture design method. At this point, the aggregate gradation or structure has been selected, the optimum asphalt binder content has been determined, and this combination has been checked if it is susceptible to moisture damage. The properly designed mixture is now ready for production. In summary, the SuperPave design method consists of the following:

- The asphalt binder is selected by the climatic conditions, traffic level, and application of the pavement.
- The aggregates are selected through the use of consensus aggregate properties that have select criteria based on various traffic levels.

- The aggregates are combined together with the aid of the 0.45 power chart, gradation control points, and a restricted zone. This combination of aggregates is usually referred to as the aggregate structure or skeleton. The gradation control points and the restricted zone apply to dense graded mixtures.
- The asphalt binder and aggregate are combined together to form a mixture that is compacted by the SuperPave gyratory compactor (SGC). The SGC simulates the compaction method of construction rollers and traffic.
- The compaction level or effort by the SGC is related to the expected traffic level on the project. The higher the traffic level, the greater the compactive effort required. The SGC increases compactive effort by increasing the number of gyrations on the specimen.
- The optimum asphalt binder content is determined by the volumetric properties of the SGC compacted specimens. These volumetric properties are air voids, P_a, VMA, and VFA. The proper aggregate gradation or structure is also verified through the use of these volumetric properties.
- The resistance of the designed mixture to moisture damage is determined and improved, if needed.

Both the Marshall and the SuperPave design methods can design all applications of dense graded and gap graded (SMA) mixtures. There has even been some success using these methods for the dense of OGFC or porous asphalt. The SuperPave design method is not just for high traffic freeways. Mixtures for low volume roads and low volume parking facilities can also be designed by the SuperPave method, since the method incorporates traffic levels. Determining the number of ESALs that the pavement will be subjected to is key to the proper design. If the number of ESALs cannot be determined, such as in specialty applications or some low volume parking lots, select the traffic level of less than 300,000 ESALs.

Other mixture design methods

There are other mixture design methods that are not practiced as widely as the Marshall or SuperPave methods. The Hveem mixture design method and the rolling wheel compactor method are two of these methods. Similar to the Marshall and SuperPave design methods, these other design methods are used to determine the optimum asphalt binder content and to confirm that the proper aggregate structure or skeleton has been developed. The following is a summary of these two methods.

Hveem mixture design method

Francis Hveem of the California Department of Highways (predecessor to Caltrans) worked with a mixture known as "oil mixes" during the late 1920s. An oil mix is a combination of a slow curing (SC) asphalt cutback and gravel that is generally used as a wearing course for low to medium traffic levels. During his work, Hveem noticed that there was a relationship between the gradation of the aggregate and the amount of oil or cutback required to maintain a consistent color and appearance of the mixture. Hveem refined a method developed by a Canadian engineer, L. N. Edwards, for calculating the surface area of aggregate gradations. Hveem refined the method through the knowledge that the film thickness on a particle decreases as the diameter of the particle decreases. There are three laboratory standard methods that are used in the current use of the Hveem mixture design method. These are ASTMD 5148, ASTM D1560, and ASTM D1561. A complete detailed procedure for the Hveem method can be found in The Asphalt Institute's MS-2 publication, *Mix Design Methods for Asphalt Concrete and Other Hot-Mix Types* (1993). Initial aggregate blends for the designed mixture can be performed in the same manner as previously described for the Marshall and SuperPave methods.

Centrifuge kerosene equivalent test

Hveem developed the kerosene equivalent test to take into account differences in oil or cutback requirements as the absorption and the surface texture of the aggregate varied (Roberts *et al.* 1991). This kerosene equivalent test is now known as the Centrifuge Kerosene Equivalent test (CKE test). The surface area of an aggregate or a blend of aggregates is calculated by the use of its gradation and a table of surface area factors. Using the surface area information and the amount of kerosene that is absorbed into the aggregate as determined by the CKE, an estimated asphalt cutback content can be determined. The CKE test is used in the Hveem mixture design method to assist in estimating the asphalt binder content through the use of an approximate bitumen ratio (ABR). The CKE test can be performed by following ASTM D5148, *Standard Test Method for Centrifuge Kerosene Equivalent.* The CKE test is performed first to give an approximate optimum asphalt cutback or binder content. This value is further verified through the use of a mixture stability test, a cohesion test, a swell test, and volumetric analysis similar to the Marshall and SuperPave methods.

California kneading compactor

After the initial asphalt binder content is determined by the CKE test, the mixture is blended together and prepared for laboratory compaction. Three trial asphalt mixtures are prepared in the same manner as for the Marshall method, but with one at the CKE determined content and the other two at 0.5 percent above and below the CKE content. The compaction method for the Hveem design method uses a special compaction device known as the "California kneading compactor." The specimen is compacted by applying a pressure of 34.5 kPa to the specimen by a tamper foot. The foot is raised after a specified pressure is sustained by the specimen and the specimen is rotated one-sixth of a revolution. This process is repeated 150 times to compact the specimen. The compaction procedure can be followed by using ASTM D1561, *Practice for Preparation of Bituminous Mixture Test Specimens by Means of California Kneading Compactor.*

Hveem stabilometer

Hveem also developed another test to evaluate the asphalt cutback mixture's resistance to rutting or permanent deformation. This test evaluated the stability of the mixture. The stability of the mixture in this application is the ability of the asphalt cutback mixture's ability to resist the shear forces applied by traffic. This test utilized a piece of equipment known as the Hveem stabilometer. The Hveem stabilometer applies a vertical load to a specimen, similar in dimensions to a Marshall specimen, while it is surrounded by an oil reservoir. It is actually a triaxial type-testing cell that measures the resistance of a compacted specimen to lateral displacement under vertical loading. Empirical stability values are obtained at different trial asphalt cutback or asphalt binder contents.

Hveem cohesiometer

Hveem developed a second mechanical device for the testing of oil mixtures. This device is known as the Hveem cohesiometer. It measures the cohesive strength of the mixture across the plane of the compacted specimen. The test was aimed at measuring the tensile property of the asphalt cutback mixtures that could be related to a minimum level to prevent raveling of wearing courses under traffic. This test later proved to be of little value in the design of mixtures using asphalt binders as opposed to cutback asphalt cement, since these mixtures always had cohesion values large enough to prevent raveling (Roberts 1991). Procedures for using both the stabilometer and the cohesiometer can be found in ASTM D1560, *Standard Test Methods for Resistance to Deformation and Cohesion of Bituminous Mixtures by Means of Hveem Apparatus*.

Swell test

The final laboratory test of the mixture is the swell test. The swell test measures the amount of swell the compacted specimen undergoes in 24 h. A measurement is also made on the amount of water that percolates into or through the specimen. A change in height while the specimen is partially immersed in water for the 24 h period determines the percent swell of the mixture (Wright and Paquette 1987).

Volumetric analysis

The only volumetric analysis required by the Hveem method is the determination of the specimen's density and air voids. The minimum air void requirement is 3 percent with optimum asphalt binder contents selected at 4 percent range. The current Hveem mixture design criteria consists of a minimum stabilometer value and a maximum swell amount. Table 5.32 gives the design criteria related to traffic level.

Loaded or rolling wheel design method

Loaded or rolling wheel design methods are just becoming an accepted tool to the mixture designer in the United States. There is, however, established use of these methods both in Europe and Asia, namely Japan. In most cases, the loaded wheel is currently used to verify a mixture that has been previously design by any of the mixture design methods, including SuperPave, Marshall, or Hveem. The concept of the method is to simulate in the laboratory

Table 5.32 Hveem design parameters

Test property	*Traffic level (ESALs)*		
	$<10{,}000$	10,000–1,000,000	$>1{,}000{,}000$
Stabilometer value	≥ 30	≥ 35	≥ 37
Swell (mm, 24 h)	<0.762		

Plate 5.17 Loaded wheel tester (Japan).

the application of a tire load on a pavement (Plate 5.17). This is performed through the use of a laboratory scale loaded rolling wheel on a laboratory prepared mixture. The wheel rolls back and forth over the sample for numerous cycles. The main purpose is to predict the potential of the mixture to rutting or permanent deformation. Another name for these testers is accelerated rut testers. A secondary function is in the prediction of the moisture sensitivity of the mixture. This is performed after it has been concluded that the mixture is resistant to rutting by repeating the application of the loaded rolling wheel to the mixture while it is immersed under water.

Various highway agencies have developed their own version of the load wheel or accelerated rut tester. In the United States, the most commonly known tester is the Georgia loaded wheel tester developed by the Georgia Department of Transportation. The method subjects the asphalt mixture to a loaded wheel system under repetitive loading conditions and measures the permanent deformation or rutting under the wheel path. This approach to assessing the rutting susceptibility is thought to be more representative of actual field or pavement condition than the Marshall stability test or through indirect correlation of volumetric properties (Collins *et al.* 1995). Variables that can be introduced into the testing (which can be said about all the various loaded wheel testers) by the Georgia loaded wheel tester are:

- wheel loading (200–440 N)
- tire pressure (0.5–0.7 MPa)
- tire size
- number cycles (8000 or more)
- application or mixture temperature (35–50 °C).

Several countries in Europe have used accelerated pavement or rutting testers for some time. The French rutting tester evaluates the resistance to permanent deformation on slabs that are 50 cm by 18 cm and 2–10 cm in thickness. The slabs are prepared with the Laboratoire Central des Pons et Chaussees (LCPC) plate tester. In the United States, some rolling wheel methods will require the specimen be prepared in accordance AASHTO PP-3, *Standard Practice for Preparing Hot Mix Asphalt (HMA) Specimens by Means of the Rolling Wheel Compactor*. Loading is with 5,000 N applied by a tire inflated to 0.6 MPa. Loading is typically done at 60 °C with rut depth measurements taken from 100 cycles up to 300,000 cycles. A successful test will have a rut depth that is less than 10 percent of the slab thickness. The Hamburg Wheel Tracking device was developed in Germany to measure the resistance to rutting and moisture damage simultaneously. The Hamburg device is similar to the French tester except that the slabs are immersed in a 50 °C water bath and loaded by a steel wheel. The wheel has a 705 N load applied to it. A successful test will have less than 4 mm of rutting after 20,000 cycles (Miller *et al.* 1995).

The Nottingham Asphalt tester developed at the University of Nottingham in England also influenced a rut tester developed by the Georgia Technical University. The primary differences between the Georgia Technical University (Georgia Tech) and the Georgia Department of Transportation testers is that on the Georgia Tech device, the loaded wheel is stationary and the slab moves back and forth on a steel plate and bearing apparatus (Miller *et al.* 1995). The ultimate in pavement rutting predication (at least from a laboratory replication standpoint) is the Accelerated Loading Facility (ALF) used by the United States Federal Highway Administration. It is a full-scale mockup of a pavement that can be tested by loads up to 10,000 kg to provide the most realistic testing possible.

Summary

The two principal functions of any asphalt mixture design method is to:

1 Verify that the aggregate blend and gradation selected will perform as desired.
2 Select the correct amount of asphalt binder for the aggregate blend.

The most common mixture design method throughout the world for performing these functions is the Marshall mixture design method. The Marshall design method is slowly being replaced by the SuperPave mixture design method that was developed by the SHRP in the United States. Three levels of designs are used in the SuperPave method, with the level one being the most common and currently practiced. The significant differences between the Marshall and the SuperPave level one design methods are the type of specimen compaction and the elimination of the Marshall Stability and Flow testing. SuperPave also implemented aggregate requirements, known as the aggregate consensus properties.

Most highway agencies worldwide have adopted some form of these two methods and may have made their own modifications to the procedures. Agencies such as federal or state highway departments are usually responsible for the majority of the asphalt pavements in a region. Typically they will develop their own mixture specifications based on years of experience with the local materials and traffic requirements. They will also have the staff to design tailored mixtures for specific project or region and verify that the specifications are being met.

A designer of smaller projects such as parking facilities or local roads should specify the design method that is prevalent in areas where the project will be located. In most cases, especially involving parking lots, a new mixture will not be required to be designed. An existing design that may have been used on a similar project or a highway agency project can be used. It is important, however, when adopting such designs that they meet the *specific requirements* for the particular project that you are designing. For example, an existing highway agency SuperPave design is available from the contractor that was designed for a freeway with 15,000,000 ESALs. That design would *not be appropriate* for the parking lot you are designing with less than 300,000 ESALs. That design would more than likely be too stiff and have durability problems when used for a parking lot. The appropriate design to select would have been one used for roadways with less than 300,000 ESALs. That design may then be tailored for parking applications by making its gradation slightly finer with slight increases in the asphalt binder content, if needed, to improve its durability. These variations are judgments that are made by designers with experience with the materials and applications and also can be recommended by experienced HMA producers. Advanced mixture design methods such as the SuperPave mix design method are completely capable of providing performing mixtures for parking facilities in addition to highways and roads. It is important to remember that low volume roads and typical parking lots will not have the truck traffic to further densify the mixture, so the initial amount construction air voids will be typically the amount of air voids throughout the pavement. These types of applications require mixtures that are highly durable and do not need the harsh rut resistant mixtures that are used for expressways. This is why it is important to follow the proper traffic levels when selecting the design criteria. Past experience with the mixture and good engineering judgment should also always be considered. A well-proven design that has met expectations on other similar projects should not be rejected just because it was not designed using the latest design methods. For all practical purposes, it may be the best choice to use.

The mixture specifications that a designer would recommend for specific projects could be adopted from the tables referenced in this chapter, from appropriate agency specifications or from a nationally recognized specification such as ASTM D3515 or AASHTO MP-2 (United States). The points to remember when creating mixture design specifications for a specific project or projects are:

- Use the mixture design method predominate in the area of the project and HMA supplier. For example, a Hveem mix design requirement from California for a project in Florida or London would cause needless difficulties.
- Try to use materials available in the region.
- Unless there is performance issues, use specifications that at least some of the materials available in the region will meet.
- Use specifications that suit the mixture's traffic loading and application.
- Mixtures with a history of satisfactory performance on similar applications in the geographical area where the project is located should not be eliminated and can usually be accepted with field or construction verification.

Chapter 6

Construction

Constructing an asphalt pavement involves three general focal points: preparation of the asphalt mixture, placing the asphalt mixture, and compaction of the asphalt mixture. This chapter deals with the construction of dense graded asphalt pavements. This type of mixture is the most popular mixture used for parking facilities and for highways. Specialty mixtures and maintenance activities such as chip seals or slurry seals are addressed in later chapters. The pavement also needs to be installed on a suitable sub-base or subgrade. Some asphalt pavements are placed as an overlay on an existing pavement that is either made of Portland cement concrete (PCC) or asphalt. Particular attention needs to be placed on both subsurface and surface drainage of the pavement. Modern day construction of asphalt pavements involves the use of complex and expensive equipment. For example, most hot mix asphalt (HMA) mixing plants must now also meet extensive environmental regulations in addition to producing an asphalt mixture to specification. Plate 6.1 illustrates a modern high

Plate 6.1 High production HMA mixing plant.

production HMA mixing plant. This particular plant can produce upward of 800 tons per hour and can incorporate the use of recycled asphalt pavement (RAP). In the left side of the illustration is the covered storage pile of RAP. This mixing plant is a counter flow drum mix plant equipped with multiple storage silos, multiple aggregate bins, and a dust collection system known as a baghouse. In order to meet the high mixture production demands of this plant, its location is actually in the aggregate quarry.

This chapter will focus on the aspects of constructing an asphalt pavement made of a dense graded hot asphalt mixture or here referred to as "hot mix asphalt." Pavements consisting of other types of mixes, such as porous asphalt, stone mastic asphalt (SMA), Delugrip, cold mix, rolled asphalt, and other specialty asphalt mixtures can also be constructed involving most of the techniques used for HMA. The three main areas of HMA construction involve mixing, placing, and compacting. Subgrade preparation and subsurface drainage is usually considered a part of site preparation. Surface drainage involves removing the water from the pavement surface, which is generally facilitated through the use of pavement slopes and crowns or through the use of porous asphalt, open graded asphalt mixtures, or an open graded friction course (OGFC). The purpose of this chapter is not to be an equipment operations manual, but to give the architect or engineer enough familiarity with what is involved with the production and placement of an asphalt pavement. Key areas of concern are operational aspects that affect the quality of the pavement.

Mixing

The production of an asphalt mixture involves mixing the aggregate and asphalt binder together according to the design or "recipe." The production of HMA requires that the aggregate be dry and hot. The asphalt binder also requires heat so that it will flow enough to give a uniform coating on the aggregate. HMA is prepared through the use of a mixing plant. The basic operations of a mixing plant are:

1 Proper storage and handling of the aggregate and asphalt binder at the mixing facility.
2 Accurate proportioning and feeding of the cold aggregate to the dryer.
3 Effective drying and heating of the aggregate to the proper temperature.
4 Efficient control and collection of the aggregate dust from the dryer.
5 Proper proportioning, feeding, and mixing of the asphalt binder with the heated aggregate.
6 Proper storage, dispensing, or discharging, weighing and handling of the finished HMA or other asphalt mixture.

(The Asphalt Institute 1998b)

A mixing plant can be permanently set up as a stationary plant or as a portable one. Portable mixing plants are usually used for larger highway projects, mostly in rural areas. Plate 6.2 illustrates a stationary batch plant. There have been three types of HMA mixing plants developed over the years: batch, continuous, and drum mix. The continuous mixing plant is no longer in commercial production. The flow of aggregate and asphalt through the various types of plants are similar except for the method of proportioning and mixing. The basic differences are in how they mix the aggregate after it has been dried and heated with the asphalt binder. The following is a brief description of each type of plant.

- *Batch*: the batch mix plant separates and screens the aggregate after it has been dried into separate bins, known as "hot bins." This separated aggregate is then recombined and mixed with the asphalt binder, one *batch*, at a time in a separate twin shafted mixer, known as a pugmill. The screens are located above the hot bins and the hot bins are above the pugmill. This vertical setup is known as the batch tower (Plate 6.3). The asphalt mixture is then discharged from the pugmill, one batch at a time, into a waiting truck or into a conveyer for transport to a storage silo. Batch plants are usually stationary plants and usually found in urban areas. Due to their ability to screen the aggregate, they are capable of making various adjustments to the proportions of a mixture.

- *Continuous:* the continuous mix plant also separates and screens the aggregate into hot bins. The separated aggregate is then recombined through the use of controlled feeders into *continuously* mixing pugmill and mixed with the asphalt binder. The asphalt mixture is discharged from the pugmill in a continuous flow without any planned interruption. The continuous mixing plant is no longer manufactured and has been replaced by the simpler and higher capacity drum mix plant.
- *Drum*: the drum mix plant differentiates from the batch and the continuous plants in that the aggregate is not separated after it is dried. In fact, the aggregate is immediately mixed after it is dried with the asphalt binder in the same dryer drum. The asphalt mixture is then discharged into a storage silo. Of all the types of plants, drum mix plants have the highest capacity or mixture output. Drum mix plants are either stationary or portable.

All HMA mix plants have three pieces of equipment in common:

1. Aggregate cold feed system that proportions the aggregate blend prior to drying and feeds the aggregate to the drier (Plate 6.4).
2. Drier for drying and heating the aggregate prior to mixing.
3. Dust collection system for collecting the aggregate dust created by the drier.

Plate 6.2 Small stationary batch plant.

Plate 6.3 Batch tower.

Most mixing plants also incorporate the use of storage silos (Plate 6.5) for storage of the completed asphalt mixture prior to transportation to the project site. The storage silo in a drum mix plant is an essential piece of equipment. Many drum mix plants are equipped with large multiple storage silos (Plate 6.6) in order that multiple trucks can be loaded at the same time or different mix types, such as a wearing course and a base course, may be loaded. The storage silo is capable of keeping the mixture hot for 24 hours or longer. A surge bin is a small type of a storage silo that only holds a few tonnes. Plate 6.7 is a surge bin that is being used for the production of a patching mixture.

Batch mixing plant operation

The major components of a batch mixing plant are the cold aggregate feed system, an asphalt binder supply system, the aggregate dryer, the batch tower, and a dust collection or

Plate 6.4 Multiple bin cold feed system.

Plate 6.5 HMA storage silo.

Plate 6.6 Multiple storage silos at a drum mix plant.

Plate 6.7 Large surge bin.

control system. The vertical batch tower consists of a hot aggregate conveyor or elevator, a screen deck, the hot bins, a weigh hopper, an asphalt binder weigh bucket, and the mixing chamber or pugmill. The operation of the batch mixing plant consists of the following:

1 Aggregate required for the mixture is stockpiled at the mixing plant.
2 The aggregate is charged into the individual cold feed bins (Plate 6.4).
3 The aggregates are proportioned out of each cold feed bin onto a feeder belt according to the percentages given by the mixture design. These percentages must be based on a total percent by aggregate basis, instead of a weight by total mixture. For example, if the mixture design calls for 60 percent coarse aggregate, 35 percent fine aggregate, and 5 percent asphalt binder, all by weight of total mixture; these percentages would then become 63.2 percent coarse aggregate and 36.8 percent fine aggregate.
4 The feeder belt transports the proportioned aggregate to the aggregate dryer.
5 The aggregate is charged into the dryer. The dryer is a horizontal rotating drum that operates on an incline in a counter flow manner. The aggregate is introduced into the drum at the upper end of the incline and is moved down the drum by gravity and by the flights inside the rotating drum (Figure 6.1). The burner is located at the lower end of the drum with the exhaust gases from the combustion and drying process moving toward the upper end of the drum, against or *counter* to the flow of aggregate. As the aggregate is tumbled through the exhaust gases, it is heated and dried. Moisture is removed and carried out of the dryer as part of the exhaust stream. The hot dry aggregate is discharged at the lower end of the drum (TRB 2000).
6 The aggregate is discharged from the lower end of the drier into a bucket elevator, which is also known as the "hot elevator." The aggregate is transported to the top of the batch tower and is discharged onto the screen deck.
7 The screen deck generally consists of four vibrating screens, placed in a vertical order of the largest size to the smallest right above the hot bins (Figure 6.2). On the first screen, the oversized material is scalped off and wasted. The coarse aggregate is retained between the top and the second screen and moves into hot bin number four. The material retained between screens number two and number three is the medium coarse aggregate which vibrates through the screen into bin number three. Bin number two receives the intermediate fine material that is retained between screen three and the bottom screen, four. Bin number one receives the finest aggregate particles (Roberts *et al.* 1991). The screen sizes can be changed and the maximum size sieve should be selected based on the largest nominal maximum size mixture being produced.
8 Up until this, point the operation of the batch mixing plant has been continuous. Aggregate is continually being heated, dried, and fed into the screen deck and hot bins. The heated and resized aggregate is stored in the hot bins until being discharged from

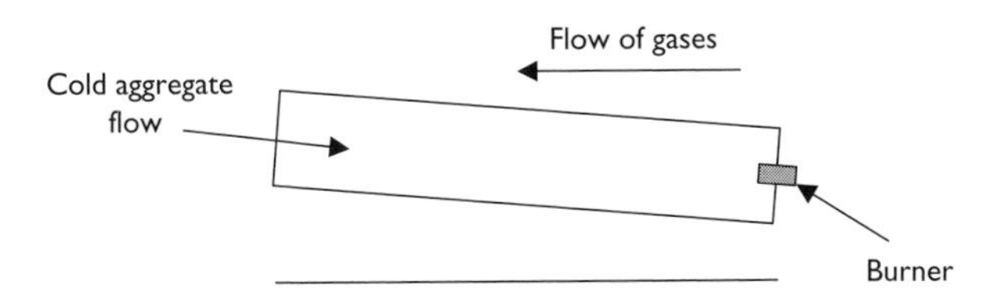

Figure 6.1 Counter flow drier for batch plants.

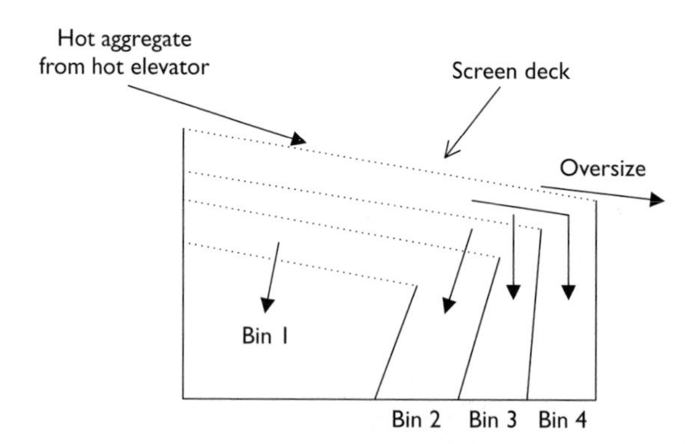

Figure 6.2 Batch plant screen deck operation.

a gate at the bottom of each hot bin into a weigh hopper that is immediately below the hot bins. The correct proportion of each aggregate from each bin for the batch is determined by weight. Hot bin number four is usually weighed first, followed by number three, and so on until the finest materials in hot bin number one is weighed. This sequence is designed to place the finest material at the top of the weigh hopper, where they cannot leak out through the gate at the bottom of the weigh hopper. This system also allows the most efficient utilization of the available volume in the weigh hopper, since the finer particles will penetrate the voids between the coarser particles (The Asphalt Institute 1998b).

9 While the aggregate is being weighed, the required amount of asphalt binder for each batch is being weighed separately in a heated "weigh bucket" located just above the pugmill. The asphalt binder is held in the weigh bucket prior to being discharged into the pugmill.

10 The aggregate in the weigh hopper is emptied into the pugmill mixer and is dry mixed for up to five seconds. After the "dry mix" time, the asphalt binder in the weigh bucket is discharged into the pugmill. The mixture is then "wet mixed" for 20–35 s and then discharged into a waiting truck or onto a conveyor or bucket elevator for storage in a surge bin or silo. The total mixing time (dry mix time plus the wet mix time) for each batch is typically 30–40 s. This process is repeated over and over until the required amount of mixture is in the truck or silo.

11 If RAP is required by the mixture design, the RAP is added from a separate cold feed bin (Plate 6.8). In a batch mixing plant, RAP is usually weighed separately from the hot aggregates into the weigh hopper. The RAP bypasses the drier in a batch plant and relies on the convection heat from the hot aggregate to bring itself up to the mixing temperature. Not nearly as common is when the RAP is added at the bottom of the hot elevator or directly into one of the hot bins (TRB 2000).

A batch mixing plant is rated or has its capacity determined by the amount of material that the pugmill will hold. The other variables that affect production rate include the cold feed capacity, capacity of the drier, screen deck efficiency, and hot bin capacity. During

Plate 6.8 RAP feed bin.

actual mixture production, the pugmill capacity and the availability or cycling of trucks are the usual limiting variables in production rates. Pugmill capacity is determined by the size of the pugmill's "live zone" or mixing zone. The live zone is the area that the paddles in the pugmill will give adequate mixing of the aggregate and the asphalt binder. Batch plant pugmill capacity ranges from 1,000 to 10,000 kg with 3,000 to 5,000 kg being typical. The production rate per hour is controlled by the pugmill capacity, the total mixing time, and any down time.

The dust collection system collects the dust that is created by the drier as it is heating the aggregate. The dust is collected through a primary and secondary collection system. A dry collector such as a knockout box (Plate 6.9) or a cyclone is used as the primary collector. The material collected here is generally returned back to the number one hot bin. The secondary collection system can be a wet scrubber or a dry fabric filter, known as a "baghouse." The baghouse is similar to a large household vacuum cleaner. The dust slurry collected from a wet scrubber is pumped to a settling pond to allow the dust to settle out. The dust collected from a baghouse is either metered back into the mixture, either at the number one hot bin, weighed separately in the weigh hopper or added at the bottom (boot) of the hot elevator (the least preferable method). Adding at the boot of the hot elevator has the potential for the dust to enter the mixture as variable "slugs" of material, leading to some batches having extremely high amount of minus 75 μm material. The dust can also be stored in a silo and hauled away from the plant.

Plate 6.9 Knockout box primary dust collector (upper right).

Batch plant proportioning

Cold feed blending

Cold aggregate feed blending is an important step in the proper production of the asphalt mixture. Improper setup of the cold feeders can cause an incorrect gradation in the hot bins. Proper setting of the cold feeders balances the overall mixture production. The cold aggregate feed rate will also need to match the capacity of the drier and may need to adjust to compensate for cold or wetter aggregate than usual. Plate 6.10 illustrates an eight-bin cold feed system that keeps the individual aggregate types separated until being fed into the drier. This particular cold feed system is for a large drum mix plant, but batch plants use a similar system. The individual cold aggregate feed rate is adjusted by varying the individual cold bin belt speed, apron feeder speed, or the size of the bin gate opening. Cold aggregate feeders are either constant speed or a variable speed.

Almost all (~90 percent) of the fine aggregate for the asphalt mixture will be screened into hot bin number one. It is necessary to properly set the fine aggregate cold feeds so that the combined fine aggregate gradation will match what is required by the mixture design. Constant speed aggregate feeders are set in the following manner:

1. Check the mixture design for the correct blend ratio of fine aggregates, for example two to one or three to one, etc.
2. Ensure feeder belts are running at the same rpm.
3. Adjust the bin of the lowest percentage fine aggregate to the desired gate opening.
4. Adjust the other fine aggregate cold bin to reflect the blend.

Plate 6.10 Large cold feed system for a drum mix plant.

For example, if the mix design requires a two to one blend of fine aggregate A to fine aggregate B, set fine aggregate B first, say at a gate opening of 75 mm, and then follow the fine aggregate A gate opening of 150 mm. Variable speed feeders are set in a similar fashion, but with the gate opening constant and the belt speeds are proportionally varied. Coarse aggregate blends can be set using the same method as for the fine aggregate blends. The overall proportions coming from each cold feed needs to add to 100 percent. The material percentages determined at the mixture design stage can be used for determining the ratios to be obtained from each cold feed bin.

Hot bin proportioning

Hot bin proportioning is the determination of the amount of material by weight that will be discharged or "pulled" from each hot bin for each batch. This material is weighed in a cumulative fashion in the weigh hopper. Depending on the nominal maximum top size of the mixture being produced, not all of hot bins will be used. Bin number one is the smallest size material, usually passing the 2.36 mm sieve. Bin number two is for the next largest size, usually passing the 4.75 mm sieve and retained on the 2.36 mm sieve. Bin number three is material, usually passing the 12.5 mm sieve and retained on the 4.75 mm sieve. Bin number four controls the nominal maximum top size of the mixture being produced and is usually for material retained on the 12.5 mm sieve. There is one more screen that is used in the batch tower and is used as the scalping screen to remove any oversize aggregate.

The first step in hot bin proportioning is to complete a sieve analysis on a representative sample of each of the hot bins. These gradations will be used to determine the percentage of aggregate from each hot bin for the mixture being produced. These samples should be taken during the actual production of the mixture. They are taken from sampling ports in each hot bin and capture the stream of falling aggregate as the aggregate is screened from the screen deck. The mixture being produced during the initial hot bin sampling is a trial production mixture for the determination or verification of the proper hot bin weights and is usually wasted. The initial hot bin weights for this trial mixture can be estimated from previous hot bin samples on the same aggregate or estimated based on calculations using the screen deck sieve sizes and laboratory prepared aggregate gradations of the duplicate of the hot bin gradations. If there is no initial hot bin samples available, the percent required from each hot bin is determined by using the screen deck sieve sizes and assuming that the screen deck has 100 percent screening efficiency.

To determine the initial hot bin percentages, obtain the target combined aggregate gradation is obtained from the mixture design. Also obtain the batch plant's screen sizes or what size fraction material is in each hot bin. Starting with hot bin number one, the small fraction material, determine what percentage of material from the bin will satisfy the same size material on the mixture design or job mix formula (JMF). For example if hot bin number one contains all the size fractions smaller than 2.36 mm, and the JMF requires 34 percent passing the 2.36 mm, the amount of material to pull from hot bin number one is 34 percent. This procedure is carried out for the remaining hot bins. Example 6.1 illustrates a step by step proportioning of the hot bins for producing a dense graded wearing course.

Example 6.1

Determine the hot bin proportions for a 3,600 kg batch plant producing a wearing course with a 5.5 percent asphalt binder content with the target JMF given in Table 6.1. The size of the material in each hot bin is also shown in Table 6.1.

Table 6.1 Example 6.1 job mix formula and hot bin sieve sizes

Hot bin sieve sizes	*Hot bin 4*	*Hot bin 3*	*Hot bin 2*	*Hot bin 1*	*Mineral filler*	*Target JMF (%)*
	+12.5	12.5–4.75	4.75–2.36		−75 µm	
Sieve sizes			Passing each bin (%)			
12.5 mm						100
9.5 mm						90
4.75 mm						59
2.36 mm						34
1.18 mm						24
600 µm						17
300 µm						10
150 µm						6
75 µm						3.5
			Asphalt binder content (%)			5.5

Solution

First determine the amount of material required from hot bin number one. Hot bin number one is all of the material passing the 2.36 mm sieve and the JMF requires 34 percent passing the 2.36 mm sieve. The initial bin number one amount becomes 34 percent.

For hot bin number two (4.75–2.36 mm), the JMF requires 59 percent passing the 4.75 mm sieve and 34 percent passing the 2.36 mm sieve. The amount needed from hot bin number two is the difference between 59 percent and 34 percent or 25 percent.

The same calculation is done for hot bin number three (12.5–4.75 mm), where the JMF requires 100 percent passing the 12.5 mm sieve and 59 percent passing the 4.75 mm sieve. The amount needed from hot bin number three is the difference between 100 and 59 percent or 41 percent.

Since the JMF requires 100 percent passing on the 12.5 sieve, no material is needed from hot bin number four, since it contains material greater that 12.5 mm.

After these initial percentages are determined, a trial mixture is produced without the addition of mineral filler, in order to sample the hot bins and determine the gradation on each bin (Table 6.2).

The next step is to calculate the individual hot bin sieve size percent by taking the initial hot bin percent and multiplying it by each percent passing sieve calculation from the hot bin samples given in Table 6.2. This step confirms the initial target bin percent and also determines the amount of mineral filler required (Table 6.3).

From Table 6.3, it has been determined that the actual combined gradation from the hot bins has 3.6 percent passing the 75 μm sieve, which is within an accepted tolerance of the JMF requirement of 3.5 percent, so no mineral filler will be required. Many mixture designs may require the addition of mineral filler at the laboratory testing stage, but during actual mixture production it is not required. This phenomenon is known as "plant generated dust" and is dependent on how easily the aggregate breaks down.

The next step is to convert the hot bin percentages into individual batch weights. The percent aggregate values are converted to percent mixture values, which include the asphalt binder. The total of percent aggregate and percent asphalt binder must add up to 100 percent. First

Table 6.2 Example 6.1 hot bin gradations taken from trial batch

Sieve sizes	*Passing each bin (%)*			
	Hot bin 4	*Hot bin 3*	*Hot bin 2*	*Hot bin 1*
19.0 mm	0	100	100	100
12.5 mm	0	100	100	100
9.5 mm	0	70.0	100	100
4.75 mm	0	6.0	90.1	100
2.36 mm	0	1.0	6.9	98.0
1.18 mm	0	1.0	1.0	75.6
600 μm	0	1.0	1.0	43.7
300 μm	0	1.0	1.0	30.0
150 μm	0	1.0	1.0	14.0
75 μm	0	0.4	0.3	9.7

Table 6.3 Example 6.1 job mix formula and hot bin sieve sizes

	Hot bin 3	*Hot bin 2*	*Hot bin 1*	*Mineral filler*	*Actual combined gradation (%)*	*Target JMF (%)*
Hot bin sieve sizes	12.5–4.75	4.75–2.36	22.36	275 μm		
	Passing each bin (%)					
Bin (%)	41.0	25.0	34.0			
sieve sizes						
12.5 mm	41	25	34		100	100
9.5 mm	28.7	25.0	34.0		88	90
4.75 mm	2.5	22.5	34.0		59	59
2.36 mm	0.4	1.7	33.3		35	34
1.18 mm	0.4	0.3	25.7		26	24
600 μm	0.4	0.3	14.9		16	17
300 μm	0.4	0.3	10.2		11	10
150 μm	0.4	0.3	4.8		6	6
75 μm	0.2	0.1	3.3		3.6	3.5
		Asphalt binder content (%)				5.5

Table 6.4 Example 6.1 batch weights

	Bin 3	*Bin 2*	*Bin 1*	*Asphalt binder*	*Total*
Mixture (%)	38.7	23.6	32.1	5.5	99.9
Batch weight (kg)	1,393	850	1,156	198	3,597
Aggregate (%)	41.0	25.0	34.0	—	100

subtract the percent of asphalt binder on the mixture design from 100 percent and then multiply this percentage by each aggregate percent.

$$100\% - 5.5\% = 94.5\%. \qquad \begin{aligned} \text{Bin one} &= (0.945)(34\%) = 32.1\% \\ \text{Bin two} &= (0.945)(25\%) = 23.6\% \\ \text{Bin three} &= (0.945)(41\%) = 38.7\% \end{aligned}$$

The final step is to determine the final batch weights, which are given in Table 6.4. Rounding of the figures gave a batch size slightly less than 3,600 kg.

In Example 6.1 the initial trial hot bin percentages also were the final values selected. If there are large discrepancies, the hot bin sampling should be repeated or the possibility of screen or bin damage should be considered.

Batch plant operation considerations

The designer of an asphalt pavement is generally not the individual responsible for running a HMA mixing plant; however, there are aspects of the plant operation that can have an impact on the quality of the mixture being produced. On large paving projects, the owner of the project may have a representative overseeing the operations of the mix plant, but with the advent of producer quality management systems, this may not be necessary. Operations of the batch mixing plant that can have an impact on the quality of the mixture are:

- Aggregate stockpiles should be prepared and handled in a manner that reduces the occurrence of segregation.
- The cold feed bins should have dividers to reduce the occurrence of contamination between the different bins.
- The moisture content of the aggregate when discharged from the drier should be less than 0.5 percent. Too much moisture in the aggregate can reduce the bonding of the asphalt binder to the aggregate, which will lead to durability problems.
- The screen decks should be checked for holes and tears.
- The amount of aggregate carryover from one hot bin to the next should be relatively constant and less than 10 percent. Carryover is the depositing of aggregate that is finer than what is required for the hot bin. Excess carryover will be apparent from the sieve analyses made on each individual hot bin. Carryover will be experienced most often coming from the number one hot bin into the number two hot bin, usually when the fine screening area is limited (The Asphalt Institute 1986a). Carryover greater than 10 percent can alter the job mix gradation and the blend of the different aggregates outside of accepted tolerances.
- The pugmill should be operated at nominal capacity. Both overloading and underloading of the pugmill will decrease its mixing efficiency and coating ability. Batch size should be consistent from batch to batch. The paddle tips and the pugmill should be checked periodically for damage and wear.
- The aggregate dry mix time should be five seconds or less. Longer dry mix times just reduce production rates, increase wear on the pugmill and may lead to some aggregate breakdown. Aggregate breakdown can increase the mixture's minus 75 μm content beyond what the JMF required.
- The wet mix time for blending the asphalt binder and the aggregate should be no longer than needed to coat the aggregate completely. This wet mix time is usually less than 30 s. Increasing the wet time to greater than 30 s, or longer than is needed for proper aggregate coating increases the likelihood of asphalt binder oxidation and increases the wear on the pugmill. The production rate of the plant is also needlessly reduced. Total mix time, including mixture discharge should be around 30–40 s.
- When the plant is waiting for trucks to load, there should be no material in the pugmill.
- If RAP is introduced into the plant at the bottom of the hot elevator, it should be placed on top of the hot aggregate and not in the bottom of the buckets.
- If RAP is introduced into the plant at the weigh hopper, it should be placed in the center of the weigh hopper, so that the hopper is balanced and an accurate weight can be determined.
- When using RAP, the temperature of the new or "virgin" aggregate must be heated enough so that adequate heat transfer occurs between it and the RAP. This temperature is a function of the amount of RAP in the mixture, the amount of moisture in the RAP, and the desired mixture discharge temperature. To prevent potential damage to the aggregate dryer, the virgin aggregate should not be heated beyond 260 °C. If it is determined that the temperature needed is greater than 260 °C for adequate heat transfer to the RAP, the amount of RAP in the mixture will need to be reduced. Since a batch plant heats RAP only through heat transfer from the virgin aggregate, batch plants rarely can use more than 30 percent RAP in the mixture (TRB 2000).

Drum mix plants

Drum mix plants have replaced almost all of the continuous mix plants and are gradually replacing batch mix plants. Almost all new mixing plants produced today are drum mix plants. There are two types of drum mix plants, parallel flow and counter flow (Figure 6.3). Drum plants do all the mixing in the same drum that is used to dry and heat the aggregate. Drum plants do not resize the material or use a screen deck, hot bins, and a pugmill. Drum plant advantages over batch or continuous plants are higher production rates, less moving parts, lower maintenance, and the ability to use higher percentages of RAP. By eliminating the screening process and the batch time sequence, production rates have become greater. When RAP is introduced into a drum mix plant, it is heated both by aggregate heat transfer and by the exhaust gases of the burner. This dual heating action allows the drum mix plant to run at higher RAP percentages than batch mix plants. It is not uncommon to have drum mix plants producing HMA with 50 percent RAP or greater. RAP is usually introduced by a conveyer near the center of the drum mixer (except for the double barrel style plant). A drum mix plant consists of five major components, the cold aggregate feeds, asphalt binder supply, combination drum drier and mixer, surge or storage silos, and a dust collection system. The cold feeds are similar to those in a batch plant with the additional function of proportioning the aggregate for the mixture. Since there is no hot bin or weigh hopper, the cold feeds must be able to accurately feed and control the blend of aggregates. Also, since there is no weigh hopper in a drum plant, the aggregate must be weighed prior to its introduction into the drier. This is accomplished by equipping the conveyor that charges the aggregate into the drier with a weighbridge and a belt speed sensor. The weighbridge is placed under the conveyer belt and measures the weight of the aggregate passing over it. This weight and the speed of the belt are used to determine the wet weight of aggregate entering the drum per hour. Using the aggregate wet weight per hour and its moisture content, the correct proportion of asphalt binder can be mixed with the aggregate. A weighbridge is also used on the RAP conveyor. The aggregate is mixed with the asphalt binder in the drier and the mixture

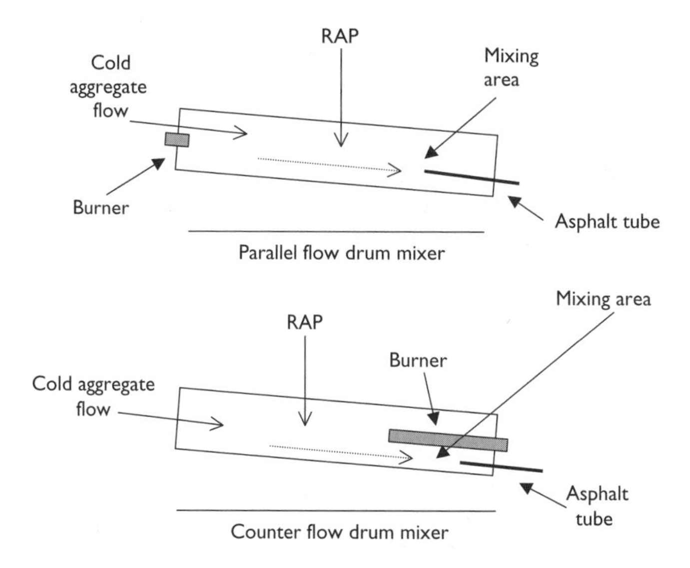

Figure 6.3 Differences between counter flow and parallel flow drum mixers.

is discharged onto a conveyor or bucket elevator for storage in a surge bin or silo. The asphalt binder is stored at the plant in the same manner as at the batch mix plants, either in vertical or horizontal tanks. Plate 6.11 illustrates vertical storage tanks at a batch mix plant and Plate 6.12 illustrates the large capacity that may be required at a stationary drum mix plant. Plate 6.13 is the RAP conveyer for the same large stationary drum mix plant. Horizontal tanks are common on portable plants, since they can be transported easily in that position.

The original drum mix plant design is a parallel flow system. Parallel flow drum mix plants are the most common, however newer designs are counter flow systems and are slowly replacing older parallel flow plants. A parallel flow drier or drum mixer has the aggregate flow in the direction of the exhaust gases or away from the burner. A counter flow drier or drum mixer has the aggregate flow against the direction of the exhaust gases or towards the burner. The parallel flow drum mixer mixes the aggregate with the asphalt binder at the opposite end of the drier from the burner. There are two types of counter flow drum mixers. The original method is illustrated in Figure 6.3 and uses a mixing chamber at

Plate 6.11 Vertical asphalt storage tanks at a batch plant.

Plate 6.12 Drum mix asphalt storage tanks.

Plate 6.13 RAP conveyor on a drum mix plant.

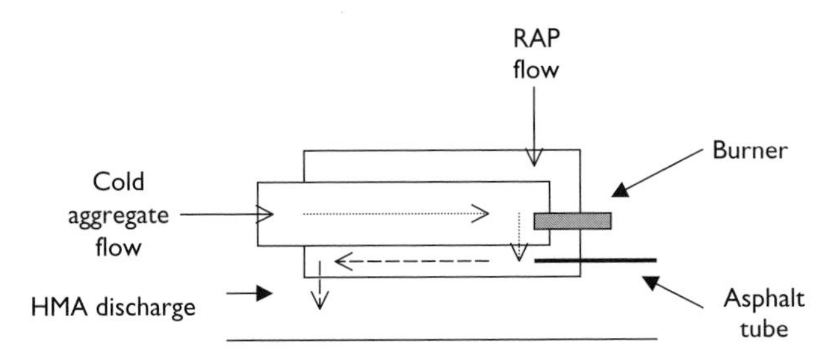

Figure 6.4 Double barrel drum mixer operation.

the burner end of the drier, but extends the burner flame away from the asphalt binder. The second type is a double barrel or double drum design (Figure 6.4), which is basically two drums, one inside of the other. The inner drum dries the aggregate and rotates inside the outer drum. The aggregate is discharged into the stationary outer drum where it is mixed with the asphalt binder by paddles or blades attached to the outside of the inner drum. Double barrel mixing plants are becoming popular because they give out fewer emissions and have better heat transfer ability. Because of these two advantages, double barrel plants can use extremely high levels of RAP. The gradation requirements of the mixture design are usually the limiting factors in the amount of RAP they can use.

After the drum mixer portion, all types of drum plants are identical in operation. Since these plants operate in a continuous manner and do not use hot bins, the asphalt mixture must be stored while awaiting the arrival of transports. The mixture is transported from the mixing portion of the drum mixer by a bucket elevator (similar to a hot elevator in a batch plant) into a surge bin or silo. Multiple silos allow for high production capacities or the ability to ship different mixture types at the same time. The dust control systems are the same as they are on a batch mix plant. These systems must be matched to the production capacity of the plant. The mineral filler or dust that is collected by the baghouse is introduced back into the mixture at the mixing portion of the drum mixer by a tube. This tube is similar to the asphalt binder supply tube.

Drum plant proportioning

A modern day drum mix plant is a highly automated and computer controlled manufacturing facility. However, by design it is a very simple operation. The cold aggregate feeders are what control the proportion and final gradation of the HMA being produced. There is no sieving or batching of materials. The key concept to remember regarding proportioning a drum mix plant is what goes in the cold feeders and how it is metered determines what comes out of the drum mix plant. Proper aggregate stockpile handling and charging of the cold feeders is mandatory for the production of a consistent and quality asphalt mixture. A good end loader operator is a key employee in the operation of a drum mix plant. An example procedure for proportioning the drum plant consists of the following:

1 Determine or obtain the starting production rate of the plant, usually in tonnes per hour (TPH).
2 Obtain from the mixture design, the aggregate blend percentages and the asphalt binder content.

3 Enter into the computer drum of the mix plant computer, the asphalt binder content, and the percentage of moisture in the aggregate as a either the total moisture in all the aggregates or the individual moisture content in each aggregate.
4 Enter that percentage of mineral filler (if required) into the computer.
5 Determine the percent of dry aggregate required from each cold feeder to meet the expected production rate (see Example 6.2).

Example 6.2

Using the mixture design information in Table 6.5, the production rate of the plant is 200 TPH. Determine the TPH of dry aggregate required from each cold feeder to meet the expected production rate.

Solution

First enter into the computer the asphalt binder content of 4.8 percent and the mineral filler content of 2 percent. Most plant computers will convert the mineral filler to a percentage of mixture, since it is added in the drum after the aggregate weighbridge. This is the same method for the asphalt binder.

The dry aggregate production rate = (plant production rate) − (asphalt binder production rate) − (mineral filler production rate).

Dry aggregate production rate = 200 − (200 × 4.8%) − (200 × 2%) = 186.4 TPH.
Adjust the percentage of aggregate to 100 percent minus the mineral filler:

Coarse aggregate 1 = 50%/0.98 = 51.0%
Coarse aggregate 2 = 8%/0.98 = 8.2%
Fine aggregate 1 = 20%/0.98 = 20.4%
Fine aggregate 2 = 20%/0.98 = 20.4%

Multiply the dry aggregate production rate by the percentage of each aggregate:

Coarse aggregate 1 = 186.4 TPH × 51.0% = <u>95.1 TPH</u> (dry)
Coarse aggregate 2 = 186.4 TPH × 8.2% = <u>15.3 TPH</u> (dry)
Fine aggregate 1 = 186.4 TPH × 20.4% = <u>38.0 TPH</u> (dry)
Fine aggregate 2 = 186.4 TPH × 20.4% = <u>38.0 TPH</u> (dry)

6 Many plant computers will complete these calculations.
7 The dry production rate is increased to compensate for the moisture in the aggregate. This is accomplished by dividing the dry production rate by (100 − moisture

Table 6.5 Example 6.2 HMA mixture design

Material	*Blend (%)*
Coarse aggregate no. 1	50
Coarse aggregate no. 2	8
Fine aggregate no. 1	20
Fine aggregate no. 2	20
Mineral filler	2
Total	100
Asphalt binder content	4.8

content)/100. For example, if coarse aggregate 1 has 2.7 percent moisture, the wet production rate becomes 95.1 TPH/[(100 − 2.7)/100] = 97.7 TPH. The plant computer also does this calculation.

8 The wet aggregate production rate is translated to the aggregate cold feeders through the variable feed control. Feed rates are increased or decreased by changing the speed or required revolutions per minute of the belt feeders.
9 After production begins, obtain samples of the combined aggregate from the weigh belt for gradation compliance. Individual aggregate cold feed samples are also useful in making cold feed adjustments, if the combined gradation is not in compliance with the mixture design.

Drum plant operation considerations

The following operations can have an impact on the quality of HMA being produced by a drum mix plant:

- The proper operation of the drum burner is important to the quality of the HMA. Unburned fuel can contaminate the mixture and reduce the asphalt binder film thickness on the aggregate.
- A proper aggregate veil in the rotating drum is important to the efficient operation of the drum. It is also important to the proper asphalt binder coating of the fine aggregate. The veil should be completely across the center of the drum at its midpoint.
- Moisture content of both the virgin aggregate and RAP are the controlling factors in the production rate of the plant. Increases in the moisture content will decrease the production rate and also limit the amount of RAP that can be put in the mixture (TRB 2000).

Mixing temperature

The mixing temperature of the HMA is dictated by the mixing temperature properties of the asphalt binder. The mixing temperature is the temperature of the asphalt binder that provides a kinematic viscosity of 170 ±20 centistokes. This is the same temperature that is used as the mixing temperature for the laboratory mixture design. However for all practical purposes, the discharging mixture temperature from a mixing plant will not be the same as the laboratory mixing temperature. The asphalt binder's storage tank temperature will usually be kept at its mixing temperature or between 150 °C and 200 °C. The actual mixture's discharge temperature is a variable of the asphalt binder temperature, the moisture in the aggregate, the plant production rate and the amount of RAP incorporated in the mixture. The *desired* mixture discharge temperature will be dependent on the distance to the paving site from the plant, any storage time and how difficult the mixture is to compact. Mixing plant mixture temperatures are generally in the range of 150 °C. Excessive mixture temperatures will cause the asphalt binder to smoke and oxidize, leading to premature pavement cracking. Generally the temperature of a typical asphalt mixture should not exceed 190 °C.

ASTM D995, *Standard Specification for Mixing Plants for Hot-Mixed, Hot-Laid Bituminous Paving Mixtures,* is a general specification that can be used to determine if mixing plants confirm to minimum requirements.

Transporting the asphalt mixture

The asphalt mixture is typically hauled to the paving project site in dump trucks. The size of the paving project, the distance to the paving project from the mixing plant and the production capacity will determine the size of the truck and how many are used. The most common type of truck used in hauling asphalt mixtures is a non-articulated end dump truck (Plate 6.14). These trucks are used on both small paving projects such as parking lots and on large scale highway projects. Typically end dump trucks, discharge the asphalt mixture directly into the paver or spreader (Plate 6.15) or sometimes into a device known as a "material transfer vehicle" (MTV).

A MTV is basically a moving surge bin that transfers the mixture into the paver. An MTV can be small (Plate 6.16) or a large machine that can hold up to 70,000 kg. The purpose of the MTV is to:

- Allow a continuous paving operation without the paver stopping and waiting on trucks. A continuous operation will give a smoother pavement.
- Reduces the occurrence of mixture segregation during the paving process. An MTV will also remix the mixture during the loading process into the paver.
- Reduces the occurrence of trucks "bumping" or striking the paver, with the end result of imperfections in the pavement surface.

The use of an MTV is only practical on a large paving project such as highways and is rarely used on parking facilities or on local road construction.

Plate 6.14 Non-articulated dump truck.

Plate 6.15 Discharging into paver.

Plate 6.16 Small material transfer vehicle.

Trucks hauling HMA should be equipped with tarps that will cover the load. The tarps provide two functions:

1 Protect the mixture from inclement weather. Large amounts of water striking a hot asphalt mixture may lead to premature stripping and will also significantly decrease the temperature of the mixture.
2 Retains heat on the asphalt mixture and slows down the cooling effects of the wind.

Trucks may also be equipped with insulation on the truck bed to provide additional heat retention. Truck bed insulation is not as beneficial as tarps are for heat retention of the asphalt mixture.

The pavement designer or specifier's concern with the vehicles hauling the asphalt mixture should only be with what impacts the quality of the mixture or the pavement. These concerns consist of the following:

- The type of truck bed release agent used. Truck bed release agents allow the mixture to slide out of the truck bed easily and provide easy cleanup of the bed. Petroleum based solvents, such as diesel fuel, should not be used as release agents. These solvents can damage the mixture by removing the asphalt film from the hot mixture. There are many commercially available release agents that are biodegradable and are usually water based. These contain no petroleum solvents and are blends of various surfactants that makes a soap or a foam film on the truck bed, which provides the releasing action.
- The method in which the mixture is loaded into the truck bed. If the mixture is dumped into the bed in a single drop, the mixture will have the tendency to segregate, especially

Plate 6.17 Separate batches in truck.

larger stone mixtures. If the mixture is dumped in a single drop in the center of the truck bed, it will build up to a conical pile. The sides of the dump bed restrict the conical pile and the larger particles will roll towards both the front and back of the truck bed. It is preferable to place the load of asphalt mixture in the truck by dividing the load into three or more separate batches or charges (Plate 6.17). In other words, if the truck capacity were 18,000 kg, the shipment would be loaded in three 6,000 kg batches. This is easily done when loading from a surge bin or silo, but if directly loading from the pugmill of a batch mixing plant, the size of the batch will need to be adjusted to match the pugmill capacity. The multiple individual batches should be loaded in the following sequence:

1 The first drop is in the front of the truck bed.
2 The second drop is in the rear of the truck bed.
3 The third or last drop is in the center of the truck bed, between the other drops.

This type of loading causes the truck to make back and forth movements under the silo or mixing plant, which may cause some complaints from the truck drivers. However, reduction of mixture segregation will lead to the longevity of the pavement. Particle segregation in the truck will transfer to the paver, and thus the pavement, giving reduced pavement durability.

Pavement site preparation

The preparation of the site for a new asphalt pavement can include land clearing, grading, compaction, ground water removal or de-watering, subsurface drainage, and surface drainage system installations. A properly prepared pavement foundation includes adequate internal and external drainage and a strong compacted sub-base or base. Water removal and drainage systems are important to public safety and to the longevity of the pavement. Internal water needs to be removed to below the depth of frost penetration or to the bottom of the improved subgrade, either through lateral drainage or subdrainage (The Asphalt Institute 1987). These topics are beyond the scope of this text and are covered in engineering hydrology publications. Sub-base and base strength is considered in the design and thickness of the asphalt pavement, and attention to their preparation is addressed.

Site preparation will vary depending upon whether the asphalt pavement is a new construction or being installed over an existing pavement. When the pavement is installed over an existing pavement, it is known as an *overlay*. Overlays are placed over pavements constructed of asphalt, PCC, or brick. Overlays are commonly done on highways and streets as part of a structural strength and smoothness improvement and a maintenance program. Overlays are typically not completed on parking facilities, but nothing prevents them from being done. Parking facilities are usually private or commercial enterprises and by the time a parking lot may need an overlay, there has usually been business related changes (such as demolition) that will negate the need for a structural overlay. New construction always involves the placement of the pavement on a soil subgrade. There may be an aggregate base course between the asphalt pavement and the soil, but the soil subgrade is the foundation of the pavement. This is true whether the pavement is for an interstate highway or motorway or for a shopping center parking lot. The soil subgrade must be properly graded to line and grade and be uniformly compacted to the required density. All utilities and storm sewers, etc. are usually installed prior to the final compaction and grading of the soil subgrade. Previous engineering work involving the *in situ* soil properties and the thickness of the pavement will have determined if the soil can be used as is, or modified to increase its load bearing capacity. In extreme cases of very poor soils, such as highly plastic clays and soil in high organic content, the soil will need to be removed from a depth of 150–1,000 mm and be replaced with a quality soil or granular fill. Once the subgrade soil has been identified as acceptable or has been improved, it is necessary to compact the subgrade to the desired density. A compacted subgrade is stronger than an uncompacted subgrade and is less likely to cause a premature pavement failure.

Subgrade soil compaction

Soil strength is obtained by removing some of its air voids, thus making it denser. Compaction of the soil removes these air voids. The degree of compaction the soil should receive depends on the nature of the structure it is to support, the properties of the soil and its depth below the pavement. Pavements, including those for parking facilities need a very high degree of compaction (NAPA 1991). In the same method that an asphalt binder aids in the compaction of an asphalt mixture, water aids in the compaction of soil. The water acts as a lubricant, allowing the soil particles to slide easily past each other and further compress, increasing its density. As the water content increases, the density of the soil, under compactive effort will increase. This phenomenon will continue until the soil reaches its saturation point, or where all of its voids are filled with water. Any additional water added to the soil will start to displace the soil particles, causing the soil density to *decrease* and its subsequent strength to weaken. The optimum soil moisture content is the moisture content at its maximum density. The easiest way to determine the optimum moisture content and its density is laboratory testing. The most common tests for this are the Proctor method, that can be followed by using ASTM D698 or AASTHO T-99, *Moisture-Density Relations of Soils using a 2.5 kg Hammer and a 305 mm Drop*. As construction compaction equipment become larger and more capable of high compaction levels, another test known as the Modified Proctor method was developed. This test uses a heavier hammer and can be completed by following ASTM D1557 or AASHTO T-180, *Moisture-Density Relations of Soils using a 4.54 kg Hammer and a 457 mm Drop*. When large compaction equipment is to be used, such as on highway or large parking facilities, the Modified Proctor method is preferred. Once the optimum moisture content and its corresponding density are determined through laboratory testing, the amount of compactive effort to be applied to the soil subgrade can be determined. The compactive effort is in terms of a percentage of the maximum *dry* density of the Proctor laboratory compacted sample. The percentage of compaction is also related to the type of soil being compacted. The soil type can be identified by the Unified Soil Classification System. Table 6.6 is a compilation of recommended minimum subgrade soil densities for various Unified Soil Classification types. Appendix B gives definitions for the various soils classified by the Unified Soil Classification system.

The desired compaction or optimum moisture content is also obtained from the Proctor laboratory testing. Water is mixed with the soil to maintain the moisture content within 2 percent of the optimum. The existing moisture content of the subgrade is determined by a simple mass loss laboratory test or by the use of a nuclear soil density and moisture content gauge that is placed directly on the subgrade. The existing moisture content is used to determine how much water to add to the soil or if the soil needs to dry out. The water is usually tilled or mixed in with the soil. A common question is how deep must the soil be treated and compacted. When no other guidelines are available, a minimum of 150 mm of the soil

Table 6.6 Minimum compactive effort for subgrade soil by type

Unified soil classification	*Percent of maximum dry density (%)*			
	Proctor		*Modified proctor*	
	<1000 mm deep	*>1000 mm deep*	*<1000 mm deep*	*>1000 mm deep*
GW, GP	97	94	94	90
GM, GC	98	94	95	90
SW	97	95	94	90
SP, SM	98	95	95	90
SC	99	96	95	91
ML, CL	100	96	96	92
OL		96		92
MH	100	97		92
CH		97		92

subgrade depth needs to be compacted. When replacing very poor soil with fill, a minimum of 1,000 mm should be removed and then replaced with compacted fill.

A soil subgrade can be compacted by various pieces of compaction equipment. This compaction equipment includes pneumatic or rubber tired compactors, sheepsfoot rollers, tamping foot compactors, and vibratory compactors. Pneumatic and vibratory compactors or rollers are also used to compact asphalt pavements. A pneumatic tire compactor compacts soil or an asphalt pavement through tire pressure, ballast, and surface manipulation or kneading. The kneading action of the tires seeks out soft spots in the soil and helps seal the surface. Pneumatic tire compactors are used on small- to medium-size compaction projects, such as parking lots, or on granular base materials. A sheepsfoot roller is a cylindrical drum filled with ballast that is covered with oval or rectangular pads that resemble sheep's feet. Roman roadbuilders herded sheep back and forth over the base until it was compacted. These pads penetrate through the top lift of material and compact the lift below. As the pad comes up through the top lift, it aerates or "fluffs" it. This action is useful in drying out a very wet clay or silt soil. A tamping foot roller is similar to a sheepsfoot roller except that the pads are not shaped like feet, but are more like oval shanks or plugs. A tamping foot roller is not able to "fluff" the top lift of soil. A vibratory compactor or roller (Plate 6.18) works on the principle of particle rearrangement to decrease voids and thus increase the density. Vibratory rollers generate three compactive forces: pressure, impact, and vibration (Caterpillar 1990b). Vibratory rollers use a rotating eccentric weight within the drum, causing it to vibrate. Table 6.7 illustrates the operating characteristics of the various soil compactors and Table 6.8 illustrates which compactor is best suited for the soil type.

The desired density is determined on the project site through the use of a nuclear density gauge. A nuclear density gauge is a relatively safe device that indirectly measures the density and moisture content of either an asphalt pavement or the soil or granular subgrade. The gauge measures density by transmitting gamma rays into the pavement or subgrade and measures the amount of radiation reflected back into the gauge in a certain amount of time (usually 1 min or less). The amount of radiation returned is compared to known amounts returned from densities of known materials by the gauge and the density of the pavement or subgrade can then be determined. ASTM D2922 is a test method that can be specified for the use of the nuclear density gauge. The designer of the project should provide the

Plate 6.18 Vibratory roller.

Table 6.7 Typical operating characteristics of various soil compaction equipment

	Pneumatic	*Sheepsfoot*	*Tamping foot*	*Vibratory*
Typical compacted lift thickness (mm)	300	150	150	150–600
Typical number of passes	6–9	4–5	2–3	1–2

Table 6.8 Soil type and equipment selection guidelines

	General soil type			
	Gravel or crushed stone	*Sand*	*Sand/clay or silt*	*Clay*
Pneumatic	X	X	X	
Sheepsfoot			X	X
Tamping foot			X	X
Vibratory	X	X	X	

minimum density and moisture content required during the compaction of the subgrade and the frequency of testing. The density the designer is concerned with is the final density prior to further construction of the pavement. The amount of acceptance testing required varies but is usually an amount per square meter of compacted subgrade. For example, one test per 1,000 m^2 of compacted subgrade could be required. The moisture content during compaction should range from 3 percent below the optimum laboratory moisture content to 2 percent above the optimum.

Laboratory testing and extensive field testing is not always available and not practical on very small projects. Proper compaction of the subgrade is still important and can be achieved. Past experience with similar soils would be helpful, and a set number of passes with a certain type of compactor could be specified. For example, seven passes with a 20 tonne rubber tired roller may be adequate for the project. Rubber tired rollers can also be used to initially seat or for proof-rolling the subgrade. These types of rollers can seek out soft spots that may exist in the soil or fill (Caterpillar 1990b). Proof rolling is one method that can be used to check the condition of the soil subgrade. The objective of proof rolling is to locate areas of slippage and cracking, which usually denotes an incorrect moisture condition sufficient to cause shear failure near the surface. The soil is observed for further densification and movement as it is rolled by the pneumatic or rubber tired roller (NAPA 1991). The rubber tired rollers are also often used to give the final surface rolling of the subgrade.

Crushed stone bases

Crushed stone or gravel bases or sub-bases are applied on top of the compacted soil subgrade. The stone is usually hauled in by trucks and spread by motor graders or bulldozers. The crushed stone is also compacted in the similar fashion as the soil subgrade. The crushed stone is spread to the desired thickness that was determined during the thickness design. However, the thickness determined during design, assumes the stone has been compacted, so the thickness to be spread by the grader needs to be thicker to allow for compaction. A general rule of thumb is to increase the design thickness value of crushed stone or gravel bases by around 30 percent to compensate for the thickness reduction due to compaction (NSA 1992). For example, if the thickness design requires 300 mm of compacted crushed stone, 390 mm of stone would be spread prior to compaction. The actual compensation value should be checked, since it will vary somewhat based on the angularity and surface texture of the stone, the moisture content and the desired final density. The motor graders will grade the stone to the final grade desired. Crushed stone or gravel bases are compacted in the same manner as the soil subgrade except sheepsfoot compactors should not be used on granular bases. When used on stone or gravel bases, the feet of these compactors tend to shove the material aside rather than compact it (Hyster Company 1986). Compaction of crushed stone is also best accomplished when it contains an optimum amount of moisture. This moisture allows the aggregate particles to easily move against each other until the base establishes particle to particle contact during compaction and reaches its maximum density. The Modified Proctor test method (ASTM D1557) is the typical laboratory method used to determine the optimum moisture content and maximum density of crushed stone and gravel. Crushed stone is usually placed at a moisture content slightly higher (1 or 2 percent) than its optimum moisture content. Stone tends to dry out faster than soil. Compaction of the crushed stone or gravel base should be completed until the density is 100 percent of the Modified Proctor laboratory dry density (NSA 1994b). The field or enplane density of the base can be measured using the same method as that used for a soil subgrade, such as a nuclear density gauge. The compaction procedure and the final grading of the stone or gravel base are similar to what would be used on the soil subgrade. Rolling or compaction should commence immediately after the spreading of the stone with a static (non-vibratory) pass of the roller. This will ensure sealing of the stone base to retain the moisture content as well as reducing additional moisture entry from the elements. Additional passes of the roller should proceed in the vibratory mode if applicable. The compaction of the base should proceed by rolling from the outside edges towards the center of the section being constructed. The roller should overlap each pass by one-third to a half of the roller width. Overlap ensures that the target density is achieved completely in all areas of the project. The surface of the completed stone base should be finished by a motor grader or other piece of equipment to produce a final surface having no deviations in excess of 10 or 15 mm when tested by a straight edge.

Subgrade and base surface preparation

A soil subgrade should be checked one final time to identify any soft areas that are too weak to support the paving equipment or haul trucks. This especially should be done if an asphalt course will be placed directly on top of the soil subgrade. Any large boulders that may interfere with the paving operation should have been removed previously. The surface should be graded smooth, since large distortions in the subgrade have the potential to be transferred to the final surface of the asphalt pavement structure. Any significant grade changes in the design of the pavement should also be applied to the soil subgrade and any base construction. This practice is done only where it is practical and is usually restrained by the project geometrics and drainage or utility installations. This procedure ensures that the design thickness of the pavement structure will be correct in all areas of the parking facility or road and also contributes to a smoother top surface of the pavement. For example, roadway requires a 2 percent crown or the parking lot requires a 2 percent slope, the surface of the subgrade should also have the same slope. This can be accomplished through final trimming or grading by the motor grader. Curb and gutters and parking islands are installed after the final trimming of the stone base or if a Full-depth®, on top of the soil subgrade. Curbs confine the asphalt mixture during placement giving support during the compaction process. The appropriate drainage apertures are also installed at this time.

The surface of a granular base or sub-base is prepared in a similar fashion as the soil subgrade, with the addition of a prime coat prior to the application of the asphalt pavement layers or lifts. A *prime coat* consists of either a slow setting or medium setting asphalt emulsion or a medium curing asphalt cutback. The prime coat is typically applied to granular bases at a rate of 0.7–2.0 l/m^2. It is used to bond the asphalt pavement layer to the granular base and acts as a temporary waterproofing layer. When a prime coat is applied to a soil subgrade, the prime coat is usually too viscous for the subgrade to absorb enough of it to make the prime coat very effective. A very dilute emulsion, consisting of a slow set emulsion diluted with water to give an asphalt binder content of 20–30 percent, are used as dust palliatives on soil subgrade. A dust palliative is especially used if there will be a significant length of time between the soil subgrade construction and the additional pavement layers being placed on top of it.

If the placement of the HMA is being put on top of an existing pavement structure, such as an asphalt or concrete pavement, it is considered in overlay. In these cases the existing pavement is considered the base or sub-base and its strength would have been considered in the thickness design stage. Its condition at that point would have been addressed and any significant remedial actions to it would have been completed prior to the construction of the overlay. The surface of the existing pavement should also be properly prepared. At a minimum failed areas should be removed and replaced; potholes properly patched; cracks cleaned out and sealed; and ruts filled in with a leveling course, or preferably planed out by cold milling. It is recommended that pavement failures due to severe load related distress be removed and replaced prior to the application of the overlay. Badly cracked pavement sections should be removed and replaced. Random cracks that are wider than 10 mm should be filled. All asphalt and granular base materials that have failed should be cold milled or excavated. New granular base material, stabilized base materials, or HMA should be placed in order to bring the strength of the pavement structure in each failed area to the same level as the surrounding good areas. Undercutting the existing pavement and then replacing with suitable fill material should repair subgrade distortion. Subsurface drainage should be installed as required (TRB 2000). Surface distortions in an existing pavement, such as rutting, shoving, or corrugations can be filled or leveled out with a leveling course. A leveling course or leveling binder is usually a dense graded wearing course placed in very thin lifts, usually 25 mm or less. Leveling courses are also called scratch coats or courses or a wedge and level course. However, it is preferable to have these distortions removed through the use of a cold planing or cold milling machine (Plate 6.19).

These same machines are what produce RAP by milling existing asphalt pavements. Milling is done by a drum (Plate 6.20) equipped with tungsten carbide and is located at the bottom of the machine (Plate 6.21). Milling machines create an extensive amount of dust,

Plate 6.19 Cold milling machine.

Plate 6.20 Milling drum.

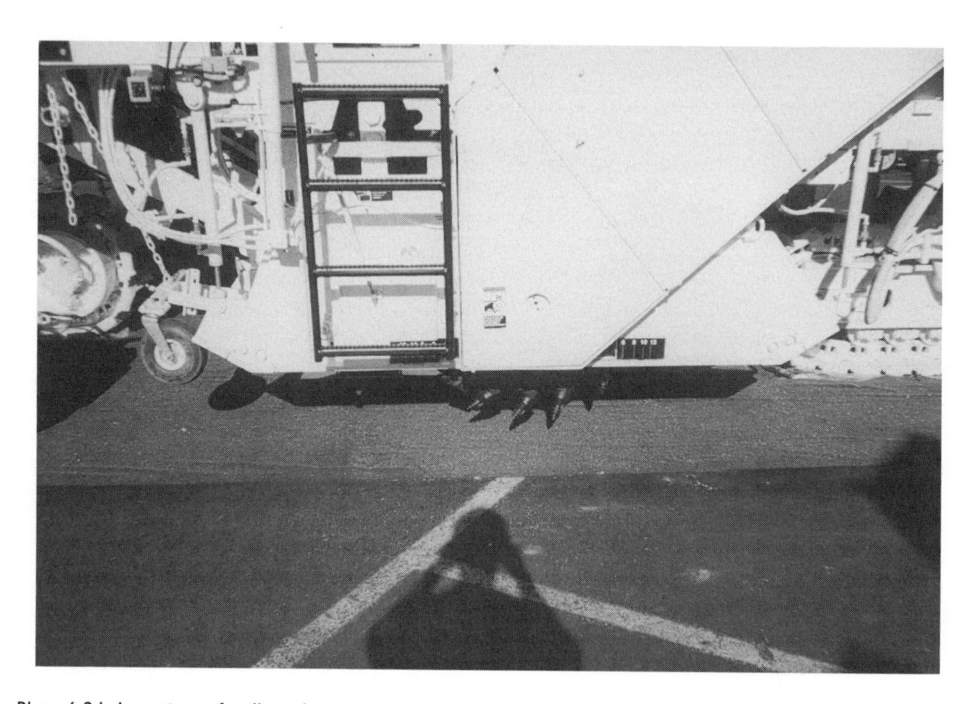

Plate 6.21 Location of milling drum.

Plate 6.22 Power broom.

which must be removed prior to the application of the overlay. This dust is generally removed by the use of a power broom (Plate 6.22).

The milling machine also leaves a grooved surface that assists in improving the bond between the existing pavement and the overlay. Overlays are bonded to existing pavements, whether they are asphalt or PCC through the use of a tack coat. A tack coat is similar to a prime coat, but is applied at a lighter application rate. Tack coats are usually applied at a rate of 0.2–1.0 l/m^2 of pavement surface area. Milled pavements, due to the grooves left by the milling machine, have a higher surface area and can require tack coats applied at 1.0 l/m^2 or more. Tack coats consist of slow setting emulsions that are diluted with water to give a final asphalt binder content of around 30 percent. Tack coats can also be straight asphalt binders such as an AC-5 or a 120–150 penetration grade or even cutback asphalt such as an RC-70. Asphalt emulsions, however are the most common material specified for use as a tack coat. Tack coats are also used to bond the various new pavement courses or lifts together. This application is not always necessary and is usually not needed if constructing the courses days or even weeks apart. If it is desirable to use a tack coat for bonding subsequent new courses together, the application rate is very light, usually less than 0.3 l/m^2. The subsequent HMA courses can be applied either directly after the tack coat has been applied or after it has "breaks" (asphalt emulsion). The telltale sign of an asphalt emulsion breaking is when it changes color from brown to black. The bond between the layers will still be created regardless of whether the asphalt emulsion broke prior to paving of the subsequent layer. Traffic, including foot traffic should be kept off the tack coat to prevent the tack being removed. A tack coat that consists of straight asphalt binder will cool very quickly. Paving can follow immediately after the application of an asphalt binder tack coat. It is acceptable for mandatory construction equipment such as haul trucks and the paver to travel over the tack coat immediately prior to paving operations. If construction delays cause either a prime coat or a tack coat to be exposed for a significant length of time, a light application of sand, 2.0–4.5 kg/m^2, will reduce any pickup by traffic. Sand can also be used to blot an excess application of a prime coat, or if it has adequately penetrated the granular base course. Excess or unabsorbed prime coat will appear as puddles, usually 24 h after application.

Paving equipment

The construction of an asphalt pavement involves spreading out the asphalt mixture to the proper thickness, grade, and slope desired. Pedestrian and vehicles also require a smooth pavement surface. An aesthetically pleasing smooth surface is a requirement for parking facilities. The best way to ensure meeting all these requirements is through the use of a mechanical spreader especially designed for laying asphalt mixtures. This type of spreader is usually referred to as an asphalt paver or asphalt finisher (Plate 6.23).

An asphalt paver can also be used to spread the aggregate during the construction of a crushed stone base course. An asphalt paver consists of two basic units, the tractor and the screed. The primary functions of the tractor are to receive, deliver, and spread the asphalt mixture or aggregate in front of the screed and to tow the screed forward. The tractor is either equipped with rubber tires (Plate 6.24) or tracks (Plate 6.25). Rubber tires provide speed of movement over smooth surfaces and tracks provide extra traction, if needed, on a soil subgrade. Both rubber tire and track tractors utilize a three-point suspension design to allow the tractor to move over irregular grades and maintain a relatively constant line of pull

Plate 6.23 Modern asphalt paver.

Plate 6.24 Rubber tire paver.

Plate 6.25 Track paver.

Plate 6.26 Screed.

on the screed. Severely irregular grade conditions will cause a change in the line of pull to the screed. The degree or amount of tow point change is averaged over the length of the wheelbase. The self-leveling action combined with the time it takes for a screed to react to the changes of the line of pull, allow the screed to place material in a constant profile (Cedarapids 1993). The primary functions of the screed (Plate 6.26) are to lay the mixture to the desired width and thickness, to level, smooth, and seal the asphalt mixture, and to provide some initial compaction (Blaw-Knox 1988). The screed also controls the thickness of the lift or course and gives a slope or grade to the pavement surface.

The asphalt mixture is supplied to the paver by haul trucks, windrow elevators, or by MTV. The most common method is to supply the paver by trucks. The haul truck backs up to the paver and dumps the load into the hopper that is attached at the front of the tractor portion of the paver. The truck is then pushed forward by the paver as the truck dumps the mixture into the hopper. The paver is equipped with push rollers (Plate 6.27) on the front of the hopper that contacts the rear tires of the truck. Some pavers also have hitches that are located by the push rollers that keep the truck in contact with the paver and prevent the truck from pulling away from the paver and dumping material on the surface in front of the paver. The hitch has arms with rollers attached on them that extend forward. The rollers are retracted into the truck tire rim preventing the truck from loosing contact with the paver during the unloading process. Once the truck has been emptied of material, the truck hitch is withdrawn, allowing the truck to pull away from the paver. It is very important that the truck does not bump the paver as it is backing up to dump the load. It is best for the truck to stop about one meter in front of the paver and allow the paver to pull up and contact the rear

Plate 6.27 Push rollers.

wheels of the truck, beginning the pushing forward process of the truck. The transmission of the truck needs to be in neutral during this process. Bumping the paver will cause forces to be transmitted to the screed causing surface irregularities in the placed mixture or mat. The hopper wings can also be folded inward (Plate 6.28) to use the material that collects in the corners. "Mat" is a term that is used to typically describe a course or lift of asphalt mixture that has recently been placed by the paver. Windrow elevators pick up the material that has been placed in front of the paver and discharge it into the hopper. This allows a continuous flow of material to the paver in addition to no stopping or contact with other vehicles. The material transfer vehicle provides the same benefits of no stopping of the paver. Two independently controlled drag slat conveyers or augers move the asphalt mixture from the hopper past flow control gates through a conveyor tunnel to a screed auger that is parallel with the screed (Plate 6.29). Control of the speed of the drag conveyers operated can be regulated manually or by an automatic feeder system (Caterpillar 1990a). The flow gates control the amount of material being moved from the hopper to the augers. They can be raised or lowered independently to control the head of material in front of the screed and match varying material demands from the left to right sides of the screed, as irregular grades are paved. The screed auger (Plate 6.30) distributes the material evenly in front of the screed. It consists of left and right augers that are operated in conjunction with the left and right slat conveyers. The augers take the asphalt mixture being delivered by the slat conveyers and move it outward across the width of the screed. The screed augers are also equipped with reversing paddles or kickback blades located by the center of the gearbox. These reversing paddles or kickback blades help fill in the void area underneath the gearbox by "tucking" or forcing some

Plate 6.28 Small paver with hopper wings folded.

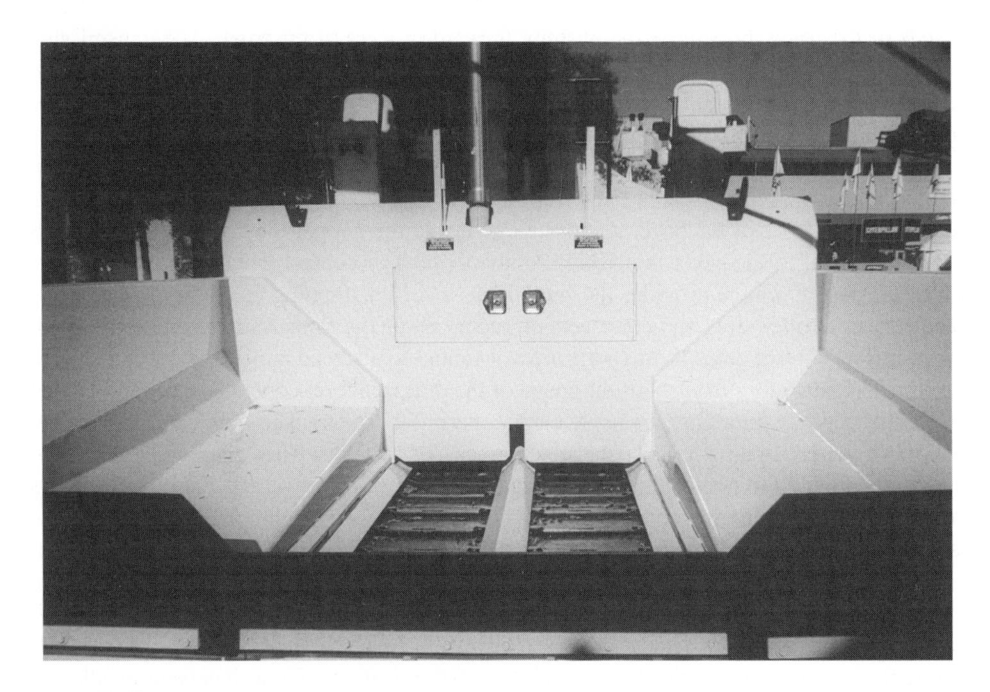

Plate 6.29 Slat conveyers and control gates.

Plate 6.30 Screed and screed auger.

material underneath the gearbox (Cedarapids 1993). If the gearbox was allowed to be void of material, larger particles would fall into this area, leaving an area of mat segregation identified as a thin strip that is approximately in the center of the mat placed by the paver. This thin strip will be short of fine material, possibly leading to pavement disintegration and cracking.

Most modern asphalt pavers, especially ones designed for highway or large parking projects, are equipped with floating screeds. Asphalt pavers equipped with floating screeds are self-leveling. Self-leveling allows the paver to compensate for irregularities in the surface of the base course or subgrade over which it is paving. The amount of self-leveling is determined by the length of the tow arms and the location of the tow points in relation to wheel base or track contact area of the tractor. A longer wheelbase or longer tow arms will increase the averaging of high and low spots in the new mat to give an overall smoother pavement surface. The floating screed works in conjunction with the three-point suspension design of the tractor. This feature allows the paver to provide a very smooth surface over large distances of paving. The floating screed is attached to the tractor by tow arms on each side of the paver. These tow arms are what allows the screed to float. The rear of the screed is attached to the tow arms by the mat thickness controls. The entire screed and tow arm assembly is free to pivot around the tow point connection. As the screed is towed forward, the asphalt mixture flowing under the screed causes vertical displacement or movement of the screed. This action causes the screed to float on top of the mixture. The rise and fall of the screed during movement establishes the mat thickness. As the screed is pulled initially through the asphalt mixture, it will initially rise or fall until it is traveling in a plane parallel to the direction of the pull. The mat thickness is maintained by controlling the rise

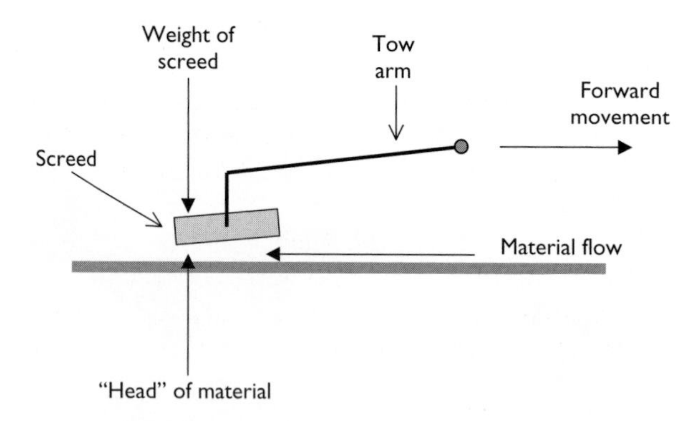

Figure 6.5 Forces acting on a floating screen.

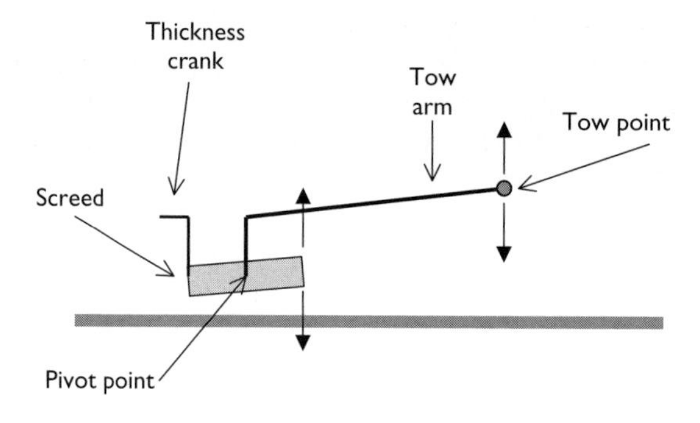

Figure 6.6 Controlling mat thickness and smoothness.

and fall of the screed. Many variables affect the vertical displacement of the screed, and if all the forces (Figure 6.5) acting on the screed are constant, the screed will ride at a constant elevation (Caterpillar 1990a).

Allowing more material to flow under the screed causes it to rise, thus increasing the mat thickness. Tilting the rear of the screed downward causes the front of the screed to rise, allowing more material to flow under the screed and increase mat thickness. Tilting the rear of the screed upward causes the front of the screed to fall, decreasing the material flow under the screed and mat thickness. The screed rises or falls gradually in about three tow arm lengths, until it is moving in a plane parallel to the grade. The tilting of the screed or "angle of attack" is controlled by three mechanisms (Figure 6.6):

- Manually by tow point elevation movement controls.
- Manually by screed depth cranks.
- Automatically through a grade control system that adjusts the tow point elevation and the screed depth.

The screed angle of attack can be changed by vertical changes in the tow point elevation. If the tow point is raised, the angle of attack is increased (the front of the screed tilts up). If the tow point is lowered the angle of attack is decreased. There is an eight to one ratio between changes in tow point elevation and angle of attack. The angle of attack can also be changed at the screed depth cranks. The controls rotate or tilt the screed at its pivot points on the tow arms. These cranks are located at each end on the rear of the screed. Turning the depth of the cranks clockwise increases the angle of attack and turning them counterclockwise decreases the screed's angle of attack. One full turn of the crank produces about a 6 mm change in mat thickness. Mat thickness will begin to change immediately, but will not be complete until the paver has moved a distance equal to three lengths of its tow arm (Caterpillar 1990a). The two point elevation controls and the screed thickness cranks both have an effect on mat thickness and smoothness.

After the correct mat thickness has been established, further adjustments using the depth cranks should not be necessary as long as the paver speed and head of material remains the same. In this regard it is best to operate the tractor at a consistent forward speed and keep the slat conveyers and augers operating at a constant speed. Adjustments to mat thickness should be made in small increments. Changes should be fully made and checked before any additional adjustments are made. Small incremental adjustments will avoid the appearance of a wavy or irregular surface in the mat. The uncompacted mat is spread at a thickness that is greater than the design thickness. The design thickness is usually based on compacted thickness. The compaction process consolidates the material down to the desired pavement thickness. Most dense graded mixtures should be placed at 1.25 times the design pavement thickness to allow for consolidation during the compaction process. Actual checks of the before and after compacted thickness should be made to confirm if any adjustments are necessary.

The screed also determines the width of the pavement. Most screeds have a fixed width with the ability to add extensions to increase the pavement width or use cut-off plates to decrease the pavement width. Small screeds will have a fixed width of 2.5 m, with larger screeds having a fixed width of 3 m. The travel lanes for roads or highways will usually range from 2.5 to 3.5 m. Additional screed auger segments can be added or removed to help spread the material across the full width of the screed. Some screeds are capable of extending their fixed width using hydraulically extendable end plates to place a mat that is up to 8 m wide. An extendable screed offers the advantage of paving the required number of lanes or sections (parking lots) in less time and in a reduction of longitudinal construction joints.

The screed imparts some initial compaction to the mat by its static weight by vibration imparted through internal tamper bars or rotating eccentric weights. The screed achieves approximately 80–85 percent of the maximum theoretical density of the mat. The rotating eccentric weights or vibrators are the most common and impart a vibrating surface to the screed. The vibrators can be adjusted for vibrations per minute (vpm) and amplitude, which is the amount of force imparted from the screed to the mat. The screed vibrations can range from 1,500 to 4,500 vpm or 25 to 75 Hz. The main screed crown control allows the screed bottom to be set to a flat profile or be bent to match any specification for various pavement profiles (Blaw-Knox 1988). The screed is also equipped with heaters that are used to warm up the screed bottom prior to the start of paving. The heaters are actually burners that are powered by the diesel fuel from the tractor. The screed should be heated prior to startup or when the screed has been extended and out of the asphalt mixture for any significant length of time. If the screed is not brought up to the same temperature of the mixture, the texture of the mat will appear open and torn. A cold screed can also cause depth control problems, since it will impart the flow of the mixture under the screed in an irregular manner. The screed heaters are not capable of heating up a cold mixture or any significant increase in mat temperature. Even if the screed is super-heated, only the surface of the mat will be heated. If the burners are left on for an extended period of time, the screed bottom could warp and the surface of the mat may burn. The screed heaters should be turned off during the actual placement of the mat and only used for preheating the screed.

Spreading process

The objective of a mechanical spreading or paving process (Plate 6.31) is to place a smooth uniform asphalt mat fairly quickly over an irregular grade.

Paving projects will not allow for more material to be placed than what was estimated or projected. Average mat thickness (or depth), material yield, profile, and slope influence the amount of material used for a project. Projects such as small parking lots and local roads normally cannot take advantage of automated screed control that controls the thickness of the mat being placed. The screed operator provides manual control in these cases. To ensure that a smooth and uniform pavement is placed, there are two methods that can be used to check and control material usage. One method is the average thickness or depth method and the other is the yield method.

The average depth method is fairly simple in that it consists of taking five depth measurements, each about 2 m apart longitudinally and then averaged. The depth measurements are made with a depth sticker that punches into the pavement surface. These can also

Plate 6.31 Paving process.

be preset to allow quick verification of the proper thickness or depth. Changes are made to the thickness cranks only after the five depth measurements are made and then averaged and checked. Changes should only be made if the average depth measurement varies more than 6 mm.

The yield method involves determining how many linear meters a truckload of material should be capable of paving and then adjusting the thickness control cranks to obtain that yield. A truckload yield is determined by the following method.

1. Determine the compacted density of the material being placed (kg/m^3).
2. Obtain the desired depth or thickness in mm and convert to meters.
3. Obtain the desired mat width in meters that will be placed by the screed.
4. Multiply the compacted density by the desired depth and desired mat width. This will give a result in kg per linear meter of travel.
5. Obtain the amount of material to be loaded in an average truck.
6. Divide the truckload by the kg per linear meter of travel result, to give a result in linear meters paved per truck.
7. Use this result to verify and adjust during construction. Keep in mind when the truck loading changes, the linear meters paved per truck will change.

Example 6.3

Determine the yield per truck that is loaded to 20,000 kg, producing a mat 3.5 m wide and 75 mm thick. The asphalt mixture has a compacted density of 2,300 kg/m^3.

Solution

The amount of material paved per linear meter is determined by:

$$\frac{75\text{ mm}}{1{,}000} \times 3.5\text{ m} \times 2{,}300\text{ kg/m}^2 = 603.75\text{ kg per liner meter paved}$$

The amount of meters paved per truck is:

$$\frac{20{,}000\text{ kg}}{603.75\text{ kg/m}} = 33.1\text{ m}$$

This value is checked during construction and adjustments are made to the screed thickness cranks if necessary.

For all practical purposes, a 100 percent identical actual yield to the calculated yield, will not occur, due to base irregularities, truck weighing, and density variations. Yield comparisons should be done over several truckloads of material. Adjusting the screed thickness cranks then compensates for any significant variations in the yield. The yield method of thickness control is better suited for large paving projects since several truckloads are required to determine and adjust the mat thickness. The average depth method is better suited for small projects.

The travel speed of the paver is ideally matched to the mixing plant and haul truck capacity and the width and depth of the material being placed. It is also matched to the compaction capacity or abilities of the rollers. The ideal speed is the speed that provides a continuous paving operation with no stopping or interruptions. The rollers must also be able to keep up with the paver and provide the desired mat or pavement density. However for all practical purposes the paver will stop, usually waiting for haul trucks. The critical or maximum paver speed will be limited by the ability of the rollers to compact the mat without allowing it to cool to below the compaction temperature. The goal is to keep from having too much uncompacted asphalt mixture spread before it becomes compacted. Equation 6.1 can be used, as a rule of thumb, in determining the maximum paver speed desired for a continuous operation with the rollers.

$$\text{Paver speed (maximum)} = \frac{\text{roller speed (m/min)} \times 0.9}{\text{number of roller passes}} \tag{6.1}$$

where roller speed is the speed of the roller that provides adequate compaction; roller passes, the number of passes that provide adequate compaction; 0.9, the time factor that allows for the stopping and reversing of the roller.

Example 6.4

Determine the maximum paver speed if the compaction roller is traveling at 75 meters per minute and requires five passes to achieve the mat density.

Solution

$$\text{Paver speed} = \frac{75 \text{ m/min} \times 0.9}{5 \text{ passes}} = 13.5 \text{ m/min}$$

The ideal operation is when the paver is moving at a fixed speed a minimum of 80–90 percent of the time and only spending 10 percent of the time at a lower speed or stopped. When a paver is stopped, the screed will have a tendency to settle causing a depth change and possibly a mark or texture change in the pavement. If a paver has to stop for an extended period of time, the temperature of the asphalt mixture in front of the screed and in the hopper drops. This change of temperature will cause a texture change to occur in the mat being placed and also a depth change. If a paver has to stop for an extended period of time, the construction of a transverse joint at this location should be considered (Cedarapids 1993).

Controlling the head of material going under the screed is important in maintaining a consistent thickness. The head of material is controlled by the slat conveyors or augers, the flow gates, if equipped and the screed augers. Variations in the head of material will cause variations in thickness and surface texture. If the flow of material is too low, the resistance by the screed to travel will decrease. The screed will now ride down as a result of the decreased pressure and the mat will start to thin. The opposite will occur if the flow or head of material is too large, causing the screed to rise and increase the mat thickness. The flow gates should be opened enough to maintain enough material to keep the screed augers half full. Most modern pavers are equipped with automatic controls to maintain the screed augers half full. The controls adjust the speed of the slat conveyers. The head of material can also change as the paver speed changes, so a constant travel speed will assist in maintaining a constant head of material. Increasing the paver speed will decrease the time the head of material spends under the screed, causing the screed to drop, decreasing mat thickness.

Decreasing the paver speed has the opposite affect, increasing the time the head of material spends under the screed, causing the screed to rise which increases the mat thickness. The rise and fall of the screed in relation to paver speed only occurs if no other changes are made in the paving system, such as changes in the tow point elevations or in the head of material. As the paver speed increases or decreases, changes must be made to the head of material flow to maintain a constant thickness. As the paver speed increases, additional material needs to be delivered by the slat conveyers and as the paver speed decreases, the material flow needs to be reduced to maintain a constant thickness. The flow of material under the screed can also be changed through adjustments in the tow point elevations or by the thickness cranks as the pavers peed changes. These adjustments change the angle of attack of the screed as the paver speed changes. Automated controls assist the paver operator in balancing material flow with paver speed.

Running the hopper of a paver empty between truckloads will also cause the head of material to change. The paver should be stopped before the level of material drops below the flow gate settings. Folding or dumping the wings of the hopper (Plate 6.32) at this point will also have some additional material to be placed back in the center of the hopper, but the hopper wings should not be dumped excessively. They should be emptied before the material that collects in the corners of the hopper has cooled to the point where chunks are formed that cannot be broken up as that material moves through the paver to the augers and under the screed. To properly dump the wings, the sides of the hopper should be slowly raised as soon as the truckload has been emptied and has pulled away from the paver. A steady forward paving speed should continue as the hopper wings begin to be folded or dumped. The wings should be fully elevated before the amount of material remaining in the hopper is lower than the top of the flow gates. The slat conveyers should never be visible at the time the wings

Plate 6.32 Hopper wings folded.

are raised. During the entire paving process the slat conveyers should be covered with material to ensure a constant mat appearance and thickness. At the end of paving, the remaining material in the hopper should be discarded. When paving is continued between truckloads, the level of material in the hopper should not be allowed to drop below the flow gate settings. If the level of material drops below the flow gate setting, the amount of material being delivered to the screed auger drops and thus the head of material is reduced. Material dropped in front of the paver by trucks or damaged hopper flashing is a common occurrence that can cause defects in the mat being placed. The material that is dropped in front of a paver adds to the volume of material that is in the screed auger area as the paver passes over it. This increased volume increases the head of material in front of the screed. If the material deposited is raked or shoveled out over a wide area in an attempt to prevent overloading or changing the head of material, a second problem is created. The material that was spread out will cool considerably and add to the elevation of the road surface. When a paver passes over this area, the mat thickness in these areas is thinner.

Paving depth or thickness

The minimum paving thickness of the mat should be at least twice or preferably three times the maximum nominal aggregate size in the asphalt mixture. This allows the vibration and weight of the screed to rearrange the aggregates into a tight uniform mat. For example, 25 mm mixture should be placed in lifts of a thickness of 50 mm or greater. Paving thickness should never be less than 1.5 times the maximum nominal aggregate size. Table 6.9 gives minimum and preferred lift or layer thicknesses based on the nominal maximum aggregate size of the mixture. If an attempt is made to pave a mixture less than this, the screed will be supported by the larger aggregates and will no longer be able to float on top of the head of material. This will cause the screed to mirror the pavement deviations below it (Cedarapids 1993). The larger aggregates will also tear out of the surface of the new mat, leaving a very poor surface texture. The maximum depth or thickest mat to be placed by the paver is limited mainly by the ability of the rollers to obtain adequate compaction. Multiple lifts of the same mixture are often placed in succession to complete the required thickness called for by the design. For example, an asphalt base course that has a design thickness of 300 mm may be placed in three lifts of 100 mm or even two lifts of 150 mm.

Pavement slopes and crowns

On roads and highways, a slope or crown to the pavement surface is provided for water runoff. A crown for a road or highway is typically at a 2 percent slope from the center of the pavement in both directions. If the road is more than three lanes in each direction, 3 percent

Table 6.9 Minimum layer thickness

Nominal maximum aggregate size (mm)	*Minimum layer thickness* (mm)	*Recommended minimum thickness range* (mm)
9.5	25.0	31.5–44.0
12.5	37.5	44.0–62.5
19.0	55.0	62.5–94.0
25.0	75.0	75.0–125.0
37.5	112.5	112.5–150.0

Plate 6.33 Floating screed, capable of variable widths and crowns.

is used on the outside lane. Parking facilities will not have a crown, but will have a slope given the surface providing water runoff to collection points. Parking facilities usually have a surface drainage slope of 1 percent. Slopes much greater than that can cause difficulty for pedestrian traffic and shopping carts. In parking areas around buildings it is often necessary for the slope to vary in both directions. Flat spots or "bird baths" tend to occur along curbs and transition areas, where the slope may approach zero transversely and longitudinally at the same spot. The amount of fall along curb lines is recommended at 0.5 percent in the flow line along the curb (NAPA 1991). Newly constructed asphalt pavements will always drain slower and hold more water than older pavements, due to the fresh oils in the new pavement. Water will bead approximately 2 mm on a new pavement and over time this beading will drop 1–2 mm as the pavement ages. The phenomenon is similar to water beading on a freshly waxed automobile (Chellgren 2002).

Screeds (Plate 6.33) have the ability to lay crowned and sloped mats. The crowning mechanism is located in the center of the screed and consists of two connected adjustment screws. The front screw controls the leading edge of the crown and the rear screw controls the tail edge of the crown. The relationship between the leading crown and the tail crown has significance since it affects the consistency and texture of the mat. When beginning to pave, there should be 3–6 mm more lead crown than tail crown. However, too much lead crown will produce too much material flow in the center of the screed. The center of the mat will be tight and the edges will be loose. This condition is characterized by a bright streak in the center of the mat. Too little lead crown reduces material flow at the center of the screed. The edges of the mat will be tight and the material in the center of the mat will be loose. Bright streaks on the edges of the mat will reveal this condition. The lead crown is changed by turning the front screw in the appropriate direction one-quarter turn at a time until a uniform mat appearance is achieved (Caterpillar 1990a).

Grade control

The screed controls the thickness of the mat and its elevation in reference to the existing grade. The screed on modern pavers can be controlled manually or automatically. Automatic screed controls are used to keep the elevation of the tow points on the paver at a predetermined elevation relative to a reference point. The reference point consists of either a preset string line installed on the grade or a mobile ski attached to the paver. Deviations in the pavement surface are averaged out over the length of the reference. The primary purpose of automatic screed controls is to produce a pavement that is smoother than the paver can accomplish by itself using only the wheelbase of the tractor and the floating screed. As the tractor moves up and down over the existing grade, the elevation of the tow points moves up and down over a smaller range than they would if the tractor's wheelbase was the reference point. A string line or ski is longer than the wheelbase of the tractor. Keeping the elevation and the tow points constant in direct relation to the reference permits the screed to maintain a more consistent angle of attack, which in turn provides a smoother mat behind the screed. The automatic controls do not permit the relative position of the tow points to change, even though the tractor unit is moving up and down vertically in response to the roughness of the surface over which it is traveling (TRB 2000).

Grade sensors are used to monitor the elevation of the existing pavement surface in a longitudinal direction. The grade sensors monitor the elevation of the existing surface by monitoring the elevation of the reference point. The reference point is either an erected string line placed on the existing pavement or various size skis or shoes that are attached to the paver. A string line provides for the smoothest pavement surface, but must be erected prior to paving and can interfere with some of the paving operations. The string line provides the elevation input to electronic grade sensors located on the paver. The grade sensor consists of a wand that follows the string line and inputs grade changes from the string line to the tow points on the paver. The tow points move up or down changing the angle of attack of the screed. The string line provides the most consistent reference for the paver tow points, enabling a predetermined grade to be matched very accurately.

Skis and shoes are mobile reference points that are attached to the tractor portion of the paver. The skis (Figure 6.7) vary in length and can be up to 12 m long. Shoes perform a similar function as skis, but are significantly shorter, usually around 0.5 m. It is used only when the grade being sensed is relatively smooth, such as a previously paved layer. The ski or shoe travels along with the paver and smoothes out the variations in the existing pavement. A string or wire is installed taut on top of the ski and the grade wand follows the changes in elevation of this reference point. The greater the length of the ski, the smoother the pavement. The short shoe is used mostly to match up to an existing joint, such as a longitudinal joint on an existing adjacent lane. Since the shoe is short, it will not reduce major variations in the pavement surface. Shoes are often used in parking facility construction, so that all the longitudinal construction joints that are made are matched up and give an aesthetically pleasing surface. Extensive smoothness is not as critical in a parking facility, due to their rather low speed operation. An appearance of seamless construction is desirable for aesthetic reasons and also impairs the introduction of water through the joints. To prevent water

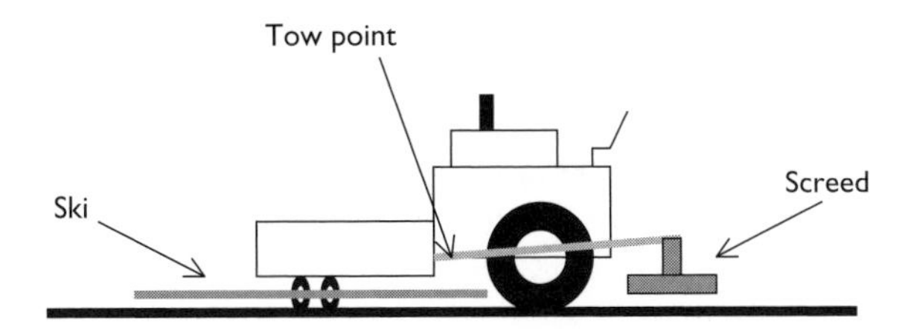

Figure 6.7 Paver with a mobile ski for grade control.

intrusion into the subgrade, joints should be offset from each other when placing multiple layers. The offset practice is similar to installing a roof on a house.

Compaction

Compaction is the mechanical process of increasing the density of a material. It compresses a material from a given volume to a smaller volume. Force and movement is exerted over a contact area, causing the particles within the material to move closer together. The voids between the particles, whether they hold air or water or both, are expelled by the combination of force and movement. During the construction of an asphalt pavement, the soil subgrade, aggregate bases, and asphalt mixtures are the materials that are compacted. The strength of all these materials increases when their density increases. An asphalt pavement must have enough density to support heavy loads; otherwise it will distort and lose its original shape when loads are placed on it. Aggregate, soil, and asphalt mixtures are made denser by reducing the voids between the particles that make them up. In time, loose materials will settle or compact themselves naturally. By applying various mechanical forces, the time to achieve this compaction is greatly reduced. Compaction is the least expensive element in extending the life of an asphalt road or parking lot (Caterpillar 1990b). Compaction also provides a smooth sealed riding surface. An asphalt mixture is compacted by one or more of the following type of forces: static pressure, impact, vibration, or manipulation. These forces are applied to an asphalt mixture by a mechanical roller or compactor.

Compaction variables

Density is the weight of a material that occupies a certain volume of space. Density is increased by reducing the volume of space while maintaining the same weight of material. Compaction reduces the volume of space in by reducing the amount of air voids in the material. The air void content of an asphalt mixture is the volume of spaces between the asphalt binder coated aggregate. The theoretical maximum density of an asphalt mixture contains no air voids. The theoretical maximum density of asphalt mixtures is usually measured by a laboratory test such as ASTM D2041, *Theoretical Maximum Specific Gravity and Density of Bituminous Paving Mixtures* or it can be calculated from the percentages and specific gravities of each component in the asphalt mixture. Since the actual air void content of an asphalt mixture cannot be measured directly, a ratio of the unit of the compacted asphalt mixture to its theoretical maximum density is used to express both air voids and the density of the mixture. Percentage of density is the compacted density of the asphalt mixture divided by its theoretical maximum density. Air voids are also expressed as a percentage by subtracting the percentage of density from 100. For example, if the compacted density of an asphalt mixture is 2,355 kg/m^3 and its maximum theoretical density is 2,467 kg/m^3, the compacted density can be expressesd as (2,355/2,467) $\times$ 100 or 95.5 percent density. The corresponding air void of the mixture content is 100 minus 95.5 percent or 4.5 percent (TRB 2000).

Several factors contribute to the ability of an asphalt mixture to densify or become compacted. These factors include:

- Mixture properties
- Mat thickness
- Mixture temperature
- Air and base temperature
- Wind speed
- Base support.

The properties of the mixture that affect compaction are the aggregate angularity, surface texture, and aggregate shape. The resistance of the asphalt mixture to internal movement defines the workability of the mixture. Cohesion and internal friction make up the internal resistance to movement and compaction of an asphalt mixture. There are three components of resistance to internal movement (Whiteoak 1991):

- The cohesion of the asphalt mixture which is influenced by the amount and type of asphalt binder and added mineral filler.
- The internal friction from the aggregate, such as grading, shape, surface area, and surface texture.
- The "viscosity" of the asphalt mixture, which is influenced by the viscosity of the asphalt binder and the internal friction provided by the aggregate and mineral filler.

Highly angular and rough surfaced aggregates are more difficult to compact than smooth and round aggregates. A cubical shaped aggregate is also more difficult to compact than round aggregate. Mixture gradations that are open graded, gap graded, or very fine (tender) are more difficult to compact than a dense graded (continuously graded) mixture. Tender mixtures are difficult to compact because they tend to move around or shove under the wheels of the roller. The asphalt binder used in the mixture also contributes to the ability of the mixture to be compacted. When compacted at the same compaction or laydown temperature, a more viscous asphalt binder is more difficult to compact than a soft asphalt binder. For example, a mixture containing an AC-20 asphalt binder will be more difficult to compact than a mixture containing an AC-10. The amount of asphalt binder in the mixture can also affect compaction. At high temperatures, the asphalt binder acts as a lubricant and allows the aggregate particles to compact easily. The asphalt binder also fills some of the air voids. However, like soil compaction, the density of the mixture will increase with increasing asphalt binder content up to the point it displaces aggregate particles and then the density will begin to decrease. The mixture will also become very tender and unstable and will shove under the roller's wheels, inhibiting the compactive effort of the roller. As a general rule, mixtures that are designed to be rut resistant or are for very high loading pavements, are more difficult to compact than mixtures intended for low volume pavements or parking facilities.

The higher the temperature asphalt mixture, the easier it is to compact. This phenomenon holds true up to the point that the temperature of the asphalt mixture is so high that the properties of the asphalt binder are entirely fluid and the mixture moves around or shoves under the roller wheel. The mixture needs enough internal resistance or cohesion to allow the particles to displace vertically (compact) instead of horizontally. As the mixture temperature decreases, the asphalt binder becomes more viscous, increasing the internal resistance or cohesion. As the mixture's temperature approaches 80 °C, density will become very difficult to achieve. Minimum mixture compaction temperatures are usually set at 80 °C, while the maximum temperature depends on the properties of the asphalt binder, but usually is in the range of 130–170 °C. The mixture temperature is measured about 15 mm deep into the pavement. Measuring just the surface temperature of the pavement will give lower temperature values that are not representative of the entire mixture temperature.

The ambient and the base temperatures affect compaction from the point in how they contribute to how quickly the mat cools in temperature. The temperature of the base or layer below the asphalt mixture will act as a heat sink and draw heat from the mixture. Wind speed can also increase the cooling effect of the ambient temperature. The greatest contributor as to how fast or slow the asphalt mixture is depends on how thick the lift or layer of asphalt mixture being placed is. Lifts that are 50 mm or less are susceptible to compaction problems during cool weather since the mixture temperature will drop very quickly. For example, for a mixture laydown temperature of 121 °C and a base temperature of 5 °C, a 25 mm thick mat will cool to 80 °C in 4 min. As the thickness of the lift being compacted increases, the time available to complete compaction increases. For a lift that is 50 mm thick under similar conditions, it will take 10 min for the mat to cool to 80 °C (TRB 2000). Paving thin lifts in cool weather will require special attention to obtaining the required density as soon as possible. Small increases in the mixture discharge temperature at the plant and increase in the compaction roller size or the number of rollers can assist in overcoming cool weather compaction problems.

Lift thickness and the nominal maximum aggregate size also influence the ability of rollers to compact the asphalt mixture to the desired density. With regard to compactive effort, the lift thickness should be at least three times the nominal maximum aggregate size. However, single lift thicknesses greater than 200 mm are difficult to compact thoroughly and have a tendency for surface irregularities (Whiteoak 1991). The base underneath also has to have enough strength to provide a stable platform for the compaction equipment without yielding under pressure.

Compaction equipment

There are four types of rollers used for compacting asphalt mixtures: static, vibratory, pneumatic, and combination. A static roller uses static pressure to compact a material. For many years, the steel wheel static roller was the standard roller used for compacting asphalt mixtures. Steel wheel static rollers are configured either in a two-wheel or three-wheel configuration. Static rollers use dead weight that is provided by ballast for compacting the asphalt mat. The steel wheels or drums are filled with ballast, usually water or sand, to provide the static pressure force along the entire length of the drum. The amount of ballast and the contact area of the drum with the asphalt mat determine the compactive effort of the static roller. Steel wheel static rollers lack the ability to adequately compact mats that are thicker that 50 or 75 mm. They also do not seek out soft spots in the mat to do their inherent bridging characteristics. On roadway projects, steel wheel rollers are used mostly to finish rolling the asphalt mixture after vibratory or pneumatic compaction has achieved the desired density. The "ironing" effect of a steel wheel roller smoothes and eliminates previous drum marks. Finish rolling with a static roller must be done while the mat is still hot enough to allow a slight movement of the particles to erase irregularities (Caterpillar 1990b). The temperature of the mat during finish rolling is usually around 50–70 °C. Steel wheel static rollers vary in size from small units (Plate 6.34) to units that weigh 10–12 tonnes (Plate 6.35). The small units are used as primary compaction rollers on small projects such as driveways or small parking lots. Changing the ballast weight and the speed of the roller are the only compaction variables that an operator can change on a steel wheel static roller. Steel wheel rollers, whether they are static or vibratory are equipped with water spray systems to keep the drums covered with a small film of water to prevent the drums from picking up material during the compaction process. Figure 6.8 illustrates the operation of a steel wheel static roller. The drive wheel or drum should always face the direction of the paving operation. The drive drum pulls or tucks the asphalt mixture under the drum, reducing shoving or displacement (Hyster Company 1986). If the non-drive wheel or drum of the roller is operated close to the paver, material shoving may occur immediately in front of the non-drive drum. This shoving will require additional passes to achieve the required density and increases the potential for hairline cracks to occur in the surface of the mat.

A pneumatic or rubber tired roller (Plate 6.36) uses static and manipulation forces to compact a material. Pneumatic rollers are used for both breakdown and initial compaction and for intermediate compaction. A pneumatic tire roller compacts an asphalt pavement through tire pressure and ballast. Depending on the number of plies in the tire's construction, tire pressures range from 400 to 800 kPa. Six ply tires are usually limited to 400 kPa,

Plate 6.34 Small steel wheel roller.

Plate 6.35 Tandem static steel wheel roller.

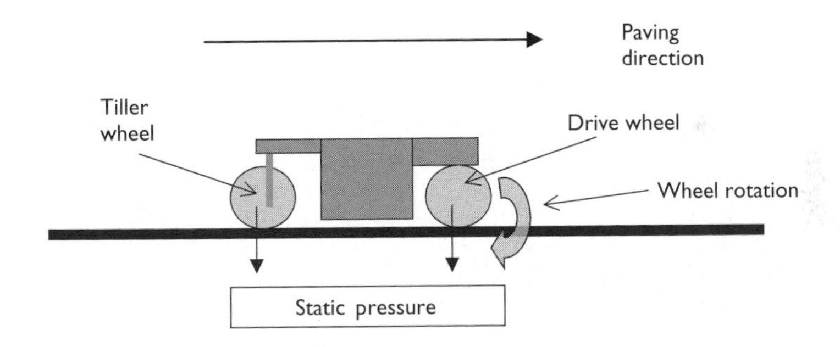

Figure 6.8 Steel wheel static roller operation.

while 12 ply tires can be inflated to 800 kPa. Instead of the wheels being filled with ballast, like the steel wheel static roller, the frame of the rubber tire roller is weighted with ballast. The total weight of rubber tire rollers is usually 13 tonnes or greater. They are also used to seal the surface of the mat through surface manipulation or kneading by the rubber tires. Rubber tire rollers usually have a total of nine tires with five in the back of the roller (Plate 6.37) and four in the front. The front set of tires and the rear set of tires are offset from each other to ensure complete coverage. The rubber tires exert a kneading action as well as varying amounts of ground pressure. These characteristics help the rubber tire roller manipulate the mat under and between its wheels in a confined manner. The result is a more stable and tighter finish to the surface of the mat.

Plate 6.36 Rubber tire roller.

Plate 6.37 Tire configuration.

The inflation pressure of the tires is a key component in the proper operation of the rubber tire roller. The higher the tire pressure, the smaller the contact area and the higher the compactive forces transmitted to the mat. During breakdown and intermediate compaction, the tire pressure is kept at a high pressure to achieve the most compactive effort possible. Overinflation, however, can lead to grooves being left in the mat. Large tires (greater than 7.50 × 15) are the most effective for use in breakdown rolling. Reducing the tire pressure causes the bottom of the tire to flatten out and thus the ground contact pressure is reduced. The compactive effort by the tire is now reduced, however the kneading or manipulation action by the tire is increased (Caterpillar 1990b). This feature is useful to seal the surface of the mat and is used when finish rolling by a rubber tire roller.

The sticking of the asphalt mixture to the rubber tires is reduced by water spray systems, scrapers, and by keeping the tires hot. Once the tire's surface becomes the same temperature as the asphalt mixture, sticking is practically eliminated. The tires are kept hot by the use of skirts (Plate 6.38) surrounding the tires. The tires will get hot by contact with the hot asphalt mixture. When the tires are cold, such as at the beginning of a paving operation, the tires can be sprayed with a truck bed release agent to keep the asphalt material from sticking during the warm up period. A water spray system for the rubber tires will also reduce the sticking of the asphalt mixture to them.

A vibratory roller (Plate 6.39) uses a combination of static and impact forces. The compactive effort caused by static forces comes from the weight of the roller frame and its drums. The compactive effort caused by impact forces is produced by a rotating eccentric weight inside each of the drums. As the eccentric weight rotates around a shaft located inside the drum, a dynamic force is produced. The dynamic force is proportional to the eccentric

Plate 6.38 Roller skirts.

Plate 6.39 Vibratory roller.

movement of the rotating weights and the speed of rotation. Changing the eccentric movement or adjusting the eccentric mass has a directly proportional effect on the dynamic force. Vibratory rollers come in several different configurations. Single drum vibratory rollers can have either a rigid or articulated frame. Single drum rollers are used on soil and granular base preparation and are not frequently used for asphalt mixture compaction. Double or tandem drum vibratory rollers can come with rigid, single articulated, or double articulated frames. Vibratory rollers can be operated in any of the three different modes of operation: static with none of the eccentric weights rotating (vibrating), one drum vibrating and one drum static and with both drums vibrating (TRB 2000).

The frequency and the amplitude of the vibrations produced by the vibratory roller are the two components of the dynamic forces used to produce the compactive effort (Figure 6.9). Both of these components are used to classify or evaluate the compactive effort of the roller. Specific elements of evaluation for the dynamic component of the vibratory roller are the magnitude of the centrifugal force of the rotating eccentric weight, its vibrating frequency, the nominal amplitude and the ratio of the vibrating and non-vibrating masses acting on the drum. The amplitude is defined as the weight of the drum divided by the eccentric moment of the rotating weight and is a function of the weight of the drum and the location of the eccentric weights. Amplitude values range from 0.25 to 1.0 mm. Some rollers have multiple amplitude settings while others will operate only at one fixed setting. The actual amplitude that enters in the material being compacted varies somewhat from the amplitude setting due to the variation in material densities. An increase in the amplitude value or setting will increase the compactive effort applied to the material. For a given vibration or frequency,

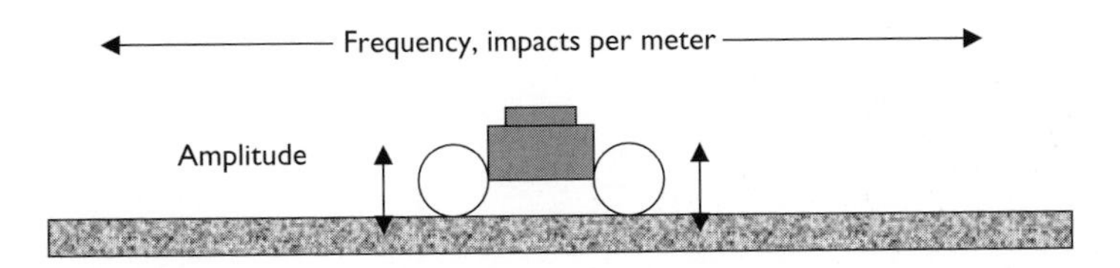

Figure 6.9 Frequency and amplitude of a vibratory roller.

changing the amplitude setting has a proportional effect on the dynamic force applied to the material. The effectiveness of the amplitude setting depends on the thickness of the material being compacted. Compacting thin mats on a high amplitude setting will cause the drums to bounce on the mat. These can lead to the mat shoving with an actual loss in density (TRB 2000). The aggregates in thin mats have the potential to fracture if compacted under a high amplitude setting. A vibratory roller compacting a thin mat on a high amplitude setting also has the potential to damage shallow utilities including water or gas lines. Mats that are 30 mm or less should be compacted either in a static (vibrators off) or on a low amplitude setting. Mats that are greater than 30 mm thick should be compacted on either a low or medium amplitude setting. Materials that are greater than 60 or 70 mm thick should be compacted on a high amplitude setting.

Vibration frequency is the number of complete rotations per minute of the eccentric weights. The faster the rotation of the weights, the greater the frequency or number of vibrations (impacts). Depending on the type of vibratory roller, vibrations can vary from 1,500 to 4,000 vpm. On asphalt mixtures, the number of vibrations or frequency should be at least 2,000 vpm. This ensures that an adequate compactive effort is applied to the mixture without introducing ripples or roughness into the surface of the mat. On asphalt mixtures, the vibratory roller should be operated at the highest frequency possible. Spacing of the impacts depends on the roller travel speed and the number of vibrations. As frequency decreases and roller speed increases, the number of impacts per meter will decrease. The largest number of impacts per meter that is attainable without severely sacrificing paving operations is most desirable. If the vibratory roller is operating at the highest frequency possible, then the roller speed will be the determining factor for the number of impacts per meter. The number of impacts per meter should be in the range of 30–40. The greater the number of impacts per meter, the greater the compactive effort provided by the vibratory roller. A range of 30–40 impacts per meter provides a balance that offers sufficient compactive effort without operating the roller at a travel speed so slow that paving productivity is reduced (TRB 2000). Impacts per meter are determined by dividing the roller speed by the frequency of vibrations (equation 6.2).

$$\text{impacts per meter} = \frac{\text{frequency (impacts/min)}}{\text{roller speed (meters/min)}} \tag{6.2}$$

The frequency of vibrations can be determined by a gauge installed on the vibratory roller or by the use of a reed tachometer. The reed tachometer can measure the vpm by placing it either on the drum of the roller or on the asphalt mat next to the roller, while it is being compacted. Using equation 6.2 and the desired number of impacts per meter can set a maximum roller speed.

Plate 6.40 Combination roller.

Example 6.5

Determine the maximum roller speed permitted if 30 impacts per meter are required and the vibratory roller is operating at a frequency of 2,100 vpm.

Solution

Using a form of equation 6.2:

$$\text{roller speed} = \frac{\text{roller frequency}}{\text{desired impacts per meter}} = \frac{2{,}100}{30} = 70\ \text{m/min}$$

A test strip is normally constructed at the beginning of the project to determine what compactive effort can be expected from the rollers and to determine the maximum roller speed, amplitude setting, and the number of passes by the roller to achieve the required density.

A combination roller has both pneumatic and vibratory steel wheels. It usually has the steel wheel or drum in the front and a set of four pneumatic tires in the rear (Plate 6.40). The concept of this roller is to provide both types of compaction, vibration and kneading, in a single pass of the roller.

The rolling pattern

The rolling pattern for the rollers includes the number of passes by each roller, the location of the first pass, the sequence of succeeding passes, and the overlapping between passes.

A pass is defined as a trip of the roller in one direction over any point on the pavement surface. The speed of the rollers is also determined during the establishment of the rolling pattern. The speed of vibratory rollers is governed by the frequency they are operating at and generally should not exceed 100 m min. Static steel wheel roller speeds should not exceed 150 m min and pneumatic roller speeds should not exceed 180 m/min. The speed of the rollers and the speed of the paver should be synchronized so that neither the paver pulls away from the rollers, nor the rollers are forced to sit idle, waiting on the paver. The number of passes of the roller required to achieve the required density is determined by the construction of a test strip at the beginning of the project. The desired breakdown roller is selected and is used for the construction of the test strip. After each pass of the breakdown roller, the density of the mat is determined and recorded. The number of passes required to achieve the desired density is then used to establish the rolling pattern. The objective of the rolling pattern is to adequately cover the entire mat with the required number of passes and to meet the required density. The mat width divided by the roller drum width determines the number of passes required to completely cover the entire mat surface.

Example 6.6

Determine the number of passes by a roller for a 5 m wide mat being paved for a section of a parking lot pavement. Four passes of a vibratory roller that is equipped with 1.8 m wide drums are required to achieve the specified density.

Solution

First determine how many passes are required to completely cover the mat. The mat width is divided by the roller drum width.

$$\frac{5\text{ m}}{1.8\text{ m}} = 2.8\text{, or three passes of the roller}$$

The mat width is divided into three sections with each section being rolled with four passes to achieve the required density. A total of 12 passes of the roller will compact the entire mat to the required density.

Breakdown rolling should achieve 91–96 percent of the final density required (The Asphalt Contractor 2001). Sometimes an intermediate roller is also used to achieve the required density in addition to the breakdown roller. The intermediate roller is usually a pneumatic roller. An intermediate roller is useful in reducing the number of passes required to be completed by the breakdown roller. The intermediate roller will generally only make one or two passes to bring the mat up to the required density. The intermediate roller needs to be included during the analysis of the test strip. The number of passes required to meet a specific density requirement depends on the mixture type and temperature, roller types, and environmental conditions. Using two rollers to achieve the required density allows the paver to operate at a higher speed and production rate.

The rolling pattern establishes the number of overlapping passes that are needed to compact the entire mat width to the required density. The roller should operate from the inner edge of the pavement surface and roll towards the outside edge. The inner edge is defined as the edge facing the median or inside of the roadway. Otherwise using nautical terms, the

Plate 6.41 Combination roller approaching paver.

roller should begin compacting on the "port" edge of the pavement and work its way towards the "starboard edge." The first movement begins on the port edge and rolls towards within 5–10 m of the paver (Plate 6.41). The second movement reverses in the same path until the roller has reached the previously compacted material. At this point the roller should swing over and move forward along path number three, again going as close as possible behind the paver. The fourth movement is reversal in the third path and a repetition of the previous operation. To maintain a constant interval between the paver and the roller, each roller reversal point is ahead of the previous one. After the entire width of the mat being placed has been covered in this manner, the roller should swing across the mat to the port side and repeat the process. With this type of pattern, the lap of the roller with succeeding passes does not need to be more than 75–100 mm (The Asphalt Institute 1992). Figure 6.10 illustrates a typical rolling pattern.

When rolling lifts greater than 75 mm thick, the roller should first start rolling 300–400 mm from the lower unsupported mat edge. This initial uncompacted edge provides some initial confinement during the first compaction pass, minimizing lateral movement of the mixture. After the center portion of the mat has been compacted, the mixture will now support the roller and allow the edge to be compacted without lateral movement.

The breakdown roller can be used to obtain all of the required density or it can be used in conjunction with an intermediate roller to obtain the final density. The intermediate roller is usually a pneumatic roller, which helps seal the mat surface through the kneading action of

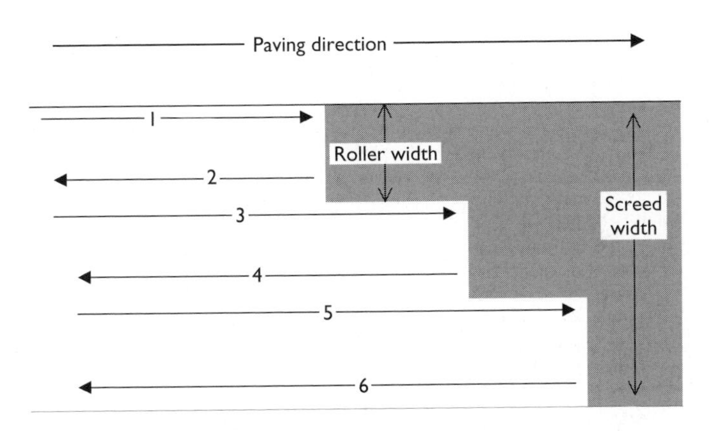

Figure 6.10 Typical rolling pattern.

the rubber tires. The pneumatic roller provides pressure and manipulation that reorients the aggregate in the mixture (Deahl 2002). A vibratory or a static roller can also be used as an intermediate roller. Usually only one or two passes of the intermediate roller (in addition to the passes of the breakdown roller) is required to finish compacting the pavement. Any further passes have the potential to damage the pavement, especially when using a vibratory roller. Intermediate rolling is usually done when the temperature of the mat is approximately 80–90 °C.

Finish rolling is used to remove any marks left by the breakdown of intermediate rollers. Finishing rolling is not used to add any additional density to the pavement. Tandem steel wheel static rollers or vibratory rollers operating in a static mode do the finish rolling of the asphalt mat. The temperature of the mat is less then 80 °C and usually in the range of 50–70 °C. The finish roller completes just enough passes to remove any marks in the mat.

Smoothness

Pavement roughness is the deviations of a pavement surface from a true planar surface. Pavement smoothness is the lack of roughness. After finish rolling the surface smoothness of the pavement is checked. A straightedge, usually in the length of 3–4 m is used. The straightedge is used to check all transverse joints and other visible surface irregularities. It is placed on the surface, parallel to the centerline or longitudinal joint, and the maximum deviation between the pavement surface and the bottom of the straightedge. A maximum deviation of 3–4 m per 3 m of straightedge is desirable. For highways or other large pavement areas, smoothness is often checked with a profilemeter or roadmeter. Two common devices for measuring pavement smoothness are the California Profilograph and the Mays Road Meter. Both devices can measure the smoothness of a pavement over large areas. In highway applications, smoothness is often given as a deviation per mile or kilometer. A maximum smoothness value of 4 inches per mile or 250 mm per km can usually be accomplished with most paving equipment.

Rolling guidelines

The following guidelines should be used during the establishment of a rolling pattern for lifts or layers that are 100 mm thick or less:

- Paving in a single lane width or full screed width, the mat should be rolled in the following order:
  1. Transverse construction joints.
  2. Outside mat edge.
  3. Breakdown rolling, beginning on the low side of the mat and progressing towards the high side.
  4. Intermediate rolling, if being used.
  5. Finish rolling.
- When paving using two pavers side by side or in echelon, or abutting a previously placed lane or other lateral restraint such as a curb and gutter, the mat should be rolled in the following order:
  1. Transverse construction joints.
  2. Longitudinal construction joints.
  3. Outside mat edge.
  4. Breakdown rolling, beginning on the low side of the mat and progressing towards the high side.
  5. Intermediate rolling, if being used.
  6. Finish rolling.

  The following guidelines should be used during the establishment of a rolling pattern for lifts or layers that are greater than 100 mm thick:
- Paving thick lifts in a single lane width or full screed width, the mat should be rolled in the following order:
  1. Transverse construction joints.
  2. Breakdown rolling, beginning 300 to 400 mm from the lower supported edge of the mat and progressing towards the high side.
  3. Outside edge: when rolling within 300 mm of the unsupported edge, the roller should advance towards the outside edge in 100 mm increments.
  4. Intermediate rolling, if being used, should begin on the low side of the mat and progress towards the high side.
  5. Finish rolling.
- When placing thick lifts in echelon or when abutting a previously placed lane or other lateral restraint, the mat should be rolled in the following order:
  1. Transverse construction joints.
  2. Longitudinal construction joints.
  3. Breakdown rolling, beginning at the longitudinal joint or lateral restraint and progressing towards the outside edge.
  4. Outside edge: when rolling within 300 mm of the unsupported edge, the roller should advance toward the edge in 100 mm increments.
  5. Intermediate rolling, if being used, should begin on the low side of the mat and progress towards the high side.
  6. Finish rolling.

(The Asphalt Institute 1987)

Roller operating techniques

Consideration of the following operating techniques during the compaction of the asphalt pavement will improve the overall quality of the pavement:

- To provide for adequate pavement density, the rollers should be used efficiently while the asphalt mixture is still above the minimum compaction temperature.
- The time available for compaction is related primarily to the mat thickness and temperature. An increase in either will substantially increase the time available for compaction.
- A decrease in the speed of the rollers, especially vibratory rollers, will increase the compactive effort applied to the material being compacted.
- The breakdown and intermediate rollers should be operated as close together as possible and as close to the paver as possible. The laydown temperature and the internal friction of the mixture will control how closely the rollers can operate to the paver. Some mixtures can behave as tender mixtures when they are in the temperature range of 95–115 °C.
- A vibratory roller should operate at the highest frequency possible when compacting an asphalt mixture.
- If the rollers can not keep up with the speed of the paver, more rollers should be used or the paver should be slowed down. Equation 6.1 is beneficial in relating paver speed to roller speed.
- The rollers should be kept operating as much as possible. If the roller is required to stop, it should stop or park on a diagonal to the mat.
- The wheels or drum of the roller should not be turned while the roller is stopped on the mat. The potential for the mat to tear or distort usually occurs while the roller is stopped.
- On vibratory rollers, always turn off the vibrators when the roller is approaching a stop or change in direction. This will prevent the drums from bouncing in one spot and distorting the mat. Some rollers are equipped with automatic shut off systems for the vibrators.
- Rollers should operate in a straight back and forth motion. Breakdown rollers should turn at a slight angle as it comes to the last meter before stopping before the paver. This procedure helps remove any distortions or marks that may have been pressed into the mat during the starting, stopping, and reversing actions of the roller.

Density requirements

The proper compaction of an asphalt mixture is a critical step in the construction of an asphalt pavement. Compaction is necessary in order to prevent the excessive intrusion of air and water into the pavement. When the inplace air voids of dense graded asphalt pavements exceed 8 percent, the voids tend to become interconnected. Interconnected air voids allow for the circulation of air and water throughout the pavement, leading to reduced pavement life. Excessive air moving through an asphalt pavement will accelerate the aging process of the asphalt binder, causing it to become brittle, reducing the overall durability of the asphalt pavement. Excessive water moving through the pavement can lead to stripping of the asphalt binder film from the aggregate. The intrusion of air and water can be greatly reduced if the inplace air voids of an asphalt pavement are less than 8 percent. Air voids are also less likely to be interconnected when dense graded pavements are compacted to less than 8 percent. Some asphalt pavements, such as OGFC or porous asphalt are purposely designed with inplace air voids in the range of 15–20 percent. These types of mixtures consist of a coarse aggregate structure coated with a thick film of asphalt binder. The thick film of asphalt binder helps overcome the effects of air and water on the OGFC, allowing it to perform satisfactorily under these harsh conditions. Sometimes OGFCs also include the use of an anti-stripping agent or additive to help resist the damaging effects of water. These types of mixtures require just a small amount of compactive effort from rollers. Only two or three passes of a roller compacting in a static mode is required to provide a finished lift or layer (The Asphalt Institute 1998b).

It would appear that the ultimate goal of compaction is to try to achieve zero air voids or compact to "refusal" density. This is not sure, because as dense graded asphalt pavements become compacted to approximately 3 percent air voids or less, they tend to become unstable under traffic loading, which leads to rutting and flushing (asphalt binder floating to the surface) of the asphalt pavement. This is especially true under high traffic applications. Just as asphalt material properties and design are tailored to the traffic loading of the pavement, the inplace air voids or compactive effort must also be tailored to traffic loading. High density (low air voids) is a desirable attribute for an asphalt pavement that will be used on a low volume parking lot, where the environment can cause more damage to the pavement than traffic loading. Mixtures that are designed for high traffic or loading conditions will be more difficult to compact than mixtures that are designed for low volume roads or the typical parking lot application.

Density of an asphalt pavement can be measured by cutting a core out of the asphalt pavement and measuring its density in the laboratory or it can be measured immediately through the use of a nuclear density gauge. The nuclear gauge measures density by transmitting gamma rays from its radioisotope source into the pavement and measures the amount that is reflected back during a set amount of time. The returning rays are proportional to the density of the material the gauge is measuring. ASTM D2950, *Standard Test Method for Density of Bituminous Concrete in Place by Nuclear Methods,* is a reference that describes the use of the nuclear density gauge. The amount of inplace air voids is determined by comparing the inplace density determined by the core or nuclear gauge with the maximum theoretical density or specific gravity of the mixture that is determined in the laboratory. The maximum theoretical specific gravity is also known as the Rice specific gravity (Chapter 5). Most construction density requirements are expressed as a percentage of the maximum theoretical density of the asphalt mixture. Typical construction density requirements are in the range of 92–94 percent of the maximum theoretical density. This specification would give an after construction inplace air void content of the pavement in a range of 6–8 percent. The application of the traffic loading over several years should give a final inplace air void content of 3–5 percent. Table 6.10 is an example of typical density requirements based on traffic level.

During the construction of a roller test strip, a nuclear gauge is used to measure the density after each pass of the roller. The rate of the density increase caused by each pass can also be determined. When no appreciable increase in density is achieved by additional roller passes, the maximum relative or maximum field density for that mixture has been achieved. Some density requirements are specified as a percentage of the maximum relative density. Using the air void content determined from the maximum relative density, a percentage of target density can be specified, ensuring an adequate range of air voids in the compacted asphalt pavement. Construction density can also be specified as a percentage of the compacted laboratory specimens. The compacted laboratory specimens are usually samples of the

Table 6.10 Density requirements as a function of traffic levels

	Traffic, million ESALs (20 years)				
	<0.3	0.3 *to* <3	3 *to* <10	10 *to* <30	≥30
	Typical minimum requirement				
Percent of maximum theoretical density	94	93	93	92	92

mixture in production and are compacted to an air void content of 3–5 percent, with a target of 4 percent. The compactive effort by the rollers is to achieve an inplace density that is about 98 percent of the laboratory compacted density.

The density value determined with a nuclear gauge is relative and will not be identical to the density value obtained from cores cut from the asphalt pavement. A correlation between the cores obtained from the pavement and the matching nuclear gauge readings can be made. A number of locations are randomly selected, usually in the range of 10–20, to perform the correlation between the cores and an individual nuclear gauge. A best-fit statistic correlation is determined and used to convert nuclear determined densities to cores densities. The significance is that some density specifications are tailored around core densities and that cores are used for referee density determination during specification disputes or interpretation. Cores have also been used in the past as the final acceptance of the density of the asphalt pavement, but are commonly being replaced by the nuclear density gauge due to its speed and ease of use.

The amount of density tests required is usually a variable in lots of the amount of tonnes or square meters of mixture placed. A length of compacted section can also be used in determining the amount of tests of required. Using the nuclear gauge allows for the immediate correction of any problems or alternations of the rolling pattern if the required density is not being achieved. A random sampling of five density spots or locations per lot is a typical requirement. The random sampling is used to determine the specific location within the lot to perform the density test. When using the nuclear gauge, five tests or readings across the mat are considered one test or spot. The five readings are averaged to give one test value. In other words, if five density tests per 150 m of pavement are specified, 25 actual nuclear readings will be taken five for each test. If cores are used for density acceptance, typically one or two cores are required per lot. If tonnes of mixture laid are used to specify the frequency of tests, lot sizes are usually between 200 and 500 tonnes, with one test being required per lot. Once the rolling pattern has been established and the mixing plant is producing a consistent asphalt mixture, the mat density should not vary significantly. A density specification should state the minimum required density and the number of lots that require testing. The lot size will also be included in the specification.

Joint construction

Pavement joints are the seams between previously placed pavements. The construction of asphalt pavements usually requires the use of two types of joints. A transverse joint is constructed whenever the paving operation is interrupted for a period of time; anywhere from an hour to several weeks or more. The transverse joint can also be constructed at the beginning and end of the paving project. The longitudinal joint is constructed when a lane or section of asphalt pavement is placed adjacent to a previously placed asphalt pavement. Joints can allow for the penetration of water and air into the pavement, so they must be constructed and compacted with care. Joints will also cause deviations in the pavement smoothness, so the minimum amount of joints should be constructed and they should be compacted flush with the adjacent pavement.

Transverse joints

A transverse joint can occur at the beginning or end of any point in the paving operation. It is perpendicular to the direction of the mat being placed. Two types of transverse joints are used in the construction of asphalt pavements (Figure 6.11):

- Butt joints are constructed if traffic will not be passing over the end of the pavement before paving is resumed. Butt joints are joints with a vertical face. Butt joints are commonly used in new construction, especially parking lots.
- Tapered joints are used if traffic will be traveling on the pavement before paving is resumed. The taper allows for the transition of traffic to the layer below. The tapered portion is removed when paving resumes, leaving a vertical joint for the new lift to butt up against.

All transverse joints used in asphalt pavement construction are actually a type of butt joint. More specifically however, a butt joint is a joint where traffic will not be allowed to travel over the unsupported edge of the unfinished joint. A taper joint allows traffic to travel over it through the use of a temporary ramp that is formed to protect the vertical edge of the unfinished joint (NAPA 1990). The temporary ramp is the taper in the taper joint. The taper is then removed prior to the resumption of paving, leaving a vertical face or butt joint.

Transverse joint construction

The paver should be run right up to the point where the transverse joint will be constructed. The head of material in front of the screed should remain as consistent as possible up to the location of the joint. The forces acting on the screed should be constant so that a consistent

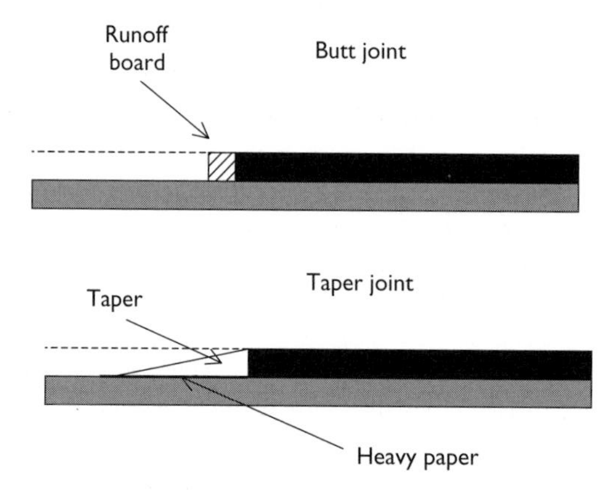

Figure 6.11 Transverse joints.

angle of attack will be maintained for the screed. The goal is to have the same thickness at the joint as has been previously laid, otherwise a surface deviation will occur at the joint. If at all possible, the transverse joint should be located while the head of material is constant or normal. In other words, the transverse joint should not be located where the hopper would empty out at the end of the day. If this occurs, the head of material in front of the screed will decrease, causing the screed angle of attack to decrease, resulting in a thinner mat at the location of the transverse joint. The thinner mat will translate into a surface deviation or dip in the asphalt pavement at the joint. If the joint is made where the head of material is constant, the paver screed is raised up at the point where the transverse joint will be constructed. All of the mixture in front of the joint will need to be removed, except for the taper, if constructing a tapered joint. A vertical edge is formed by temporarily pushing aside the asphalt mixture in front of the intended location of the joint. Placing a strip of heavy paper, such as kraft paper, down in front of the paver will facilitate the removal of the taper when paving is resumed. The width of the paper is the width of the layer being paved while the length should be about 1 m. The asphalt mixture that was previously removed is then shoveled back on top of the paper and a ramp is formed with a lute (a type of construction rake). When constructing the ramp, the required lift thickness should be continued for about 150 mm past the location of the joint to ensure that the full compacted lift thickness will be attained at the joint (NAPA 1990). What should be left is gradual taper leading down to the layer below. The taper allows traffic to be allowed on the pavement until paving is ready to resume. Taper joints are mostly used on highways and roads and sometimes on parking facilities undergoing rehabilitation. Taper joints are also used when placing the mainline portion of the parking lot prior to the construction of the entrances and approach aprons. This allows vehicles to use the facility during construction. When paving is resumed, the taper is removed mechanically with the assistance of the kraft paper, leaving a vertical or butt joint. An application of a tack coat or rubberized crack filler is applied to the face of the vertical joint to help ensure a tight joint. The paver is brought on the previously placed layer close to the joint. The screed is placed up on starting blocks that have a thickness that is 1.25 times the desired compacted lift thickness. For example, if the compacted design lift thickness is 100 mm, the blocks should be about 125 mm thick. Placing the screed up on these blocks prior to the commencement of paving helps ensure that there will not be a dip at the joint. This procedure should be applied during the construction of any type of transverse joint. During the startup of asphalt paving, when there is no joint to place the lift up against, the starting blocks should be the same thickness as the uncompacted lift thickness being placed. A proper head of asphalt mixture is fed to the screed prior to being moved off the starting blocks. The paver then begins moving and brought up to the correct paver speed (NAPA 1990).

Butt joint construction follows the same method as the taper joint construction, with the exception of the taper or ramp construction. Butt joint construction also involves the placement of boards or planks, known as runoff boards, on the unpaved surface immediately adjacent to the joint location. One board is placed perpendicular to the pavement at the joint location to form a vertical face for the butt joint. The thickness of the boards should be about the same thickness as the desired thickness of the compacted lift. The other boards must be wide and long enough to support the full length of the roller being used. These boards allow the roller to pass completely over the asphalt mixture at the joint and onto the boards before reversing the rolling direction and performing the normal rolling pattern (NAPA 1990). This procedure ensures the proper compaction is applied to the joint. The boards are then removed and when paving is resumed the same starting procedures of the paver used for taper joints should be applied. When possible the roller should operate transversely across the width of the mat at the transverse joint. The roller has about 150 mm of its drum width on the newly laid lift. This operation should be repeated with successive passes each covering an additional 150–200 mm of the new lift until the entire drum width is on the new lift (The Asphalt Institute 1998b). This practice helps ensure proper compaction at the joint. For most practical purposes this operation is difficult to complete due to traffic on the adjacent pavement or other similar constraints. This procedure can be applied easily to new construction, especially parking lots. Usually compaction at the transverse joint occurs normally with the roller operating parallel to the joint. Breakdown rolling should begin as soon as the paver has moved away from the joint, allowing sufficient room for the operation of the rollers.

Longitudinal joints

Longitudinal joints are seams between lanes or screed widths of material being placed. In highways or roads they occur next to the previously paved lane or existing pavement. When paving large spans of pavement, such as runways or parking lots, the joint is between each screed width of pavement being placed. The joint is parallel or longitudinal to the direction of paving. When constructing a pavement, the longitudinal joint of successive lifts should not be on top of each other. The longitudinal joint formed should alternate or overlap about 25–50 mm from the longitudinal joint below (Figure 6.12). This practice is similar to overlapping joints during the construction of a roof and helps ensure that water does not directly flow through the joint to the subgrade below. This practice also provides the required amount of material on top of the joint to allow for proper compaction of the joint with minimal handwork such as luting or raking. During roadway construction, the joints are staggered between layers such that the final longitudinal joint in the surface course is in the center of the roadway. The mixture located at the longitudinal joint almost always is lower in density than the surrounding mat. It is typically 2 percent less than the density in the surrounding mat. This defect is due mostly to the unconfined edge that occurs during the placement of the first mat or lane. The unconfined edge has a tendency to move laterally during compaction and not provide enough internal resistance for proper compaction. Low densities at longitudinal joints can lead to raveling of the mixture at the joint and water intrusion into the pavement. The low density can cause excessive oxidation of the asphalt binder, leading to cracking at the longitudinal joint.

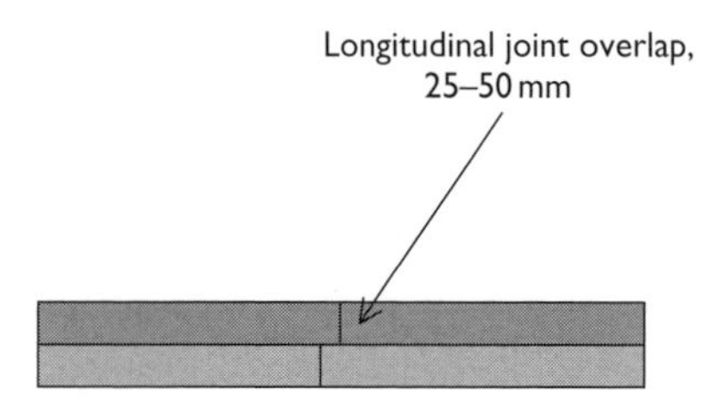

Figure 6.12 Longitudinal joint overlap.

There are two major types of longitudinal joints:

- Hot joints that occur when two or more pavers are paving in echelon or parallel to each other at the same time. The pavement being placed is hot on both sides of the newly formed longitudinal joint.
- Cold joints occur when the newly placed mixture is placed next to a previously placed lane, layer, curb, or other restraint. The pavement is hot on one side of the longitudinal joint and cold on the other side.

The construction of the cold longitudinal joint is the most common. The first lane or pavement section is placed, compacted, and allowed to cool. At some later time the companion lane or section is placed against it and compacted.

Hot longitudinal joints are usually made during the construction of large spans of asphalt pavements, such as in runways, automobile raceways, and large parking lots. Brand new highway construction or when the entire roadway can be shutdown to traffic can also utilize hot longitudinal joint construction.

Cold longitudinal joint construction

The construction of a tightly sealed cold joint first requires the first paved layer to be placed in a straight or uniform line. There are three types of cold joints (Figure 6.13):

- Vertical joint
- Wedge joint
- Notched wedge joint.

The vertical joint is formed during the normal paving process. It is the unsupported edge that is on either side of the mat. As the adjacent mat is paved next to the unsupported edge,

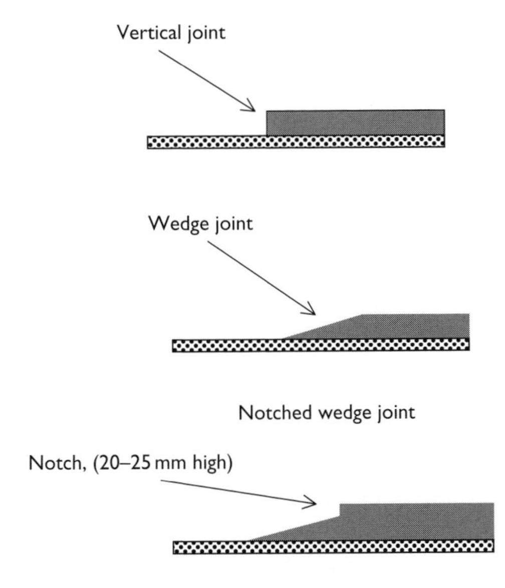

Figure 6.13 Longitudinal joint types.

the longitudinal joint is formed. It is essential that the paver operates in as straight or consistent line as possible. This provides a crisp edge for the adjacent mat to be placed against. An application of a rubberized crack sealer or an emulsion tack coat can be applied to the vertical edge of the joint prior to the placement of the adjacent mat. This helps ensure a waterproof joint. Proper compaction of the unsupported edge is provided by beginning compaction of the hot side of the joint about 100–200 mm away from the edge. The roller then progresses towards the unsupported edge. During the placing of the adjacent mat, the level of the asphalt mixture at the longitudinal joint must be above that of the previously compacted (cold) mat by an amount equal to approximately 6 mm for each 25 mm of thickness of the compacted mat. When the adjacent mat is placed against the now cold joint, most of the compactive effort should be on the hot side of the joint, with only 100–200 mm of the roller drum on the cold side of the joint.

The wedge joint is thought to provide good density and with the tapered edge, is somewhat safer for vehicular traffic than the drop off of the vertical joint (Wallace 2000). Differentials in density on both sides of the longitudinal joint can lead to cracking and/or raveling of the asphalt mixture at the joint. The wedge joint is formed during the first pass of the paver by attaching a metal form to the end gate of the screed. The degree of the slope of the wedge is typically from 6:1 to 12:1. The wedge should taper down to a thickness that is about the equivalent to the nominal maximum aggregate size of the mixture being placed. The New Jersey Department of Transportation has developed that wedge joint with a slope that is 3:1 with the thought that the steeper slope will reduce the occurrence of raveling of the joint. This type of steeper slope joint is also known as the New Jersey wedge (Kandhal *et al.* 2002). The wedge is usually compacted by the use of a small roller that trails the screed and is just wide enough to cover the wedge. The wedge is then overlapped when the adjacent mat is placed. Heat from the newly placed mat transfers down to the wedge below, allowing a slight amount of further compaction of the wedge when the adjacent mat is compacted. Due to the taper or slope of the wedge, the wedge joint may not be suitable for thin lifts unless the nominal aggregate size in the asphalt mixture is very small. Tearing of the wedge needs to be avoided.

The notched wedge joint is similar to the wedge joint, with the addition of a vertical notch placed in the wedge joint. The notch is 18–25 mm^2 in depth from the top of the lift into the beginning of the taper of the wedge joint (Figure 6.13). The notched wedge joint is constructed in a similar manner as the wedge joint, using a 150×300 mm steel plate attached to the paver screed and is dragged through the freshly placed mat, creating the notch and wedge. Due to the vertical face or notch in the notched wedge joint, the joint will tie into the adjacent mat without having to “feather” the mixture into the adjacent mat. The notched wedge joint does not work well with thin lifts. A clearly defined notch and a uniform taper are difficult to achieve with small lift thicknesses. The notched wedge joint should be used on lifts that are 50 mm or greater in thickness (Wallace 2000).

Cul-de-sac paving

A cul-de-sac is a street or roadway that is closed at the end. A cul-de-sac is shaped as a circle placed at the end of the street. It is used in residential subdivisions and smaller commercial properties. It also has select usage in parking facilities. Paving operations generally consist of paving roadways or parking lots in straight lines. The layout is rectangular and when curves are encountered, they consist of gradual arcs or transitions. The paving of

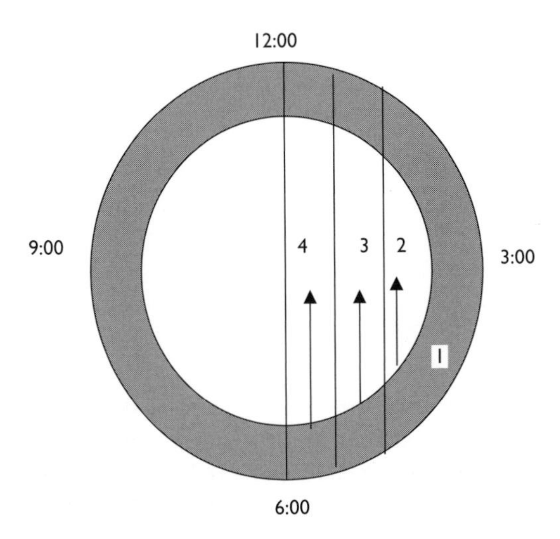

Figure 6.14 Cul-de-sac paving layout.

a cul-de-sac, for all practical purposes consists of paving of a circle. Particular attention should be placed properly mapping out the movements of the paver prior to construction. The cul-de-sac should be mapped out in a pattern of a clock.

Since a cul-de-sac is shaped as a circle, the circumference of the circle must be determined. The circumference measurement determines the number of paver passes and the width of each pass. Half of the circumference is the diameter and that is where the paver should begin, on one half of the cul-de-sac. The paver should begin the first pass on the outside of the circle or cul-de-sac and begin its way towards the center of the circle. In other words, the first half of the circle is paved from the outside in, followed by the second half of the circle, also paved from the outside in. Laying out the circle or cul-de-sac as a clock (Figure 6.14) assists in keeping the tonnage required for the cul-de-sac correct. The cul-de-sac is first marked with a line from noon to 6 o'clock. Paving will begin at the 6 o'clock mark, depending on the location of drainage structures (inlets, grates, etc.). The 6 o'clock mark can be rotated away from a drainage structure if necessary. A longitudinal line is then marked every 3 m (or use the actual paver screed width) dividing up the circle. To begin paving, the paver operator begins paving at 6 o'clock, facing 3 o'clock. The first pass is laid counterclockwise along the outside edge of the circle. The paver stops at 2 o'clock for the finish of the first pass. The paver then moves down (in transport mode, not paving) to 4 o'clock, again facing the top of the clock. This second pass goes in a straight line up to about the 1:30 area of the clock. The third pass with paver then sets down on the next marked line, pulling away from around the 5 o'clock area up to the 1 o'clock area. The fourth pass then sets on the next marked line, pulling around the 5 o'clock area up to the 12:30 area. The final pass of this side of the cul-de-sac begins back down at the 6 o'clock area and paves up towards noon. Once the first half of the cul-de-sac is paved, the process is repeated on the second half (Ball 2001). The operation of the roller follows a similar operation as the paver, with repeat passes for compaction completed in the same fashion as for the construction of a roadway.

Chapter 7

Specialty asphalt mixtures

Dense graded mixtures make up the greatest percentage of all the type of mixtures used for asphalt pavements. They also are the greatest tonnage of all the mixture types produced. Dense graded mixtures are also generally known as hot mix asphalt (HMA). Dense graded mixtures are very versatile and can be used both for the structural component of the pavement and for the wearing course. They can also be used in the maintenance of asphalt pavements as thin overlays, also known as "scratch coats" or thin surface courses. Various specialty types of mixtures have been developed for specific pavement applications and can provide better performance than dense graded mixtures for those specific applications. There are many types of specialty mixtures and not every one is described in this chapter. The more common and relevant mixtures that a designer for a parking lot or city or county road may consider are described. The distribution of aggregate sizes in the asphalt mixture typically defines the type of asphalt mixture. Dense graded mixtures contain a uniform distribution of aggregate particles of different sizes throughout the mixture. Open graded friction courses (OGFCs), popcorn mixtures, or porous mixtures consist mostly of larger aggregate sizes with hardly any of the smaller particles. There are also various types of gap graded mixtures with several different names. Gap graded mixtures are somewhat similar to dense graded mixtures with the exception that several sizes of the aggregate particles are missing, leading to a "gap" in the gradation of the mixture. Various components that have been incorporated into asphalt mixtures for specific purposes are for enhancements in the engineering properties of the asphalt mixture. For example, a dense graded asphalt mixture that incorporates recycled asphalt pavement (RAP) has both economical and environmental advantages over a mixture that does not contain RAP. Or an asphalt binder may be modified with one of the several types of polymers to provide improved performance under high temperatures and loading.

Stone mastic asphalt

Stone mastic asphalt (SMA) was developed in Germany in the late 1960s as an effort to resist the detrimental effects of tires equipped with winter driving studs. In Germany it is known as "*Splittmastixasphalt.*" Studded tires have the tendency to tear the small aggregate particles out of the typical dense graded wearing course. The use of studded tires was banned in Germany, however SMA pavements proved to be very durable and rut resistant. It is the most common road surfacing material in use in Germany today. Variants of the mixture have also been adopted in other countries including Sweden, Denmark, the Netherlands, Belgium, France, Switzerland, Japan, Australia, and the United States (Department of

Transport 1999). In the United States, SMA mixtures are named "Stone Matrix Asphalt" to prevent it from being confused with mastic. In the United States, mastic is typically used for roofing applications and may contain coal tar or other derivatives (EAPA 1998). SMA is a high stone content mixture that has a gap gradation. An SMA mixture consists of coarse aggregates and a mastic of fine aggregate, mineral filler, and asphalt binder. The mastic fills the voids between the coarse aggregates. The coarse aggregates interlock to form a stone skeleton. The asphalt binder is often reinforced with stabilizing agents such as small fibers or polymers. The stabilizing agents help keep the SMA mixture homogenous and prevent the drain down of the asphalt binder during transport or laydown. The gap grading and high asphalt binder content contributes to the problem of drain down in SMA mixtures. SMA has very high strength due to its stone skeleton. Because of the stone skeleton, an SMA mixture can provide a very stable and durable rut-resistant wearing course for highway applications. The mastic provides a thick film around the coarse aggregate particles. This film thickness provides help to ensure a durable mixture. The coarse aggregate must be durable and hard to provide the SMA's high wearing resistance property. The stone skeleton in the SMA mixtures causes the surface of the mixture to have a significant "negative" texture. A negative texture is a texture with voids in the surface and no exposed aggregate protruding from the surface. A protrusion from the surface lends to the surface having a "positive" texture. A negative texture is beneficial from a wearing course standpoint in that it is skid resistant and generates lower tire noise than a wearing course with a positive or neutral texture. Proper aggregate selection is key to the proper performance of SMA pavements.

Stone material asphalt mixtures are typically not used for parking facility pavements. The high wearing resistance is generally not required for parking lots and it is not as aesthetically pleasing as a dense graded wearing course. It can also cause some difficulty

Plate 7.1 Open negative texture.

for pedestrian traffic, especially in high-heeled shoes. SMA mixtures are more expensive to produce than dense graded mixtures, due to their high asphalt binder content, highly processed aggregates, and additional ingredients such as fibers or polymers. Parking facilities that undergo heavy loading such as truck terminals or industrial traffic such as forklifts can benefit from the use of an SMA mixture for the wearing course.

The cross-section of an SMA wearing course will show almost all single sized coarse aggregate particles bonded together by the mastic. The single sized aggregate forms the stone skeleton. At the bottom and middle of the wearing course or layer, the voids in the coarse aggregate are almost entirely filled with the mastic. At the top or surface of the layer, the voids are only partially filled with the mastic, giving the surface of the layer a coarse or open texture (Plate 7.1). This coarse texture provides for excellent skid resistance properties for highway applications. The texture is one of the properties that usually exclude SMA from use in parking facilities or other low speed applications.

SMA design

Stone mastic asphalt pavements are designed such that the interlocking stone skeleton carries traffic loading. An SMA mixture is proportioned such that when it is fully compacted, the voids in the stone skeleton exceed the volume of the mastic used by 3–5 percent. The mastic alone is almost voidless. The stone skeleton makes up over 60 percent of the volume of the SMA mixture. About 35 percent of the remainder volume is the mastic, leaving between 3 and 6 percent air voids. The mastic itself is composed by volume of approximately 37 percent sand, 26 percent mineral filler, and 37 percent asphalt binder (Nicholls 1998). The aggregate grading for SMA mixtures is gap graded. Table 7.1 and Figure 7.1 illustrate a typical SMA grading in the United States.

Table 7.1 Typical SMA grading (United States)

Sieve size	*Total passing (%)*
12.5 mm	>95
4.75 mm	<30
75 μm	8–10

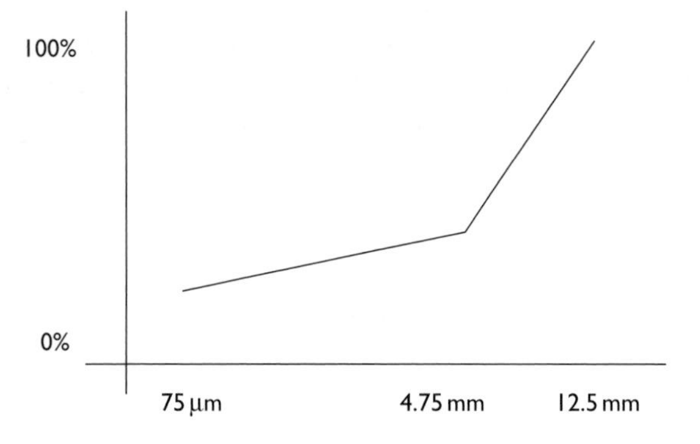

Figure 7.1 Typical gap graded SMA mixture.

Stone mastic asphalt mixtures are mostly used for wearing course applications. The texture of an SMA mixture provides excellent skid resistance for high speed traffic applications. The stone on stone contact provides resistance to distortion and deformation. An SMA mixture's high resistance to deformation is also a benefit at traffic intersections and parking facilities that may experience high surface stresses such as ship loading terminals, truck terminals, and warehouses with large amount of forklift traffic. The higher expense of SMA mixtures is usually not justified for use as base or binder courses, where not all of the benefits of an SMA mixture will be utilized. SMA mixtures have been beneficial as base courses on facilities that undergo heavy slow moving traffic. A truck or ship loading terminal would be one such facility. Since an SMA mixture is a wearing course mixture, the nominal maximum top size of the mixture is usually 19 mm or less and typically 12.5 mm or less.

Typically, there are five steps involved in designing an SMA mixture:

1 Select the coarse aggregate, fine aggregate, mineral filler, asphalt binder, and stabilizing agents.
2 Determine the proper gradations or blends of aggregates that will achieve stone on stone contact that creates the stone skeleton.
3 Validate which gradations will meet any required voids in the mineral aggregate (VMA) requirements.
4 Determine the optimum asphalt binder content for the desired air void content.
5 Evaluate the drain down potential and moisture susceptibility of the final mixture.

Most SMA mixtures are designed using the Marshall mix design method, while an increasing amount of designs are now incorporating the SuperPave® gyratory compactor instead of the Marshall compaction hammer. The mixture design method used for SMA mixtures is no different than what method may be used for any other gap graded mixture. An SMA mixture is designed to have a void content level of 3–6 percent. Most dense graded or continuously graded mixtures are designed to have a void level content of 3–5 percent. The quality of the coarse aggregate is very important to the performance of the stone skeleton in the SMA mixture. The shape of the coarse aggregate is preferred to be cubical with a limit on flat and elongated particles. The coarse aggregate should be crushed with at least one fractured face. Table 7.2 gives some typical quality criteria for the coarse aggregate in SMA mixtures.

Aggregates are typically blended to meet the gradation requirements for SMA mixtures. The gradation requirements will be for the entire SMA mixture, the coarse aggregate, and the mastic. Aggregate blending is accomplished in the same manner that is used to blend aggregates for dense graded mixtures. The mastic is a fine aggregate, usually a crushed manufactured sand, and mineral filler. An SMA mixture will consist of a least one coarse aggregate, a crushed sand, and some mineral filler to meet the required minus 75 μm

Table 7.2 Coarse aggregate quality for SMA mixtures

Test criteria	*Typical specification*
Flat and Elongated Particles, ASTM D4791	≤5%
Toughness, Los Angeles Abrasion, ASTM C131	≤30%
Angularity, ASTM D5821	≥90%, two fractured faces
Durability, Soundness, ASTM C88	≤15%

Table 7.3 Typical SMA gradation specification

Sieve size	*Total passing (%)*
19.0 mm	100
12.5 mm	85–95
9.5 mm	≤75
4.75 mm	20–28
2.36 mm	16–24
600 μm	12–21
300 μm	11–17
150 μm	9–15
75 μm	8–12

content. Mineral filler should meet the same requirements as that used for dense graded mixtures. Table 7.3 is a typical gradation specification for SMA mixtures.

The asphalt binder grade selected for SMA mixtures can follow the same method used for dense graded mixtures. The grade selected should match the applicable traffic and climate conditions. Any of the methods discussed in Chapter 3 for selecting asphalt binder grades would be acceptable.

Stabilizing agents would be incorporated into the mixture also at the time of the SMA mixture design. The most common stabilizing agents used for SMA mixtures is cellulose fiber. Mineral fiber can also be used instead of cellulose fiber. A dosage rate of fiber of 0.3 percent by weight of mixture should provide for acceptable drain down protection. The length of the fiber should be less than 6 mm. There are several commercially available fibers that are specifically made for application in SMA mixtures. Polymers are also used as SMA stabilizing agents, but are not as common as fibers. Polymers that have been used are styrene–butadiene–styrene (SBS), styrene–butadiene–rubber (SBR), ethylene vinyl acetate (EVA), and natural rubber. The polymers are not as effective as fibers for reducing drain down in SMA mixtures. The polymers are generally used to modify the performance of the asphalt binder and have secondary use as a stabilizing agent (Nicholls 1998). A polymer modified asphalt binder may be used in conjunction with fibers in an SMA mixture.

An SMA mixture design can be completed by following the same principles and procedures used in either the Marshall or SuperPave mixture design method. Sample preparation can also follow the same methods. The mixture design criteria used for design verification and determining the optimum asphalt binder content will differ. Table 7.4 can be used for the optimum asphalt binder selection using either the Marshall compaction hammer or the SuperPave gyratory compactor. Various highway agencies have already developed guidelines and specifications for SMA mixture design and those specifications should be used if relevant to the geographical location and the application.

There is doubt if the Marshall stability and flow tests have significant relevance for determining the optimum asphalt binder content. Volumetric properties determined during the design stage should be the ultimate determining factor for the asphalt binder content, while Marshall stability and flow tests are useful for limited performance verification of the SMA mixture. The amount and composition of the mastic in the SMA mixture determines the majority of its Marshall stability and flow characteristics. In some countries, a Marshall quotient is used for performance verification. The Marshall quotient is the Marshall stability divided by the Marshall flow. A Marshall quotient specification can be easily developed

Table 7.4 Guidelines for SMA mixture design using the Marshall procedure

Design parameter	*Design traffic level, ESALs*	
	10,000–1,000,000	>1,000,000
Number of hammer compaction blows per side of specimen	50	75
Mixture air voids, P_a (%)	2–4	3–4
Stability, kN	4,500	6,000
Flow, 0.25 mm	8–16	8–14
VMA (%)	16	17
VFA (%)	70–90	70–85

from Marshall stability and flow specifications by using the maximum and minimum values of each and calculating the Marshall quotient specification range. The Marshall quotient can be increased by any of the following:

- Reducing the amount of mastic in the mixture
- Increasing the stiffness of the asphalt binder in the mastic
- Increasing the stiffness of the mastic by increasing the mineral filler content
- Increasing the stiffness of the mastic by increasing the angularity of the fine aggregate.

Completing any of these changes will also change the volumetric properties of the mixture. The asphalt binder content should again be verified, ensuring that the mixture meets the requirement volumetric properties.

Some specifications include a minimum asphalt binder content for the SMA mixture. A common minimum content specification is 6 percent or more, depending on the gradation. A high asphalt binder content is required for SMA mixtures to be durable. VMA, is another method of ensuring a minimum asphalt binder content. By increasing the VMA content and keeping the air void level, the compaction and the gradation constant, an increase in the asphalt binder content will occur. The higher asphalt binder content will give the aggregate a thicker film thickness, ensuring a more durable pavement.

The susceptibility of the SMA mixture to drain down its asphalt binder needs to be verified. The final design selected is proportioned in the laboratory for testing. Several procedures have developed by various highway agencies for predicting the potential of the asphalt binder to drain away from the aggregate during transport and laydown. These procedures can usually be applied to any type of mixture including SMA and open graded or porous mixtures. ASTM D6390, *Standard Test Method for Determination of Draindown Characteristics in Uncompacted Asphalt Mixtures,* is one test that involves placing a heated SMA mixture into a basket made of 6 mm sieve cloth. The basket and mixture are then placed in an oven for 1 h at the specified plant production mixing temperature. At the end of 1 h, the amount of asphalt binder draining from the mixture is determined. A typical maximum amount of drain down permitted during this test is 0.3 percent. Another drain down test is the Schellenberg/Von Weppen test. This test involves storing 1 kg samples of the SMA mixture in 800 ml beakers for 1 h at 170 °C. At the end of 1 h; the SMA mixture is tipped out by rapidly unending the beaker. The loss in mass resulting from the SMA mixture adhering to the beaker is determined. A typical maximum amount of loss permitted during this test is 0.3 percent. The fiber stabilizing agents generally prevent any type of asphalt binder drain down (Nicholls 1998). The moisture sensitivity of the SMA mixture can be determined by the same methods used for dense graded mixtures. One of the more popular methods is ASTM D4867 or AASHTO T-283. These methods involve comparing the indirect tensile strengths of unconditioned and conditioned samples through the use of a tensile strength ratio.

SMA construction

Construction practices used for dense graded mixtures will perform satisfactorily with SMA mixtures with a few exceptions. The mixing of the SMA mixture is conventional with the exception of the use of large amounts of mineral filler and the addition of the fiber. The large dust or mineral filler content (over 8 percent) of SMA mixtures requires a higher addition rate of the mineral filler than what would be used for dense graded mixtures. This consideration should be made when determining the size of the mineral filler silo and the source of the mineral filler. The addition of the fiber into the mixture at the mixing plant must also be considered. In batch mixing plants, the fiber is generally added through a separate inlet directly into the weigh hopper above the pugmill. The addition of the fiber should be timed to occur during the hot aggregate charging of the weigh hopper. Pre-weighed fiber packaged in melting bags can also be added through the use of a conveyor system going into the weigh hopper. Additional dry mixing time may be required to ensure a homogenous dispersion of the fibers throughout the mixture. The dry mixing time is typically increased for an additional 5–15 s. The wet mixing time should also be increased by an additional 5 s to ensure that the fiber is adequately blended with the asphalt binder. In a drum mixing plant, a separate fiber feed system is used to add loose fiber by blowing it into the drum at the required production rate (NAPA 1994). Fiber that has formed into pellets through the use of a small amount of asphalt binder is also commercially available. The pellets tend to handle easier when used with a pneumatic feeder for a drum mixing plant.

Transporting the SMA mixture to the project site requires no additional procedures or equipment than what would be needed for dense graded mixtures. There is some minimum layer thickness guidelines to prevent tearing of the SMA mixture by the paver screed and to allow room in the layer for the stone on stone contact of the mixture. Table 7.5 gives some minimum thickness guidelines for SMA mixtures. An SMA mixture should be delivered to the paver at a temperature of around 150 °C. This high temperature assists the paver screed in providing almost all of the compactive effort required for the SMA mixture. Rolling is accomplished immediately after placing using static rollers that have a weight of 9–12 tonnes.

Table 7.5 Minimum layer thickness for SMA mixtures

	Nominal maximum top size (mm)		
	19.0	12.5	9.5
Minimum thickness range (mm)	50–75	37.5–50	25–37.5

Due to the stone on stone contact in an SMA mixture, vibratory compaction should not be used. Pneumatic tire rollers can have excessive pickup of the mastic out of the mixture and should be used with caution. The SMA mixture should be compacted to a minimum of 94 percent of its maximum theoretical density. Once the SMA mixture is compacted, stone skeleton is interlocked and provides a high resistance to deformation or tearing.

Open graded or porous asphalt

Porous asphalt is another type of gap graded asphalt mixture. The dominant use of porous asphalt is, as a wearing course and is usually placed in layers that are 25 mm thick or less. Porous asphalt also has application as asphalt treated permeable base mixtures. Porous asphalt differs from dense graded and SMA mixtures in that it contains no fine material, only single size, usually large aggregate pieces. This allows for a high percentage of air voids in the mixture. The amount of air voids can be in the range of 17–22 percent of the volume of the mixture. The air voids in the mixture are designed to be interconnected, allowing the water to drain freely. The lack of fine material in porous asphalt allows the voids to become interconnected. The lack of fine material also creates a negative texture on the surface of a porous asphalt wearing course (Plate 7.2). This negative texture provides for a highly skid resistant pavement. Porous asphalt is usually known as OGFCs in the United States and some other countries. Porous asphalt is the common terminology in Europe and Asia. The terms "porous asphalt" and "open graded friction course" can be used interchangeably. Porous asphalt is typically used as a high performance, skid resistant wearing course and has been in use since the 1950s.

Plate 7.2 Porous asphalt texture.

Porous asphalt was developed as a special purpose wearing course that drains water away from the pavement surface and absorbs tire noise. The porous property of the mixture is its dominant feature in Europe (NAPA 1993). In the United States, OGFC, involved from experimentation with mixing plant produced seal coats in some western states during the 1960s. The cover aggregate used for seal coats or chip seals is typically 9.5–12.5 mm maximum nominal size and is mixed with a relatively high amount of asphalt binder in an HMA mixing plant. It is placed 15–20 mm thick. This plant mix seal coat, also known as a "popcorn mixture" offered the same benefits as a conventional chip seal with an additional benefit in a reduction in chip or stone loss. This plant produced seal coats or popcorn mixtures also offered a high degree of skid resistance. The term OGFC was adopted in the United States (NCAT 2001). The advantages of porous asphalt or OGFC are as follows:

- The presence of air voids in the mixture allows surface water to quickly drain below the pavement surface, offering a significant reduction in water spray and an increase in visibility. The water drains vertically through the porous asphalt down to usually an impermeable dense graded mixture below the wearing course. The water then travels laterally (Figure 7.2) to the exposed or "daylighted" edges of the porous asphalt wearing course (Plate 7.3, Plate 7.4).
- The mixture offers improved road safety over dense graded wearing courses by reducing the occurrence of hydroplaning.
- Headlight glare is reduced due to lack of water on the pavement surface. The open or negative texture of porous asphalt also reduces headlight glare.
- Porous asphalt or OGFC has a negative surface texture, reducing high-speed tire noise.

Similar to SMA mixtures, the largest value of porous asphalt is on high-speed roads and is typically not used as a wearing course on low volume roads or parking facilities. Porous asphalt typically does not have enough stability or cohesive strength to withstand significant amounts of tire turning movements, which are typical to parking facilities. The disadvantages of porous asphalt include the following:

- The large amount of air voids in the mixture can become filled with dirt and debris and cannot be easily cleaned out. Due to dirt filling these air voids, the predicted life for porous asphalt is about 8–10 years.
- In the wintertime, the porous asphalt surface temperature may be 1–2°C less than pavements without a porous asphalt wearing course. This lower temperature is due to the porous asphalt containing a large amount of air voids. A larger amount of deicing chemicals is required on porous asphalt due to its quicker freezing.
- A dangerous situation known as "mushrooming" can occur when freezing conditions follow rain. Water in the voids of the porous asphalt expands when it freezes, pushing up the pavement surface. Deicing chemicals are not effective on mushrooming. Prevention is by ensuring the water completely drains away from the mixture.
- Porous asphalt or OGFC has no significant structural strength like dense graded or SMA mixtures. During thickness design, it should not be considered as any contribution for the overall thickness required for the pavement structure. Additional thickness that would be contributed by a dense graded or SMA wearing course would need to be added to the lower layers or lifts.
- Porous asphalt generally has a higher cost than dense graded mixtures.

(Rushmoor Borough Council 1998)

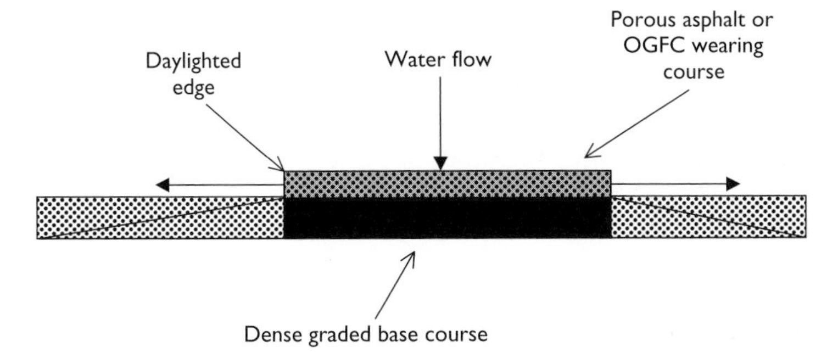

Figure 7.2 Water flowing through porous asphalt.

Plate 7.3 Porous asphalt wearing course with "daylighted" edges. Also note the rumble strip that is pressed into the shoulder.

Plate 7.4 Daylighted edges.

For high-speed applications and in climates with heavy rainfall, the advantages of porous asphalt usually exceed the disadvantages of the mixture. Porous asphalt should be placed on pavement that is sound and structurally sufficient.

Porous asphalt design

Similar to other types of asphalt mixtures, the selection of suitable materials and their proper proportioning is key to the performance and life expectancy of porous asphalt or OGFC. Porous asphalt is designed to have a significantly higher amount of air voids than a dense graded or SMA mixture. The high amount of air voids ensures that they are interconnected throughout the pavement structure and allow water to rapidly pass through. A lack of fines or mastic in porous asphalt allows room for the interconnected voids.

The compactive effort applied to porous asphalt is just enough to "seat" the mixture into the pavement layer below. The mixture design method for porous asphalt is to select the asphalt binder content that provides some desirable film thickness. The lack of fine aggregate or mastic in porous asphalt can lead to durability problems for the mixture and this is compensated for by large asphalt binder film thicknesses on the aggregate. The drain down of the asphalt binder during construction is also a concern for porous asphalt. This is compensated for in a similar fashion as for SMA mixtures by either using a stiffer asphalt binder or adding fibers. Fibers are added to porous asphalt at a rate of 0.15–0.5 percent by weight of mixture. Polymers or ground tire rubber can be added to the asphalt binder at a rate of 5–7 percent by weight of asphalt binder to reduce the occurrence of drain down during mixture transporting or construction.

The aggregate selected for porous asphalt depends on the application of the mixture. Porous asphalt is typically used as a wearing course on high-speed traffic facilities so the aggregate will possess some skid resistant properties and be highly durable to wear. The aggregate should have a maximum Los Angeles abrasion loss (ASTM C131) of 40 percent. Very high traffic facilities may require a maximum abrasion loss of 25 percent. A crushed aggregate with a least one crushed face should be used for porous asphalt. The crushed faces create some interlock among the aggregate particles, giving the mixture some resistance to deformation. Porous asphalt generally does not provide any structural strength to the pavement, especially when used as a wearing course. The gradation of the aggregate provides the water draining characteristics of the mixture. Porous asphalt is a gap graded or single particle sized mixture (Figure 7.3). There is no significant amount of fine aggregate or mastic used in the mixture. Table 7.6 illustrates a typical gradation for porous asphalt when used as a wearing course.

The gradation of porous asphalt can be tailored to the wearing course thickness and is illustrated in Table 7.7. A small amount of minus 75 μm material or mineral filler is included in the porous asphalt gradation. This gives a small amount of cohesive strength to the mixture, helping with placement and compaction and providing some stability under loading. Being that the mixture is gap graded, there is usually stone on stone contact throughout the gradation.

The asphalt binder grade selected for porous asphalt usually corresponds to a similar grade that would be used for dense graded mixtures in a similar loading and temperature environment. Due to concerns of drain down of the asphalt binder the grade of the binder is

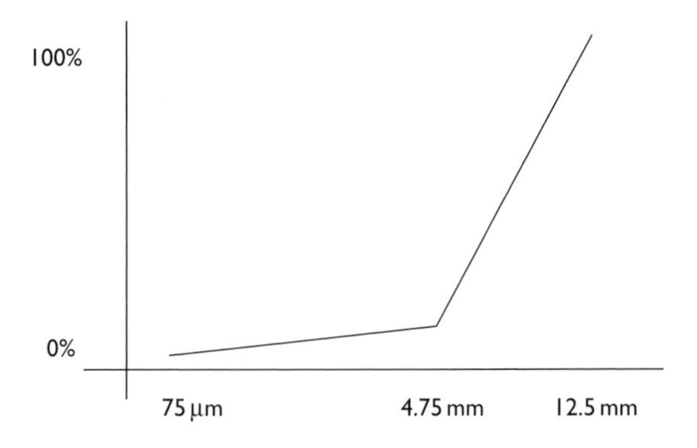

Figure 7.3 Open graded or porous asphalt mixture.

Table 7.6 Typical porous asphalt gradation

Sieve size	*Total passing (%)*
12.5 mm	100
9.5 mm	95–100
4.75 mm	30–50
2.36 mm	5–15
75 μm	2–5

Table 7.7 Porous asphalt gradation specification

Sieve size	*Total passing (%)*	
	19 mm thick wearing course	*25 mm thick wearing course*
19.0 mm	100	100
12.5 mm	100	70–100
9.5 mm	80–100	35–75
4.75 mm	25–40	25–40
2.36 mm	10–20	10–20
600 μm	3–10	3–10
75 μm	2–5	2–5

"bumped" or increased one grade or the binder is polymer modified. Fibers can also be used to reduce the potential of drain down. The cohesive strength of the porous asphalt can be increased by using polymer or tire rubber modified asphalt binders. Under the SHRP grading protocol, a PG70-xx or PG76-xx is usually specified for use in porous asphalt.

There are several methods available to the designer for determining the optimum asphalt binder content in porous asphalt. The determination of the proper asphalt binder content is a balance of having enough asphalt binder to provide adequate film thickness, but not so much that there is excessive drain down during construction and an unstable mixture. One method is simply a comparison of the asphalt binder "blots" or stains left by mixtures with varying asphalt binder contents. A visual comparison is made to standard blots of mixtures that have performed satisfactorily in the past. Another method takes steps from the Hveem dense graded mixture design method by determining the proper asphalt binder content on the basis of the surface capacity of the aggregate. The surface capacity, or K_c value, of the aggregate retained on the 4.75 mm sieve is determined from the percent of SAE 10 engine lubricating oil retained by the aggregate. The determination of K_c is described in the centrifuge kerosene equivalent (CKE) test method and can be completed by following ASTM D5148. A surface constant is derived and an optimum asphalt binder content is calculated. This test is valid for aggregates with an apparent specific gravity in the range of 2.60–2.80 and water absorption of less than 2.5 percent. The asphalt binder content determined by this method is inversely proportional to the specific gravity of the aggregate used (United States Army Corps of Engineers 2001).

A more common method of determining the optimum asphalt binder content in porous asphalt is to use the same methods for asphalt binder content determination in dense graded or SMA mixtures. Either the Marshall or the SuperPave method will give satisfactory results. The Marshall stability and flow values are not used and the asphalt binder content is determined solely on the volumetric properties of the porous asphalt mixture. The amount of compactive effort applied on the mixture samples is the equivalent of 20–50 blows of the Marshall hammer, with 35 being typical. The compactive effort by the SuperPave gyratory compactor is 50 or less. Establishing the acceptable compaction values is mostly trial and error, with a balance of the required asphalt binder film thickness and the desired air void content. The porous asphalt design and in-place air void content should be in the range of 17–22 percent, with 18 percent considered somewhat optimum. Sample preparation would follow the similar manner as that used for dense graded mixtures, with varying asphalt binder contents used and the optimum asphalt binder content selected that corresponds with an air void content of 18 percent.

Once the desired asphalt binder content is determined, a sample of the porous asphalt is made to determine if the mixture is susceptible to drain down. The same tests or methods for determining the drain down potential in SMA mixtures can be used for porous asphalt mixtures. The maximum drain down of the asphalt binder permitted is 0.3 percent by weight of the mixture. The drain down test is not used to adjust the asphalt binder content but rather to determine if fibers or polymers are required to adjust the porous asphalt mixing temperature (NAPA 1993). The mixing temperature should correspond to the temperature that gives the selected asphalt binder viscosity in the range of 170–190 centistokes.

Porous asphalt should be tested for moisture susceptibility due to its high air void content (USDOT 1990). Tests used for dense graded mixtures, such as ASTM D4867 can also be used for porous asphalt. Porous asphalt mixtures generally include the use of an antistripping additive as additional protection against moisture damage. The Cantabro abrasion test can be used to confirm if the porous asphalt mixture is durable and resistant to moisture damage. The test involves tumbling compacted 100 mm diameter porous asphalt specimens in a steel drum for 5 min. The test is similar to the Los Angeles abrasion test without the use of the steel shot. The loss in mass due to abrasion is measured. The test can also be completed after the specimens have been immersed in water for a period of time. The purpose of this procedure is to assess the ability of the porous asphalt mixture to resist the effects of water and traffic.

Porous asphalt construction

The construction methods of pavements using porous asphalt or OGFCs is not very different from the methods used for dense graded mixtures with a few exceptions:

- Since porous asphalt is subject to drain down of the asphalt binder, silo storage or any other storage of the mixture is not recommended. Any storage of the mixture should not exceed 15 min.
- There are no compaction density requirements for porous asphalt. The mixture is just seated to the pavement underneath. Two passes with a 10 tonne steel wheel static roller should be satisfactory to properly seat the porous asphalt (United States Army Corps of Engineers 2001).
- Porous asphalt is designed to have a high in-place air void content. The air voids typically are in the range of 17–22 percent.

- The thin layer, high air void content, and lack of fine aggregate in porous asphalt cause the mixture to cool rapidly. All placing and rolling of porous asphalt should occur prior to the cooling down of the mixture to 80 °C or below. Tearing of the mixture will occur if placing or compaction is attempted below 80 °C.
- Since porous asphalt cools so rapidly, some highway agencies do not permit the construction of porous asphalt when the ambient air temperature falls below 15 °C.

Porous asphalt is mostly used as a wearing course and is placed and compacted in a single thin lift or layer. Table 7.8 gives some minimum thickness ranges for porous asphalt when it is used as a wearing course. Porous asphalt should not be placed next to a curb and gutter (Plate 7.5) and should be placed adjacent to another porous asphalt course or left day-lighted. Obstructing the edges of a porous asphalt layer will not allow lateral flow of water.

Porous asphalt or OGFCs have the most value when used on high-speed traffic facilities, usually with traffic speeds that are greater than 70 km/h. The skid resistance benefits of

Plate 7.5 Porous asphalt next to curb.

Table 7.8 Minimum layer thickness for porous asphalt wearing courses

	Nominal maximum top size (mm)		
	19.0	12.5	9.5
Minimum thickness range (mm)	25–50	25–37.5	19–25

Plate 7.6 Transition between porous asphalt and a dense graded wearing course.

porous asphalt will generally not occur on low speed pavements. A dense graded wearing course is usually used on lower speed pavements with a transition occurring between the porous asphalt and the dense graded wearing course (Plate 7.6). A less expensive dense graded mixture will adequately perform under such applications. Porous asphalt generally will not withstand the continuous steering movements that occur on parking lots. The open, negative texture of porous asphalt is not considered as aesthetically pleasing as dense graded mixtures for parking applications.

Asphalt treated permeable bases

Asphalt treated permeable bases (ATPB) for all practical purposes are porous asphalt mixtures used as a freely draining base courses. Other names for ATPB are "open graded base mixes" and "asphalt treated permeable material" (ATPM). ATPB mixes, OGFCs, are highly permeable and provide for rapid drainage of water. ATPB are used to improve the subsurface or base drainage of either flexible or rigid pavements. The permeability of ATPB can exceed 3,000 m/day. It can be used as a drainage layer between a concrete pavement and a dense graded asphalt overlay. Whether the ATPM is used as a base course or overlay drainage layer, its edges must be daylighted or connected to edge drains to allow for the rapid removal of water. When the ATPM is placed between the asphalt overlay and the rigid concrete pavement, the ATPM absorbs some of the temperature related stresses and reduces reflective cracking in the overlay (Roberts *et al.* 1991). When ATPM is used solely for the function of reducing reflective cracking in overlays, they are sometimes known as a "stress absorbing membrane interlayer" (SAMI). When ATPM is used as a base course, it is placed directly on the soil subgrade or on a granular sub-base (Figure 7.4). If the ATPM is placed on a soil subgrade, a filter fabric should be placed on top of the soil to prevent the upward movement of soil particles into the ATPB and clogging it.

The highly durable and skid resistant properties of the aggregate used for porous asphalt or OGFCs is not required when used as a base course. The coarse aggregate should be angular to provide an interlocking aggregate skeleton. The aggregate gradation is similar to that for porous asphalt with the exception that the aggregate is usually larger. Table 7.9 gives some typical gradations for ATPM. The optimum asphalt binder content of ATPM can be determined using the same methods used for porous asphalt. Depending on the

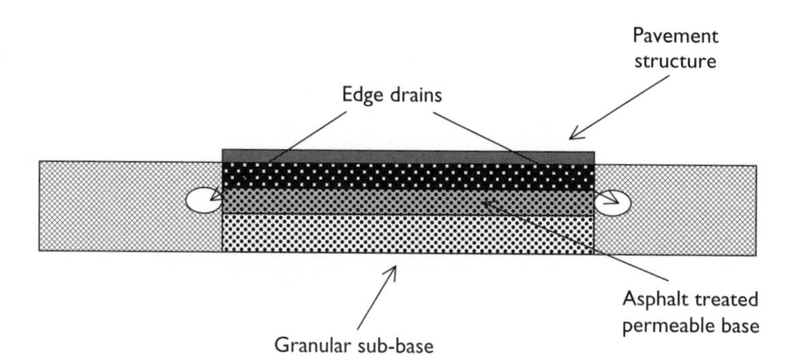

Figure 7.4 Pavement using asphalt treated base (ATPB).

Table 7.9 Typical asphalt treated permeable base gradations

Sieve size	*Total passing (%)*	
	19.0 mm nominal	*25.0 mm nominal*
37.5 mm		100
25.0 mm	100	90–100
19.0 mm	85–100	
12.5 mm	35–65	25–60
4.75 mm	0–20	2–10
2.36 mm	0–10	0–5
75 μm	0–4	0–2

Table 7.10 Minimum layer thickness for asphalt treated permeable bases

	Nominal maximum top size (mm)		
	12.5	19.0	25.0
Minimum thickness range (mm)	25–50	37.5–75	50–100

aggregate gradation and absorption, the asphalt binder content of ATPM can be as low as 2–3 percent.

The asphalt binder used for ATPM is typically the same binder that has been selected for the structural component of the asphalt pavement. When selecting a performance grade (PG) graded asphalt binder, it is not necessary to bump the binder grade due to traffic variables. Since ATPM is typically near the bottom of the pavement structure, it does not receive the effects of oxidation that an asphalt binder used for a wearing course may receive. The addition of an anti-strip additive is usually necessary to reduce the occurrence of moisture induced damage to the ATPB. The structural strength coefficient of ATPM is similar to that of a treated base, typically in the range of 0.012–0.013 (1/mm) or 0.30–0.34 (1/inch). Minimum thickness values are given in Table 7.10.

The same construction methods used for porous asphalt can be used for the construction of ATPB. Attention should be given to preventing the asphalt binder from draining down during transport or placement. Just seating the ATPM into the sub-base completes compaction. Two to three passes of a 10 tonne roller in a static or non-vibratory mode accomplish this. ATPM has a tendency to cool down quickly, so paving should be completed when the ambient temperature is 15 °C or higher.

Asphalt treated permeable bases have application on all types of pavements, including low volume roads and parking facilities. Pavements being placed in low-lying areas or other water saturated areas can benefit from the use of ATPBs. The installation of any ATPB requires additional subdrainage structures or day lighting to remove the water that travels laterally through the ATPB.

Polymer modified asphalt

Polymer modified asphalt (PMA) refers to any asphalt mixture that uses an asphalt binder that has been modified with a polymer. Some references include PMA as describing the entire mixture, asphalt cement, polymer, and aggregate, but more specifically PMA refers to an asphalt binder that is polymer modified. The chemical definition of a polymer is a natural or synthetic compound of a usually high molecular weight consisting of many, repeating, linked molecules. Natural rubber and various types of plastics are examples of polymers. The purpose of adding the polymer is to physically or chemically alter the asphalt cement in order to make an asphalt binder that ultimately improves the durability and service life of the asphalt binder. The polymer is dissolved in certain fractions of the asphalt cement, creating an interconnecting matrix of the polymer throughout the asphalt cement, now known as an asphalt binder. Previously described in Chapter 1 is the difference in terminology between asphalt cement and asphalt binder. The term "asphalt binder" is more general, including all the materials; the petroleum derived asphalt cement and any additional added materials, such as polymers.

One of the primary purposes of adding polymer to the asphalt binder is to increase the resistance of the asphalt mixture to permanent deformation, especially at high temperatures without significantly affecting the properties of the asphalt binder at lower temperatures. The asphalt binder is stiffened so that its total viscoelastic response is reduced with a corresponding reduction in permanent strain. Permanent strain is reduced by increasing the elastic component of the asphalt binder and reducing its viscous component. Increasing the stiffness of the asphalt binder also increases its dynamic stiffness. When the dynamic stiffness of a pavement is increased, its structural strength increases, thus increasing the load carrying ability of the pavement (Whiteoak 1991). Some studies have suggested that for the same traffic loading a PMA pavement can be thinner than an unmodified asphalt pavement.

The polymer can be specifically requested to be added to the asphalt binder or the asphalt binder may be modified with the polymer to meet certain test criteria. The most common use of polymers is for the asphalt binder to meet specific Strategic Highway Research Program (SHRP) PG grading requirements. For example, typically there are no straight or unmodified asphalt cements that will meet a PG76-xx grade or higher. Polymer is added to the asphalt cement in the range of 2–6 percent to meet the upper PG grade requirements. A very general rule of thumb is that if there is a 90 °C or greater difference between the top and bottom temperature grades of a PG binder, a polymer will be required to be added to the asphalt cement.

Polymers can be grouped into two classes: thermoplastic and thermoset. Thermoplastic materials soften when heated and become hard when cooled and can repeat the process. Thermoset materials become hard when cooled and cannot repeat the process and become soft when heated. Thermoplastic materials have the most utility for asphalt cement. For paving applications, thermoplastic polymers can be further divided into two categories, elastomers and plastomers. Elastomer polymers consist of natural rubber or latex, synthetic rubber, styrene butadiene (SB), SBR, SBS, polybutadiene, polychloroprene, and polyisoprene. Plastomer polymers are also known as crystalline polymers, consisting of polyethylene, polypropylene, EVA, ethylene methyl acrylate (EMA), and polyvinyl chloride (PVC). By far the largest use of any polymers in asphalt paving are elastomers. Of elastomers, SBS is the most commonly used.

Thermoplastic polymers derive their strength and elasticity from a physical cross-linking of the molecules into a three-dimensional network. The polystyrene end blocks of the three dimensional network provide the strength to the polymer. The polystyrene middle block provides for the elasticity of the polymer. At temperatures above the grass transition temperature, about 100 °C, the three-dimensional networks dissociate and weaken, allowing the polystyrene type polymers to soften or melt. This softening enables the polymer to be blended into the asphalt binder. Upon cooling the three-dimensional networks reassociate and its strength and elasticity is recovered (Whiteoak 1991).

Not all asphalt cements are compatible with all types of polymers. The asphalt cement and polymer are carefully chosen from both a compatibility and performance standpoint. The addition of thermoplastic polymers to asphalt cement disturbs its balance or phase equilibrium of asphaltenes, resins, and maltenes. In asphalt cement, the asphaltenes and resins are dispersed in the maltene phase. Asphaltenes require maltenes to be present; otherwise the asphaltenes will precipitate causing the asphalt cement to become gelled and unworkable (Morgan and Mulder 1995). When a thermoplastic polymer is added, the polymer and the asphaltenes (the resins are generally attached to the asphaltenes) compete for the solvency power of the maltene phase. The thermoplastic polymer absorbs proportionately the maltene portion of the asphalt cement. If not enough maltenes are available in the selected asphalt cement, phase separation will occur and the polymer will generally drop out during storage. The main factors influencing the compatibility or storage stability of an asphalt cement and polymer combination are:

- The amount and molecular weight of the asphaltenes.
- The aromatic or solvency power of the maltene phase.
- The amount of polymer in the asphalt cement.
- The molecular weight and structure of the polymer.
- The storage temperature.

(Whiteoak 1991)

When phase separation occurs in a PMA binder, the polymer generally rises to the surface and the asphaltenes precipitate to the bottom. The top portion of the asphalt binder becomes very soft and elastic while the bottom portion becomes brittle and hard. Due to this phenomena, simple tests involving testing the top and bottom portion of an asphalt binder sample and determining their penetration, ring and ball softening point, or complex modulus (dynamic shear rheometer) can determine if the asphalt binder has the potential for phase separation.

Ethylene vinyl acetate is another type of thermoplastic polymer. EVA increases the viscosity or stiffness of the asphalt binder but does not typically increase its elasticity. EVA is polymer with random structure produced by the copolymerization of ethylene and vinyl acetate. The performance properties of EVA are controlled by molecular weight and vinyl acetate content. One physical property of EVA that is measured is its melt flow index. The melt flow index is a viscosity test that is inversely related to molecular weight. The higher the melt flow index, the lower the molecular weight and viscosity. The vinyl acetate content is significant in providing the thermoplastic properties of the asphalt binder. EVA polymers typically blend easily with the asphalt binder and have a low probability of phase separation.

Polymer mixing

Mixing or blending a thermoplastic polymer into asphalt cement either involves low shear or high shear mixing. Low shear mixing utilizes a blend tank and a low-speed paddle type mixer. The polymer, in either pellet or powder form, is slowly added to the asphalt cement in the mixing tank until a homogenous blend is created.

In high shear mixing, the thermoplastic polymer particles are physically reduced in size by a high shear rotor-stator mixer in a blending tank with the asphalt cement. At the same time, the polymer is dispersed through the asphalt cement.

The addition of heat in the range of 170–190 °C is usually required to facilitate the blending of the polymer into the asphalt binder. During the addition of the thermoplastic polymer to the asphalt cement, the polymer swells to about 6–9 percent of its volume. The polymer absorbs the maltene portion of the asphalt binder resulting in the swelling of the polymer. High shear equipment will provide for faster dispersion of the polymer and is usually the mixer of choice when producing PMA. The blending of the polymer into the asphalt cement consists of three stages:

- Pre dispersion stage
- Disintegration stage
- Swelling stage.

In the first stage, the asphalt cement is heated to about 180 °C in the blending tank. The thermoplastic polymer is added to the asphalt cement as quickly as possible without causing the polymer to agglomerate. The polymer is mixed throughout the asphalt cement. The time for mixing depends on the capability of the mixer and the size of the blending tank. A 10 tonne batch of asphalt binder may require 30 min of premixing time.

In the second stage, the premixed batch of asphalt binder is passed through the high shear mixer to disintegrate the thermoplastic polymer particles. The milling stage can be completed in a single pass or recirculated through the high shear mixer. The temperature of the asphalt binder should rise due to the energy being dissipated throughout the blend.

The final swelling stage of the process may be carried out under low shear mixing while maintaining the asphalt binder temperature at 180 °C until there are no visible particles of polymer left. The final blended asphalt binder is then checked for specification compliance (Morgan and Mulder 1995).

PMA design and construction

The mixture design methods used for PMA can be the same methods used for unmodified asphalt binders. The polymer in the asphalt binder typically has no effect on the volumetric properties of an asphalt mixture. Mixtures with PMA can exhibit increased asphalt binder film thickness around the aggregate, leading to a more durable mixture. PMA will also increase the cohesive strength of the mixture, giving higher Marshall stability values. One of the purposes of using PMA is to provide a mixture that is more resistant to deformation, especially at higher temperatures, than a mixture that does not use PMA. Depending on which SHRP PG binder grade is specified, the asphalt binder may already come with a polymer added to it. For example, if the mixture designer requires a PG 76-22 for the asphalt mixture, more than likely the mixture will be using PMA. In the asphalt industry, there are three approaches to specifying a PMA. One approach is to specify that an asphalt binder contain a specific amount of a specific type of polymer, another approach is to require that the asphalt binder meet certain performance requirements with no specific mention that the asphalt binder contain polymer or any other modifier. The current SHRP PG binder specifications are an example of the second approach. The third approach is a combination of the two previous approaches, with no specifically mentioning dosages or polymer types, but requiring specific test parameters, that will almost certainly require a specific type of PMA to meet the test criteria. The American Society for Testing and Materials (ASTM) has developed standard specifications for the third approach. Table 7.11 outlines the various ASTM specifications. These specifications are ways to specify a type of PMA, but are not performance specifications. Future trends are expected to continue to expand the use of performance specifications, such as the SHRP PG binder specification, with the gradual elimination of method and material specific specifications.

The construction practices for pavements using mixtures containing PMA binders are similar to methods that would be used for other mixtures with two important exceptions. Polymer modified binders are stiffer and more sticky than mixtures containing no polymer. Mixing and compaction temperatures are increased in account for the stiffer mixtures. A 10 °C or higher increase over conventional compaction temperatures maybe required. The asphalt binder supplier can provide guidelines for the proper mixing and compaction temperatures or they can be determined in the laboratory. Theoretically, the mixing temperature is the temperature of the asphalt binder that provides a kinematic

Table 7.11 Various PMA specifications

Specification designation	*Specification name*	*Polymer controlling test requirements*	*Typical polymers that meet the:*
ASTM D5976	Type I PMA	Elastic recovery separation test	SB or SBS
ASTM D5840	Type II PMA	Ductility toughness and tenacity	SB rubber latex polychloroprene latex
ASTM D5841	Type III PMA	Softening point	EVA
ASTM D5892	Type IV PMA	Elastic recovery	Non-cross-linked SBS

viscosity of 170 ± 20 centistokes. The compaction temperature is the temperature of the asphalt binder that provides a kinematic viscosity of 280 ± 20 centistokes. For practical purposes, it is essential that a compaction test strip be completed so that the optimum compaction temperatures and the effort of compaction can be determined. It is not unusual for PMA to be mixed and placed at temperatures in the range of 160–170 °C. Overheating of PMA is a concern and excessive temperatures can cause oxidation of the asphalt binder, potentially causing cracking in the pavement. The supplier of the PMA binder can provide the proper mixing and compaction temperatures. Constructing porous asphalt pavements containing polymer during cool weather can be difficult and lead to tearing of the porous asphalt surface. The open gradation of porous asphalt increases the cooling rate of the mixture. Ambient temperature paving restrictions should be observed, especially with porous asphalt.

The use of specially formulated release agents will overcome most of the problems of the stickiness of the PMA mixture. The mixture tends to stick to the bottom of truck beds, in the hopper of the paver, and on the pneumatic tire rollers. As the temperature of the mixture increases, the sticking problems are usually reduced. The release agent is applied to the surfaces that come in contact with the asphalt mixture. Petroleum solvents, such as diesel fuel are not recommended as release agents, since they have the potential to damage and remove the hot asphalt binder from the aggregate. Petroleum solvents can also have damaging effects on the environment. There are many commercially available asphalt release agents that work well with PMA. These agents are usually water based and biodegradable. The water based release agents are similar in chemistry to detergents with additional compounds such as silicone. There are also asphalt release agents that are made from natural ingredients such as soybean oil. These products can also be formulated into biodegradable solvents, such as soybean derived methyl esters. The excessive use of methyl esters can also remove the hot asphalt binder from the aggregate.

Application

Adding any additional ingredients to an asphalt mixture will increase the cost of the mixture. PMA is no exception. A designer considering the use of PMA should consider the following two points before requiring the use of PMA:

- Does the use of PMA extend the life of the asphalt pavement?
- Does the use of PMA provide additional pavement performance that an unmodified asphalt binder would not provide?

These questions can apply when considering the use of any additional or supplemental paving material. PMA is considered a premium paving mixture and has the potential for providing additional service life to the pavement under select conditions. Pavements that incur high traffic conditions and high ambient temperatures can benefit from the use of PMA. A fast food restaurant parking lot in a moderate or cold climate such as Wisconsin or the United Kingdom will probably neither benefit nor require the use of PMA. The additional cost of the PMA would not be justified. A parking facility that receives heavy truck traffic such as a shipping terminal would benefit from the use of PMA.

The use of an economic analysis technique for pavements should be used when determining if an additional cost incurred during pavement construction will provide an overall lower pavement life cost. One such technique is known as "life cycle costing." In life cycle costing, all the pavement costs are considered in terms of discounted costs and benefits over the period of time the pavement is required. The period of time for a parking lot may be considerably shorter than the pavement service time for an interstate highway or motorway. For example, if a fast food restaurant has an economic life of 8 years, after which it would be torn down, the parking lot for the restaurant should not have a pavement service life longer than 8 years. Any additional costs incurred at the beginning of construction may not be recovered.

Pavement recycling

The use of recycled materials in asphalt pavements has been occurring with varying degrees of success for the past 20 years. RAP reclaimed concrete pavement, coal fly ash, and blast furnace slags are the most common materials that are recycled back into an asphalt pavement (USDOT 2000). Asphalt pavement recycling is the recycling or reusing an existing asphalt pavement into a new and structurally sound asphalt pavement. There are four common methods used in asphalt pavement recycling:

- Cold in-place recycling
- Hot in-place recycling
- Full depth reclamation
- Hot mix recycling.

Cold in-place recycling involves the recycling of the existing asphalt pavement *in situ* without the use of heat. An asphalt emulsion is typically used as recycling agent. The process includes pulverizing or tilling the existing pavement, the application of the recycling agent, placement, and compaction. The use of a recycling train is often used for large or long roadway projects. The recycling train consists of pulverizing, screening, crushing and mixing units. The processed roadway is deposited in a windrow from the mixing machine, where it is then picked up, placed, and compacted with conventional HMA paving equipment. Advantages of cold in-place recycling include significant remedial corrections of most pavement distresses, environmentally friendly, and the complete reuse of the existing pavement.

Hot in-place recycling involves heating and softening the existing asphalt pavement and then scarifying it. A recycling agent, usually an asphalt emulsion, is added to the scarified RAP. Sometimes a new asphalt mixture is also added to the RAP. The depth of recycling can vary from 20 to 50 mm. Hot in-place recycling can be performed in either a single pass or a multiple pass operation. In a single pass operation, the scarified RAP is combined with a new asphalt mixture if desired, and then compacted. In a multiple pass operation, the RAP and recycling agent is compacted first and then a new wearing course is added. The advantages of hot in-place recycling are that pavement distresses, including surface cracks can be corrected and the existing oxidized asphalt binder can be rejuvenated. If needed, the aggregate gradation and asphalt content can be modified by the addition of new HMA.

Full depth reclamation is where the entire asphalt pavement and a predetermined amount of the underlying base course are treated to produce a stabilized base course. The existing asphalt pavement becomes part of the new pavement's base course. Full depth reclamation is a cold mix recycling process using asphalt emulsions, calcium chloride, fly ash, and possibly other additives to stabilize the base course. The process consists of pavement pulverization, mixing with the additives, compaction, and the construction of a wearing course. If the existing pavement material is not adequate to provide the desired thickness of stabilized base, new aggregates may be added. Full depth reclamation is typically performed to depths of 100–300 mm. Full depth reclamation can remove most of the pavement distresses and upgrade the structural strength of the pavement.

Hot mix recycling involves removing or milling up the existing asphalt pavement, crushing it if necessary, and adding it to HMA at a mixing plant. RAP can be added at both batch mixing and drum mixing plants (Plate 7.7). The HMA containing RAP is constructed using the same methods for conventional asphalt mixtures (USDOT 1997).

The most common method of pavement recycling is the hot mix recycling method. The use of hot mix recycling is prevalent to geographical areas with some areas using RAP in

Plate 7.7 RAP stockpile at a HMA mixing plant.

all the HMA being produced and some areas with no RAP usage at all. The availability of RAP and landfills usually determine how many tonnes of HMA being produced contain RAP. Urban areas generally see more RAP usage than rural areas. The use of RAP in the mixture should not preclude the pavement designer or architect from accepting the mixture for any particular project. When properly designed and manufactured, mixtures containing RAP will perform to the same satisfaction as mixtures containing no RAP.

The incorporation of RAP reduces the demand for both new aggregate and asphalt binder in the mixture being produced. RAP is generated through pavement restoration projects. The pavement has been milled to either remove pavement defects or distresses or to maintain utility heights during overlay projects. The RAP can then be disposed off at a landfill or recycled into a new pavement. The economic value of RAP consists of the following:

- No longer a disposal issue
- The value of the aggregate in the RAP
- The value of the asphalt binder in the RAP.

The disposal issue and the value of the asphalt binder in the RAP are the main economic incentives for the incorporation of RAP into HMA. As the cost of the asphalt binder increases, the value of the RAP increases. The contribution of the asphalt binder in the RAP allows for a reduction in the amount of new or virgin asphalt binder in the mixture being manufactured. The gradation of the RAP and the heat transfer ability of the mixing plant limit the amount of RAP that can be utilized in an asphalt mixture. The gradation of the RAP needs to be accounted for in the final gradation of the asphalt mixture. During the milling or crushing process of the existing asphalt pavement, a significant amount of fine material can be generated, which can limit the amount of RAP used in the new asphalt mixture. The heating of the RAP in the mixing plant extracts most of the asphalt binder from it and allows it to be blended through out the new asphalt mixture. The ability of the mixing plant to transfer enough heat to dry the RAP and extract the asphalt binder from it limits the total amount of RAP that can be incorporated in the HMA. Some modern drum mixing plants, such as double drum plants, have been designed to incorporate up to 70 percent RAP or more, but typically the maximum is 50 percent. Batch mixing plants can usually only incorporate up to 30 percent RAP in the HMA.

In batch mixing plants the heat transfer method is a conduction method, which consists of the use of the heat from the new aggregate to increase the temperature of the RAP. The new aggregate temperature is increased to account for heat transfer to the RAP. Drum mixing plants use one or more of the following approaches to heat the RAP:

- Indirect heating of the RAP from the exhaust gases of the burner.
- Convection heating by containing the entire combustion process and circulating air throughout the RAP.
- A combination of convection and conduction heating by introducing the new aggregate in the burner end of the drum, introducing the RAP in the center of the drum and mixing the two together at the end of the drum.

The amount of moisture in the new aggregate and the RAP, along with the desired discharge temperature restricts the amount of RAP any type of mixing plant can adequately handle.

RAP mixture design

When an asphalt mixture containing RAP is properly designed it will perform just as well as a mixture that is all new aggregate and asphalt binder. A mixture containing RAP should be required to meet the same performance characteristics as a mixture containing no RAP. In the RAP mixture design and manufacturing process, RAP is treated as another aggregate. Sampling of the RAP stockpile or source is handled in the same manner as sampling of an aggregate source. It is important that the RAP sample be representative of the RAP that will be used by the mixing plant. The gradation of the RAP is determined and the asphalt binder is extracted from the RAP using one the laboratory test methods listed in ASTM D2172, *Standard Test Methods for Quantitative Extraction of Bitumen from Bituminous Paving Mixtures*. The asphalt binder content of the RAP can then be determined. The extracted asphalt binder is reclaimed from the extraction solvent One method to reclaim the asphalt binder is ASTM D1856, *Standard Test Method for Recovery of Asphalt From Solution by Abson Method*. The viscosity or PG of the recovered asphalt binder is then determined. It is an accepted practice not to grading tests on the recovered asphalt binder with the mixture is expected to contain 15 percent or less RAP. At amounts of 15 percent or less, generally there is not enough influence from the asphalt binder in the RAP to significantly alter the desired asphalt binder grade for the entire mixture. For example, if the designer requires an AC-20 or PG64-22 for the asphalt mixture being purchased, the aged asphalt binder in the RAP should not alter those grades when used at levels of 15 percent or less. In mixtures containing large amounts of RAP it may be necessary to use a softer grade asphalt binder to counter the stiffening effects of the aged asphalt binder in the RAP. The amount of asphalt binder in the RAP and the RAP gradation is always required to be determined regardless of the amount of RAP being used in the mixture.

The following steps provide an overview of the mixture design procedure for mixtures containing RAP:

1 Determine the gradation; asphalt binder content and recovered asphalt binder grading properties of the RAP. Determine the $G^*/\sin\delta$ of the recovered asphalt binder, if applicable.
2 Determine the gradation of the aggregate in the RAP after having its asphalt binder removed (solvent or ignition oven method). This material is now known as reclaimed aggregate material or RAM.
3 Calculate the combined aggregates, including the RAM, using the same methods for mixtures containing no RAP.
4 Approximate the required total asphalt binder content for the combined aggregates.
5 Estimate the amount of new asphalt binder that will be required for the mixture. The amount of new asphalt binder plus the amount of asphalt binder contributed by the RAP should equal the required total asphalt binder content for the combined aggregates.
6 Select the grade of new asphalt binder required for the mixture. Determine the $G^*/\sin\delta$ of the new asphalt binder, if applicable. It may be necessary to select a softer grade to counter the effects of the aged asphalt binder from the RAP.
7 Complete mixture design such as the Marshall or SHRP SuperPave methods.
8 Determine the overall optimum asphalt binder content and combined aggregate blend or job mix formula.

There are various specific equations for the mixture design process that can supplement the equations provided in Chapter 5.

Estimated percentage of new asphalt binder in the mixture, the quantity of new asphalt binder to be added to the trial mixes of the recycled HMA mixture, is expressed as a percentage by weight of the total mixture. Estimated new asphalt binder, P_{nb}, is determined using equation 7.1.

$$P_{nb}=\frac{(100^2-rP_{sb})P_b}{100(100-P_{sb})}-\frac{(100-r)P_{sb}}{100-P_{sb}} \tag{7.1}$$

where P_{nb} is the percentage of new asphalt binder in mixture containing RAP; r, the new aggregate expressed as a percentage of the total aggregate in the mixture containing RAP; P_b, the estimated total asphalt binder percentage for mixture containing RAP; P_{sb}, the asphalt binder percentage in the RAP.

Percentage of new asphalt binder to total asphalt binder content, the percentage of the new asphalt binder, P_{nb}, to the total mixture asphalt binder content, P_b, is given as R in equation 7.2.

$$R=\frac{100(P_{ab})}{P_b} \tag{7.2}$$

Percentage of RAP in the mixture, the percentage of RAP in the mixture, P_{sm}, expressed as a percentage of the total mixture is given in equation 7.3.

$$P_{sm}=\frac{100(100-r)}{100-P_{sb}}-\frac{(100-r)P_{sb}}{100-P_{sb}} \tag{7.3}$$

The total percentage of new asphalt binder, recovered asphalt binder, recovered asphalt material or aggregate and new aggregate should equal 100, when all percentages are expressed as a percentage of total mixture.

Percentage of new asphalt binder, by weight of aggregate, the percentage of new asphalt binder, P_{nb}, can also be expressed as a percentage of the total weight of aggregate by equation 7.4.

$$P_{nb}=P_b=\frac{(100-r)P_{sb}}{100} \tag{7.4}$$

When using a percentage of RAP greater than 15 percent, the stiffness of the recovered asphalt binder from the RAP must be considered. A target viscosity of the combined asphalt binders is selected. The target viscosity is usually specified or provided by the designer and is the desired asphalt binder grade for the total mixture. Under the viscosity grading system, if using greater than 15 percent RAP in the mixture use one grade softer than the normally selected for a 100 percent new or virgin asphalt mixture. For example, if the pavement designer requests a mixture with AC-20 and the mixing plant is providing a mixture with 25 percent RAP, the grade for the new asphalt binder should be an AC-10. The percentage of RAP allowed in the mixture due to the effects of the aged asphalt binder can also be determined. Figure 7.5 is a viscosity blending chart that uses a log–log viscosity on the

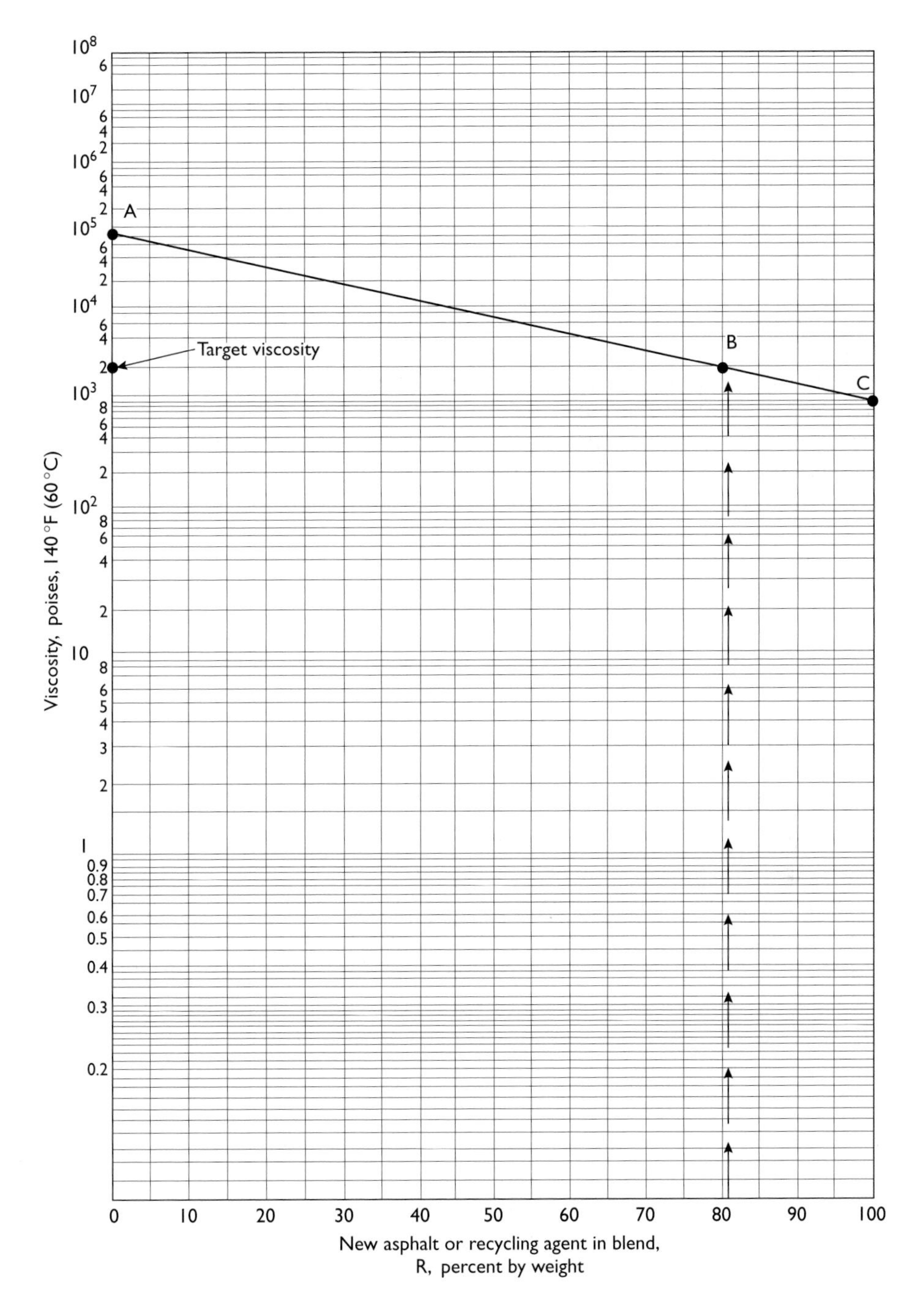

Figure 7.5 Asphalt viscosity blending chart.
Source: © The Asphalt Institute; reprinted with permission.

y-axis versus percent new asphalt binder on the x-axis. A target viscosity for the blend of reclaimed asphalt binder and the new asphalt binder is selected, this value is usually already determined through specification or selected by the pavement designer. The viscosity of the aged asphalt binder is plotted on the left hand vertical scale as point A. A vertical line is drawn representing the percentage of new asphalt binder and is drawn to the horizontal line representing the target viscosity. This point is point B. A straight line is drawn from point A to point B and is extended to the right-hand vertical scale. The point on the right-hand scale is point C. Point C is the viscosity at 60 °C of the new asphalt binder required to blend with the asphalt binder in the RAP to obtain the target viscosity in the blend. Select the grade that has the grade of the new asphalt binder that has a viscosity range that is closest to viscosity at point C. As previously suggested one softer grade of asphalt binder will usually provide satisfactory results with most mixtures containing more than 15 percent RAP. Large amounts of RAP in a mixture will require the use of very soft asphalt binders or recycling agents. Large amounts of RAP in a mixture are usually limited by RAP availability, plant heating capacity, and the mixture gradation requirements.

The same chart can be used for asphalt binders that are graded using the SHRP PG grading system. The left-hand vertical or y-axis is replaced by the desired binder grade's original G*/sin δ values at the appropriate temperature. The appropriate temperature is the upper PG grade temperature required for the total mixture. For example, if the designer requires a PG64-22 for the mixture, point A on the left-hand axis would consist of the original (before the rolling thin film oven) G*/sin δ value of the recovered asphalt binder measured at 64 °C. The G*/sin δ of both the new and recovered asphalt binder must be determined at identical temperatures. SHRP PG blending charts have been developed based on G*/sin δ values (rutting resistance), G* sin δ values (fatigue resistance), and low temperature values.

A specific blending chart is created for each PG grade being considered. Some specific guidelines have also been developed in conjunction with the SHRP PG blending charts:

- When using 15 percent RAP or less in the mixture, the new asphalt binder grade should be the same, as the one required for the total asphalt mixture.
- When using 16–25 percent RAP, the selected PG grade of the new asphalt binder should be one grade below (both high and low temperature) the grade required for the total asphalt mixture. For example, if the designer requests a PG64-22 for the asphalt mixture, an asphalt mixture containing 21 percent RAP should use a PG58-28 as the new asphalt binder.
- When using greater than 25 percent RAP, a PG blending chart, similar to Figure 7.5, should be developed. The maximum and minimum amount of the new asphalt binder to be used is based on two horizontal lines corresponding to stiffness (G*sin δ) values of 1 and 2 kPa. The recovered and the new asphalt binder need to be tested at the same grading temperature.

Table 7.12 Viscosity of various RAP recycling agents

	Recycling agent grade					
	RA-1	*RA-5*	*RA-25*	*RA-75*	*RA-250*	*RA-500*
Viscosity at 60 °C (cSt)	50–175	176–900	901–4500	4,501–12,500	12,501–37,500	37,501–60,000

Another alternate method consists of using 2.0 kPa as the maximum allowable value of $G^*/\sin\delta$ that can be contributed from the recovered asphalt binder. Regardless of the method used to determine the proper blend of new and aged asphalt binder, the availability of the softer new asphalt binder should be verified. In some geographical areas, there are specific materials manufactured as recycling agents for hot mix recycling. These recycling agents are usually a soft asphalt binder or high flash point (>215 °C) oil that meet specific viscosity requirements. Table 7.12 gives some typical viscosity ranges for various recycling agents.

Recycling agents should be capable of performing the following functions:

- Capable of altering the viscosity of the aged asphalt binder in the RAP to the desired level.
- Be compatible with the aged asphalt binder to ensure that the extraction of the paraffins or maltenes does not occur.
- Have the ability to redisperse the asphaltenes in the aged asphalt binder.
- Be uniform in properties.
- Be resistant to smoking and flashing.

(USDOT 1997)

Once the amount of RAP, the amount of the new asphalt binder, and the grade of the new asphalt binder or recycling agent is determined, the mixture design process proceeds in the same manner as designing for virgin HMA mixture. The volumetric properties are determined along with optimum asphalt binder content. The optimum asphalt binder content includes the new and the recovered asphalt binder. The estimated new asphalt binder content is then adjusted to meet the overall optimum asphalt binder content. If the new asphalt binder content is significantly different from the estimated asphalt binder content; a trial mixture can be produced in the laboratory to verify the mixture's volumetric and other design properties.

The performance expectations of the mixture containing RAP should not be any different from a mixture containing no RAP. The same performance and volumetric guidelines for virgin mixtures should be used on mixtures containing RAP. The structural thickness design values for HMA containing RAP are the same as for mixtures containing no RAP.

Construction techniques

An asphalt mixture is designed with the assumption that all the ingredients are consistent. The RAP stockpiles at the mixing plant should be segregated by gradation and asphalt binder content. Both batch and drum mixing plants can incorporate RAP into the mixture being produced. The method in which the mixing plant incorporates the RAP can vary with various types and manufacturers of the mixing plants. Batch plants typically add the RAP directly into the weigh hopper prior to its introduction into the pugmill. It is dry mixed with the hot aggregate prior to the addition of the new asphalt binder. A drum mixing plant incorporates the RAP at various points along the drum, depending on the drum mix plant type and manufacturer. The transportation, placement, and compaction of mixtures containing RAP are the same methods used for virgin HMA mixtures.

RAP is almost exclusively used for dense graded mixtures. The gradation of RAP typically prevents its use in porous asphalt gap graded mixtures. If the quality of the aggregate in the RAP meets the specific performance criteria, it can also be used in wearing courses. The use of RAP allows the HMA manufacturer to make more economical mixtures. If these mixtures meet the same criteria as expected of virgin mixtures, nothing should prevent their use on any type of pavement, from parking lots and low volume roads to expressways.

Hot rolled asphalt

Hot rolled asphalt (HRA) or rolled asphalt is not very common or even used in the United States, it was one of or the most commonly used mixture in the United Kingdom. In the United Kingdom, Japan, and a few other countries, HRA is not considered a specialty mixture, but is a common paving mixture for highway or road applications. HRA is a wearing or surface course mixture, but can be successfully used as a base or binder course. HRA is typically not used for parking lots (car parks) since less expensive, dense graded mixtures will perform adequately, and have a smoother surface. HRA is basically a hot mastic that has single-sized aggregate rolled into the mastic. The mastic consists of fine aggregate, mineral filler, and a stiff asphalt binder. The mastic can also incorporate coarse aggregate. The asphalt binder is typically a 50 penetration asphalt binder. A corresponding SHRP PG grade, if used for HRA, would be a PG70-xx or PG76-xx. The stiff mastic contributes the majority of the strength of an HRA mixture. The mineral filler stiffens the already somewhat stiff asphalt binder. The mineral filler is very fine, with 75 percent passing the 75 μ sieve.

Hot rolled asphalt has been used mostly for the surfacing of major roads or motorways. Recently proprietary thin surfacings and porous asphalt have replaced the use of HRA on heavily traveled motorways and trunk roads in the United Kingdom. The rolled in aggregate or "*chippings*" provides the majority of macro texture for traffic wear and skid resistance. If the chippings are not rolled in the HRA at the correct temperature, they can dislodge from the HRA. The gradation of HRA is gap graded with the nominal maximum top size usually 14 mm. The smaller particle sizes are missing with the remaining particle sizes being sand and mineral filler. Most HRA can be considered as two components:

1 Gap graded mixture with a high filler content
2 Rolled in 14 or 20 mm coarse aggregate or chippings.

The coarse aggregate, part of the HRA mixture is used to extend the mastic and increases the stability of the mixture. Wearing course HRA mixtures can contain no coarse aggregate or up to 55 percent coarse aggregate. It is difficult to roll additional coarse aggregate into HRA mixtures that contain more than 45 percent coarse aggregate. Wearing courses that contain 55 percent coarse aggregate do not require additional coarse aggregate to be rolled in to provide the macro or aggressive texture (Whiteoak 1991). The HRA mixtures that contain high amounts of coarse aggregate are similar to an SMA mixture, but most HRA mixtures do not contain enough coarse aggregate to provide the stone on stone contact or skeleton structure of an SMA mixture. The coarse aggregate in most HRA mixtures is used as filler to reduce the volumetric requirements of the more expensive asphalt binder and filler.

Hot rolled asphalt comes in various standard formulations. The common variables are the coarse aggregate content in the mixture and the penetration of the asphalt binder Table 7.13 provides some of various combinations of HRA mixtures. Not all of the HRA combinations are listed in Table 7.13, but it illustrates the nomenclature in the HRA designations. The top size of the coarse aggregate is mostly determined by the thickness of the HRA being placed.

Table 7.13 Hot rolled asphalt combinations

HRA designation	*Coarse aggregate amount (%)*	*Nominal top size (mm)*
0/3	0	3
15/10	15	10
15/14	15	14
30/14	30	14
35/14	35	14
55/10	55	10
55/14	55	14

Wearing course thickness of HRA is usually 40 or 50 mm thick. HRA is further defined as being fine, Type F or coarse, Type C.

The coarse aggregate is crushed stone, gravel, or blast furnace or steel slag. The grading of the coarse aggregate is secondary in importance, since the mortar or mastic in HRA provides the majority of the mixture's stability. The minus 75 μ particle size of the coarse aggregate is usually less than 6 percent. The coarse aggregate should be crushed and have a maximum flakiness index of 45 percent of crushed stone and 50 percent for crushed gravel (Nicholls 1998).

The fine aggregate in HRA provides for the majority of the mastic. Whether the mixture is a type F or type C mixture depends on the grading of the fine aggregate. The fine aggregate for a type F mixture has 95–100 percent passing the 2.36 mm sieve and 0–8 percent passing the 75 μ sieve. The fine aggregate for a type C mixture has 90 to 100 percent passing the 2.36 mm sieve and 0–17 percent passing the 75 μ sieve. The mineral filler is typically limestone dust that has 75 percent or more passing the 75 μ sieve. The mineral filler and the asphalt binder bind the fine aggregate together to form the mastic. The mineral filler contributes to the stiffness of the mastic, providing strength to the HRA.

The standard asphalt binder used for HRA is usually 50 penetration. Other penetrations that have been successfully used in HRA are as follows:

- 35 dmm
- 50 dmm
- 85 dmm
- 125 dmm.

The traffic levels determine which asphalt binder to use. Eighty five and one hundred penetration grade asphalt are used for lower volume roadways, while 35 penetration grade asphalt is used for heavily traveled roads or motorways.

The Marshall mixture design method can be used to design HRA, somewhat in the same manner as that used for designing SMA mixtures. The Marshall stability, Marshall quotient, and air void values are used in selecting the optimum asphalt binder content for HRA. The air voids of HRA at the design stage should be about four percent. The optimum Marshall stability and quotient values depends on traffic. Marshall stability values are in the range of 3–10 kN, with 6 kN being typical. Quotient values usually vary from 0.6 to 1.1 with 1.0, a typical value. The same procedure used for selecting the optimum asphalt binder content for dense graded mixtures can be used for HRA mixtures. Once the asphalt binder content is determined from the Marshall design, an empirical factor of 0.7 percent is added to the selected asphalt binder content to become the *target* asphalt binder content. This empirical factor can vary and is used to increase the workability of the HRA mixture.

The coarse aggregate to be rolled in is usually precoated with a 50 penetration asphalt binder. *Precoat* is the term given for the coated chippings. The chippings or aggregates that are precoated with an asphalt binder will have better adhesion to the HRA mixture and will be less likely to become dislodged. The chippings are usually 14 or 20 mm coarse aggregate. 10 mm HRA mixtures use 14 mm chipping and 14 mm HRA mixtures use a 20 mm chipping. The flakiness of the chippings should be less than 25 percent. The amount of asphalt binder mixed in with the chippings is about 1.5 percent.

The chippings or precoats are spread at an approximate rate of 14 kg/m^2. The precoats are applied immediately after laying of the HRA mixture. They are spread or applied in same manner as applying aggregate or chippings for chip seals or surface dressings. The precoats are then rolled into the HRA with one or two passes of a steel wheel roller. If the HRA cools too fast, the chippings may not adhere to the mixture and will dislodge. Colored aggregate can be used (usually not precoated) as the chippings to provide various colors to the HRA wearing course.

The major problem of HRA is the loss of the rolled in aggregate or chippings. If the HRA is allowed to cool prior to the pressing or rolling in of the chippings they will dislodge and become a hazard to traffic. For a HRA containing a 50 penetration asphalt binder, the minimum temperature for rolling is 85 °C. The use of SMA mixtures and proprietary thin surfacings are replacing the use of HRA on high traffic facilities. Porous asphalt is also becoming increasingly popular due to its ability to quickly remove water from the pavement surface.

Chapter 8

Pavement distresses

An asphalt pavement, when designed and constructed properly, will provide years of service. All pavements will eventually require some type of maintenance. Asphalt pavements for parking facilities may require an application of sealer just to maintain or give a new appearance to the pavement for aesthetic reasons. Pavements continually undergo various types of stresses that induce minor defects into the pavement. Traffic loading, temperature, moisture, and subgrade movement can cause stresses. An asphalt pavement can exhibit various distresses that will eventually lead to the pavement's failure. Cracks, holes, depressions, and other types of distresses are the end result of wear on the pavement. The ability for the engineer or maintenance professional to recognize the various distresses that can occur in asphalt pavements allows for the relevant remedial corrective measures to be applied. Not all asphalt pavements will exhibit all of the distress types. The environment and the type of service the asphalt pavement is in, usually determines the type of distresses it may exhibit. For example, an asphalt pavement in service in Aruba will probably not display any low temperature cracking. High traffic, expressway, or motorway type asphalt pavements can undergo a type of distress known as rutting, while parking facilities generally do not incur rutting. A pavement can exhibit one or more distresses in various severity levels. Distresses can range from surface imperfections to distresses that lead to structural failure of the pavement. Most pavement distresses, if identified early, can be repaired and not have any significant effect on the service life of the pavement. The identification of the various types of pavement distresses allows the pavement maintenance personnel to determine what type of remedial action is necessary.

The early detection and repair of defects in the pavement will prevent minor distresses from developing into a pavement failure. The identification of the distress aids the engineer or maintenance professional in identifying what caused the distress and the required approach in repairing it. Cracks and other defects start appearing very small and are usually only detectable when walking along the pavement. Water pooling in select areas of the pavement is identification of potential problems in the pavement later on. It is important that the water quickly drains away from the pavement surface. Inspection and cleaning of drainage systems ensures that they are working properly and will eliminate some of the major causes of pavement damage. Each inspection should include all surface drainage structures, ditches, gutters, and channels to ensure that they are working as designed and not clogged or restricted. At least twice a year subsurface drains should be inspected to make sure they are working as intended. The abnormal appearance of water on the pavement surface may indicate that the surface or subsurface drains are improperly located, incorrectly designed, or restricted. The identification of drain outlets on maintenance maps will assist the inspector in not overlooking them (The Asphalt Institute 1983).

In all cases of pavement maintenance, it is best to determine the cause of the distress or defect. Determining the distress cause assists in making the proper repair and in preventing the distress from reoccurring. Identification of the distress is the one of the first steps in a pavement maintenance program. A formal program of routinely inspecting the pavement for distresses is a key part of any pavement maintenance program. A program of inspecting and repairing a pavement is known as a pavement management program or system. Asphalt pavements for parking facilities and low volume roads up to a motorway or expressway will all benefit from a pavement management program. Large highway departments have extensive pavement management programs while a routine inspection of a fast food restaurant parking lot followed by an annual application of a coal tar sealer can also be considered a pavement management program. A full-scale comprehensive pavement management program involves collecting data and assessing pavement characteristics that include roughness (ride), surface distress (condition), surface skid characteristics, and structure (pavement strength and deflection). By combining data on these pavement conditions with an economic analysis, engineers or maintenance professionals can develop short range and long range plans for a variety of budget and financial constraints (TIC 1989).

Full-scale or elaborate pavement management programs may not be needed or attainable by small agencies or owners of parking facilities. Large pavement management programs require staffing to support it and such a program is usually not financially feasible for parking lots or low volume roads. The smaller size of such pavements does not warrant an extensive management program. A pavement surface evaluation is the appropriate tool for smaller agencies and parking facility maintenance professionals. A pavement surface evaluation can be used to evaluate any type of pavement, but it is especially suited for use by existing staff and can be implemented relatively inexpensively. A visual inspection of the pavement is used to develop a condition rating for the pavement. The rating can be in terms of slight, moderate, and severe or a numerical value on a scale of 1–10 can be used to rate the pavement distress or condition. The visual inspection gives a picture of the condition of the pavement and can identify candidates for maintenance and rehabilitation. The pavement surface evaluation and rating (PASER) is one of many formal and simple pavement inspection and management programs that are available to the engineer or maintenance professional. The essential ingredients of the PASER program consists of:

1 Inventory of all local roads, streets, or parking facilities
2 Periodically evaluate the condition of all the pavements
3 Use the condition evaluations to set priorities projects
4 Select the required maintenance treatment, if necessary.

The key element to surface evaluation management programs is to identify the different types of pavement distresses and determine their cause. Knowing what caused the pavement distress allows the appropriate maintenance treatment to be applied. There are four major categories of asphalt pavement distresses:

1 Surface defects: raveling, flushing, oxidation, and polishing
2 Surface deformation: rutting, shoving, settling, and heaving
3 Cracking: transverse, longitudinal, reflective, block, and alligator
4 Potholes.

The causes of pavement distresses and deterioration are environmental and structural. Environmental induced distresses are due to weathering, moisture, and aging. Loading causes structural induced distresses. Pavement deterioration usually occurs from both loading and weathering (TIC 1989).

Surface defects consist of raveling, flushing, oxidation, and polishing. Both high volume and low volume pavement can exhibit surface defects.

Raveling

Raveling or fretting is the progressive disintegration of the asphalt mixture from the pavement surface downward caused by the loss of the asphalt binder and/or dislodged aggregate particles (Plate 8.1). Small movements of individual aggregate particles, caused by traffic or water movement, develop sufficient tensile stresses and strains which exceed the breaking strength of the asphalt binder. Slight to moderate raveling has a loss of the fine aggregate, while severe raveling includes the loss of the coarse aggregate. Raveling is caused by moisture or solvent-induced stripping of the asphalt binder film from the aggregate, oxidation of the asphalt binder, and poor or low compaction during construction or insufficient asphalt binder content in the mixture. Raveling usually occurs in cold or wet weather. Raveling caused by traffic is usually during cold weather, when the stiffness of the asphalt binder is high (Whiteoak 1991). Raveling may also be caused by the abrasive action of tires or deicing sand, especially in intersections and parking lots. Oil or fuel spillage is a common cause of isolated raveling that occurs in parking stalls (Plate 8.2). Birdbaths and other low-lying

Plate 8.1 Raveling.

Plate 8.2 Oil spillage in a parking stall.

areas in parking lots can cause water to pool, leading to stripping and raveling (Plate 8.3). Raveling is a common surface defect on parking lots. The severity of raveling can be described in the following manner:

- *Low severity*: The fine aggregate or asphalt binder has started to wear away, exposing the tops of large aggregate. The surface is starting to pit. In cases of oil or fuel spillage, the stain can be seen, but the surface is hard and cannot be penetrated with a coin.
- *Medium severity*: Erosion further exposes large aggregate with at least one-third of the large aggregate diameter exposed. The surface texture is moderately rough and pitted. In cases of oil or fuel spillage, the surface is soft and can be penetrated with a coin (Plate 8.4).
- *High severity*: The fine aggregate and asphalt binder has been considerably worn away and large losses of the coarse aggregate have occurred (Plate 8.5). The surface texture is very rough and severely pitted. In cases of oil or fuel spillage, the asphalt binder has lost its binding effect and the aggregate has become loose.

Plate 8.3 Birdbath.

Plate 8.4 Moderate oil spillage damage.

Plate 8.5 High severity raveling.

Plate 8.6 Raveling due to stripping.

Plate 8.7 Wheelpath raveling and alligator cracking.

Moisture-induced damage or stripping is the most common cause of raveling (Plate 8.6). Pavements with high in-place air voids and low asphalt binder contents are likely candidates to experience moisture induced damage and raveling. Traffic will cause raveling in wheel paths to accelerate in severity (Plate 8.7). Highly oxidized pavements also have a tendency to experience raveling in the later stages of the pavement's life.

The extent of raveling can be measured by square meter of surface area. Low severity raveling can be corrected through an application of pavement sealer, or a fog seal. The fog seal can consist of an asphalt emulsion or a commercially available rejuvenating agent. The fog seal or rejuvenating agent is also a rehabilitation technique for oxidized pavements. Medium severity raveling can be corrected by an application of a pavement sealer, a slurry seal, a chip seal, or seal coat. In high severity raveling, the pavement must receive some type of rehabilitation. On localized severe raveling, the damaged section is removed and replaced with a patch. On larger sections of severe raveling, a surface treatment such as a seal coat or slurry seal, or even an asphalt overlay may be required. The remaining surface texture will dictate whether a surface treatment or an overlay is required. If the coarse aggregate is protruding over 12 mm, an overlay should be used instead of a seal coat (Local Road Research Board 1991).

Flushing

Flushing is the presence of excess asphalt binder on the pavement surface. Flushing appears as a film of asphalt binder on the pavement surface. The film can appear shiny and will become sticky when hot. Flushing is the result of free or excessive asphalt binder migrating upward to the pavement surface. Excessive asphalt binder content in the mixture or a very low air void content can cause flushing. Soft asphalt binders used in hot climates will also contribute to flushing. Flushing typically occurs in vehicle wheel paths and can be accelerated by traffic and hot weather. Flushing can also be a symptom of moisture damage. The asphalt binder is stripped away from the aggregate by water, with the asphalt binder rising to the pavement surface. Flushing on roadways is a potential safety hazard, since it creates low skid resistance. The severity of flushing can be described in the following manner:

- *Low severity*: Flushing has occurred to a slight degree and is only noticeable for a few days a year. The asphalt binder does not stick to tires or pedestrian traffic.
- *Medium severity*: Flushing has occurred to the extent that the asphalt binder sticks to tires and shoes during a few weeks of the year. Flushing will appear as black blotches on the pavement surface (Plate 8.8).
- *High severity*: Flushing has occurred extensively and a considerable amount of asphalt binder sticks to shoes and tires during at least several weeks of the year. Almost the entire pavement or wheel path will appear covered with a light film of asphalt binder.

Flushing is measured in square meters of surface area. Flushing is not easily corrected with the cause of the flushing usually occurring lower down in the pavement structure. Low severity flushing is usually not corrected until it becomes medium severity. Low and medium severity flushing is repaired by blotting it with fine sand. High severity flushing can be corrected by the removal of the wearing course. A seal coat can also be used to cover up the flushing, however, this is somewhat of a temporary solution.

Plate 8.8 Moderate flushing.

Polishing

Polishing is caused by repeated wear on the pavement surface. This wear is usually due to traffic but even heavy pedestrian traffic can cause slight polishing on parking lot pavements. Polishing is a smooth slippery surface that was created by traffic polishing off the sharp edges of the coarse aggregate. Some aggregate has a greater tendency to polish than other types of aggregate. The tendency to polish is influenced by the geologic source of the aggregate. Limestone and Chert have a greater tendency to polish than most other types of aggregates. Polished aggregate will have the asphalt binder film removed and be smooth to the touch. It will cause the pavement's skid resistance to be reduced. Polished aggregate, as a form of surface distress appears when the skid resistance values are low or have dropped significantly from previous ratings. It can be a serious surface distress for high-speed pavements, while it is a negligible distress on most parking facilities. The severity of polished aggregate is typically not rated, but considered as part of an entire skid resistance evaluation program. The elimination of the use of aggregates that easily polish from high-speed applications is the best preventive measure for polished aggregate distresses. High polished pavement surfaces can be repaired by the application of a seal coat or an overlay. The removal and replacement of the wearing course should also be considered.

Rutting

Rutting is surface deformation that is a longitudinal surface depression in the vehicle wheel paths. Rutting is sometimes called grooving or channeling. Rutting displaces the asphalt mixture in the wheel path, creating channels. In moderate or severe rutting, the surface of the pavement may uplift along the sides of the channel. Rutting is usually very visible after a rainfall. Even very slight rutting is visible after rain, so surface visual evaluations are best completed after a rainfall. Rutting can occur on any type of pavement including high and low volume roads. When ruts fill with rainwater or ice, vehicle hydroplaning can occur leading to safety hazards. It occurs where vehicles continually drive over the same spots in the roadway. It most frequently occurs on high volume roadways such as expressways or motorways. Rutting typically does not occur on parking lots with the exception of locations such as drives or vehicle entrances or exits. The random movement of vehicles through parking stalls etc., usually does not provide enough repeated tire movement to develop the rut.

Traffic compaction or displacement of unstable asphalt mixtures causes rutting. Rutting may also be caused by base or subgrade consolidation. Displacement or plastic flow type rutting is related to the design of the asphalt mixture. Low design air voids, excessive asphalt binder, excessive sand or mineral filler, rounded aggregate particles, and low voids filled in mineral aggregate (VMA) can all contribute to displacement rutting. The higher the traffic or loading level (in terms of ESALs) the more likely these variables will contribute to displacement rutting. For example, a very tender or "sandy" mixture that might perform with good results in a restaurant parking lot, will probably exhibit displacement rutting on an expressway. Consolidation rutting is caused by consolidation of the asphalt pavement including an uncompacted base or subgrade. Consolidation rutting can also occur in the mainline or wearing course, when not enough construction compaction has been completed. Proper compaction of all components of the pavement structure will prevent consolidation rutting. Rutting can also be caused by excessive surface wear on the wearing course. A typical example is caused by the abrasion of tire chains or studded tires. Abrasion rutting will also appear as raveling in the wheel paths. Mechanical deformation is rutting in the wheel path caused by insufficient structural strength in the pavement. It is caused by overloading, insufficient pavement thickness, or a strength deficient asphalt mixture. Mechanical rutting is usually accompanied by longitudinal or alligator cracking, which typically initiates at the bottom of the asphalt pavement layer where there are excessive tensile stresses (Rosenberger and Buncher 2000).

The severity of rutting can be described in the following manner:

- *Low severity*: rut depths less than 12 mm (Plate 8.9)
- *Medium severity*: rut depths of 12–25 mm (Plate 8.10)
- *High severity*: rut depths greater than 25 mm, hydroplaning can occur.

Rutting is measured by determining the depth of each rut over a square meter of surface area. On large roadways, it is often measured in centerline distance of rutting. The severity of rutting is determined by measuring the average depth of each wheelpath rut with a straight edge. The rut depths are measured over standard interval, such as every 20 m. The average rut depth is calculated by laying the straight edge across the rut, measuring its depth, then using the measurements taken along the length or centerline of the rut to compute its depth in millimeter.

Low severity rutting is usually not addressed until it becomes medium or severe rutting. Low severity rutting can be corrected through the use of a thin overlay or patch, or through a slurry seal or microsurfacing surface treatment. Medium or moderate rutting is often corrected by an overlay or microsurfacing. Medium and severe rutting can also be corrected by

Plate 8.9 Low severity rutting.

Plate 8.10 Medium severity rutting.

milling off the existing pavement, to a depth below the rut, and then replacing the pavement. The cause of severe rutting must be determined prior to the rehabilitation of the pavement. The base or subgrade may need to be reconstructed in cases of severe consolidation rutting. Coring or transverse full depth trenching of the pavement can usually reveal which layer the rutting is occurring in and to what extent.

Shoving and surface distortions

Shoving is a surface deformation that involves the wearing course to be displaced transversely across the pavement. Shoving is a permanent displacement caused by traffic loading. Shoving that appears as transverse undulations in the pavement surface consisting of closely spaced alternate valleys and crests is also called corrugations. When traffic pushes against the pavement surface, it produces a short abrupt wave in the pavement surface (Local Roads Research Board 1991). Shoving typically occurs at intersections or entrance/exit ramps off of freeways or motorways. The braking action of trucks and even automobiles is the cause of shoving. Localized distortion such as swell heaves, frost heaves, and others can also be grouped under shoving. In parking facilities and local roads, a common distortion is caused by the tree roots (Plate 8.11). Frost heave is usually caused by a boulder pushing up through the pavement due to frost expansion of the subgrade. A well-draining and boulder-free subgrade will prevent the occurrence of frost heaves. Settling of the subgrade, utility cuts, drainage structures, etc. can cause a localized depression to occur in the surface of the pavement. Proper compaction of the subgrade and proper fill and compaction of utility cuts and structures will prevent the occurrence of settling induced depressions.

Plate 8.11 Distortion caused by a tree root.

Shoving and localized distortion can also be observed at commercial refuse containers. The refuse truck unloading the container causes the distortion. The container is moved over the cab of the truck into the refuse receiving area of the truck. This action puts high stresses under the front tires, leading to depressions into the surface of the pavement (Plate 8.12). Placing the refuse container on a concrete slab reduces the occurrence of damage caused by the small wheels of the refuse container.

Shoving, corrugations, and distortions are forms of plastic movement of the asphalt mixture. The mixture lacks stability for its intended application. The low stability can be due to low air void content, excessive asphalt binder content, excessive sand, round aggregates, or an asphalt binder that is too soft for the application. A tender mixture that exhibits excellent handwork characteristics has the potential for distortion when placed under any significant

Plate 8.12 Distortion caused by a refuse truck.

Plate 8.13 Pothole.

loading. An asphalt mixture designed for high traffic applications is also very suitable for intersections and will prevent most cases of shoving and corrugations.

The rating of shoving is fairly simple:

- *Low severity*: The shoving or corrugations are just noticeable with regard to vehicle ride quality.
- *Medium severity*: The shoving or corrugations are noticeable and have an effect on the ride quality.
- *High severity*: The shoving or corrugations are very noticeable and vehicles are slowing down when approaching them.

Shoving and corrugations are measured in square meter of surface area. Localized distortions are evaluated on case-by-case basis, with significant heaves or depressions requiring immediate attention. The repair of shoving or corrugations requires milling the corrugations and replacing the material with a patch or overlay. Localized heaves or depressions are usually repaired in the form of a patch. Localized depressions will eventually become a pothole that requires filling (Plate 8.13). If a tree root or a boulder caused the heave, the root or boulder should be removed prior to patching.

Transverse cracking

An asphalt pavement will crack when temperature or traffic generated stresses and strains exceed the fatigue or tensile strength of the compacted asphalt mixture. Transverse cracking is a crack that extends across the pavement perpendicular to the pavement centerline (Plate 8.14).

Plate 8.14 Transverse crack.

Transverse cracks are usually spaced at some routine interval. Transverse cracks will initially be widely spaced, usually 15–20 m apart. Over time more transverse cracks will appear until they are closely spaced, only a meter or two apart. Transverse cracks begin as narrow or hairline cracks and will widen with age. If wide transverse cracks (6–12 mm) are not sealed, secondary and multiple cracks will develop parallel to the initial transverse crack. The edges of the wide cracks will deteriorate further leading to raveling of the adjacent pavement. This crack edge deterioration is known as "*spalling*." Sealing cracks that are greater than 6 mm wide will prevent water from intruding into the crack and damaging the pavement structure below.

Loading or traffic does not cause *temperature induced* transverse cracks. Pavement movement due to temperature changes and aging related shrinkage of the asphalt binder causes transverse cracks. *Load induced* transverse cracks occur on overlays of pavements that have unfilled joints or gaps in them. Load induced transverse cracks are sometimes mistaken as reflective cracks. Non-load induced transverse cracking can also be called "temperature cracking." During periods of cooling, the asphalt pavement tries to contract and the pavement for most practical purposes is laterally restrained. Thermal stresses develop and exceed the tensile strength of the asphalt pavement, resulting in a transverse crack. Transverse cracks will typically appear in an asphalt pavement 2–7 years after construction. Transverse cracks or joints in concrete pavements that reflect up through an asphalt overlay are grouped or described as reflective cracking and will follow the direction of the joint below the crack. A transverse crack that occurs above a wide joint, but propagates downward, is a load induced transverse crack. As traffic traveled over the bridged joint, excessive stresses and strains developed at the joint and asphalt pavement interface. The tensile strength of the asphalt mixture is exceeded and a crack developed at the interface. The crack then propagates toward the pavement surface.

There are not many preventative measures that a designer can do to prevent the occurrence of transverse cracks. Selecting the proper asphalt binder for the climate the pavement is located will reduce the occurrence of temperature induced transverse cracks. The development of the SHRP Performance Grades for asphalt binders was a direct result in the attempt to reduce thermal induced transverse cracks. Filling and maintenance of the cracks when they occur will prevent any serious damage to the pavement structure later on.

The severity of transverse cracking can be rated in the following manner:

- *Low severity*: A non-filled crack width is less than 6 mm or a filled or repaired crack of any width is present.
- *Medium severity*: A non-filled crack width is 6–12 mm or a non-filled crack of any width up to 12 mm is surrounded by light (hairline) and random cracking or a filled crack of any width is surrounded by light and random cracking.
- *High severity*: Any crack, filled or non-filled, surrounded by medium or high severity random cracking or a non-filled crack width over 12 mm or a crack of any width where the pavement immediately adjacent to the crack is broken or raveled.

Transverse cracks are measured in linear meters of crack. The length and severity of each crack should be recorded after identification. If the crack does not have the same severity level along its entire length, each portion of the crack having a different severity level should be recorded (Local Roads Research Board 1991).

The maintenance or filling of cracks will reduce the amount of water that can penetrate the pavement structure and it will retard the deterioration of the crack edges. Low severity transverse cracking is usually not filled until it becomes 6 mm in width or greater. The application of a fog seal or pavement sealer will seal narrow or hairline transverse cracks. Any surface treatment applied to a pavement surface will seal the hairline and transverse cracks that are less than 6 mm in width. Medium severity transverse cracks are sealed with crack filler. The crack is generally routed out first prior to the application of the crack filler. Chapter 9 provides the proper guidelines for the filling of cracks. High severity cracks are routed and filled, but is usually more feasible and economical to remove the damaged section of pavement and patch it. High severity transverse cracking can degrade the pavement to a point that reconstruction will be required. Reconstruction consists of milling and replacing the pavement with an overlay.

Longitudinal cracking

Longitudinal cracking is cracking that appears parallel to the centerline or laydown direction of the pavement. Longitudinal cracks are either load induced or non-load induced. Load induced longitudinal cracks occur in the wheel path or loading area of the pavement. Non-load induced longitudinal cracks can occur anywhere throughout the pavement, but are typically in the center or at the edge of the pavement (Plate 8.15). Load induced longitudinal cracks are a form of fatigue cracking which eventually involves into "Alligator cracking." Load induced cracking is addressed under the section on alligator cracking. Longitudinal cracking can also appear as a diagonal cracking across the pavement. This is a load-induced crack that is caused by traffic loading on an unsupported or poorly supported pavement edge. The subgrade settles or is pumped out at the pavement edge, resulting in a diagonal crack.

Non-load induced cracks usually occur where any longitudinal construction joint is present. These construction joints occur when lanes or sections of asphalt pavement are constructed adjacent to each other (Plate 8.16). Inadequate bonding of the sections during construction causes the crack (Plate 8.17). A crack can also occur in a longitudinal direction at patches for utility cuts (Plate 8.18) and other pavement repairs. A longitudinal crack underneath the pavement can also reflect up to the surface of the pavement. Longitudinal cracking can occur at locations in the pavement that has segregation of the asphalt mixture. Segregation induced cracking typically is caused by segregation that has occurred between the paver screed augers. Longitudinal cracking close to the edge of the pavement is due to insufficient shoulder support, poor drainage, and frost conditions.

Longitudinal cracking at construction joints can be reduced through proper construction, compaction, and sealing of the construction joint. Reducing the occurrence of paver segregation can obviously reduce longitudinal cracking. Screed auger segregation occurs under the auger gearbox, between the two screed augers, in the center of the screed. This type of segregation occurs in the center of the mat being laid, with the appearance of a narrow dark streak behind the screed. The problem is often corrected on modern pavers with the installation of kickback blades on the auger. The kickback blades are mounted on the end of each auger by the auger gearbox. The kickback blade moves or "tucks" some of the asphalt mixture underneath the gearbox, reducing the occurrence of segregation. Filling and maintenance of the cracks when they occur will prevent any serious damage to the pavement structure.

Plate 8.15 Longitudinal cracking.

The severity of longitudinal cracking can be rated in the following manner:

- *Low severity*: A non-filled crack width of less than 6 mm or a filled or repaired crack of any width is present.
- *Medium severity*: A non-filled crack width of 6–12 mm or a non-filled crack of any width up to 12 mm is surrounded by light (hairline) and random cracking or a filled crack of any width is surrounded by light and random cracking.
- *High severity*: Any crack, filled or non-filled, surrounded by medium or high severity random cracking or a non-filled crack width over 12 mm or a crack of any width where the pavement immediately adjacent to the crack is broken or raveled. Longitudinal cracks at the pavement edge that resulted in the edge of the pavement breaking away are considered high severity.

Plate 8.16 Longitudinal crack at construction joint.

Longitudinal cracks are measured in linear meters of crack. The length and severity of each crack should be recorded after identification. If the crack does not have the same severity level along its entire length, each portion of the crack having a different severity level should be recorded (Local Roads Research Board 1991).

The maintenance or filling of cracks will reduce the amount of water that can penetrate the pavement structure and it will retard the deterioration of the crack edges. Low severity longitudinal cracking is usually not filled until it becomes 6 mm in width or greater. The application of a fog seal or pavement sealer will seal narrow or hairline longitudinal cracks. Any surface treatment applied to a pavement surface will seal the hairline and longitudinal cracks that are less than 6 mm in width. Medium severity longitudinal cracks are sealed with crack filler. The crack is generally routed out first prior to the application of the crack filler.

Plate 8.17 Water induced longitudinal crack.

High severity longitudinal cracks are routed and filled, but it is usually more feasible and economical to remove the damaged section of the pavement and patch it. High severity longitudinal cracking can degrade the pavement to a point that reconstruction will be required.

Significant longitudinal cracking occurring at the pavement edge will require the investigation of the surrounding support and the adequacy of drainage. Large loading at the pavement edge, such as truck loading, is considered fatigue related (Plate 8.19). Proper pavement thickness and edge restraints such as a curb and gutter or a shoulder will help eliminate fatigue related edge cracking. High severity longitudinal cracking that occurs at the pavement edge will require the edge of the pavement to be reconstructed.

Plate 8.18 Utility cut patch.

Reflective cracking

Reflective cracks are cracks in asphalt overlays that reflect up from the pavement underneath. The sources of reflective cracks are either joints or existing cracks in the pavement underneath. The movement of the joint crack induces high tensile strains in the material above, eventually resulting in a crack at the interface. The crack propagates upwards toward the pavement surface from the joint below. The pavement underneath can also crack after the asphalt overlay, possibly resulting in a reflective crack in the overlay later on (Plate 8.20). The most common source of reflective cracks is from joint in concrete pavements that have been overlaid. Vertical movements in the concrete slabs caused by traffic or thermal expansion or contraction will result in shear stresses induced in the overlay (Whiteoak 1991). When the reflective crack is from a concrete joint, it can appear as either a transverse or

Plate 8.19 Edge failure.

longitudinal crack (Figure 8.1). The crack will follow the direction of the crack underneath. The crack typically occurs in the overlay within 1–3 years after resurfacing.

Knowing the condition or type of pavement under the overlay assists in determining if a transverse crack actually is a reflective crack. Reflective cracks are not loading or fatigue related, but traffic loading may cause a breakdown of the surface near the crack (Plate 8.21). If the pavement is broken or fragmented along a crack, the crack is known as a *spalled* crack. Vertical or horizontal movement in the pavement structure underneath causes reflective cracks. A crack, especially a transverse crack, that occurs above a joint in a pavement below, but is not caused by movement in the pavement structure, is not a reflective crack, but a load induced transverse crack. The transverse crack is caused because the joint below is of significant width (>6 mm) and is not filled with joint or crack filler. As a load such as traffic is applied, across the asphalt surface over the joint, the overlay cracks. The crack occurs

Plate 8.20 Reflective crack.

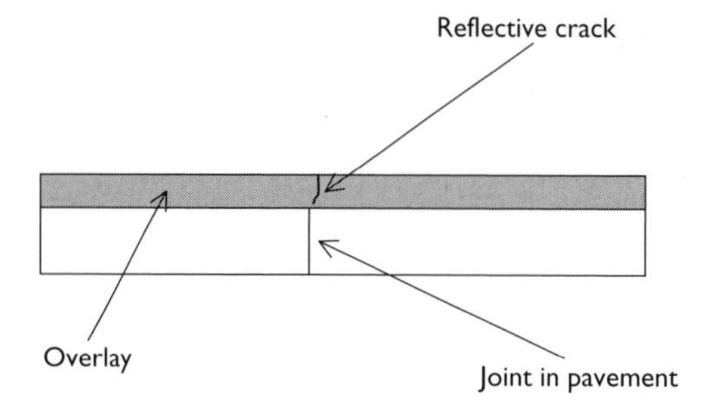

Figure 8.1 Reflective cracking.

because the overlay is attempting to bridge the gap, but does not have sufficient thickness to do so.

Preparation of the cracks or joints in pavement prior to an overlay will reduce the occurrence of reflective cracking. Such preparation is as extreme as rubblizing the pavement, eliminating all the cracks and joints by turning the pavement into what could be considered the same condition as a crushed aggregate base course. Another method is purposely cracking the pavement underneath in a very frequent pattern through the use of a mechanical or ultrasonic breaker. A very heavy roller, 20 tonnes or more follows the breaker, which seats the pavement. The methods available to most designers, especially for parking facilities or local roads is to either cover each joint with a strip of construction fabric or to use excessive thickness in the overlay to reduce the occurrence of the crack reflecting through. In that case the overlay should be at least 100 mm thick or greater. Reflective cracking is rated in the same manner as transverse and longitudinal cracking. The severity of reflective cracking can be rated in the following manner:

- *Low severity*: A non-filled crack width is less than 6 mm or a filled or repaired crack of any width is present.
- *Medium severity*: A non-filled crack width is 6–12 mm or a non-filled crack of any width up to 12 mm is surrounded by light (hairline) and random cracking or a filled crack of any width is surrounded by light and random cracking. The cracks are moderately spalled.
- *High severity*: Any crack, filled or non-filled, surrounded by medium or high severity random cracking or a non-filled crack width over 12 mm or a crack of any width where the pavement immediately adjacent to the crack is broken or raveled. The crack is severely spalled.

Reflective cracks are measured in linear meters of crack. The length and severity of each crack should be recorded after identification. If the crack does not have the same severity level along its entire length, each portion of the crack having a different severity level should be recorded. If a bump occurs at the reflective crack, it should also be recorded (Local Roads Research Board 1991).

The maintenance or filling of cracks will reduce the amount of water that can penetrate the pavement structure and it will retard the deterioration of the crack edges. Low severity reflective cracking is usually not filled until it becomes 6 mm in width or greater. Any surface treatment applied to a pavement surface will seal the hairline and reflective cracks that are less than 6 mm in width. Medium severity reflective cracks are sealed with crack filler. The crack is generally routed out first prior to the application of the crack filler. High severity reflective cracks are routed and filled, but it is usually more feasible and economical to remove the damaged section of the pavement and patch it.

Block cracking

Block cracking is a non-load related cracking that can also be called as shrinkage cracking. The pavement cracks in both longitudinal and transverse directions. Shrinkage cracks become interconnected, forming a series of large square or rectangular blocks (Plate 8.22). The size of the blocks is usually from half to three meters on each side of the block. The shrinkage cracks are usually caused by a volume change or shrinkage in either the

Plate 8.21 Spalled transverse crack.

Plate 8.22 Block crack.

asphalt pavement or the base or subgrade underneath. Asphalt mixtures that are graded on the "fine side" with an oxidized or stiff asphalt binder typically will display shrinkage or block cracking. Using an asphalt binder that is more appropriate for a high traffic or motorway application will increase the occurrence of shrinkage cracks on low volume or typical parking lots. Block cracking mostly develops on low volume roads and parking lots. The asphalt mixture that did not have enough compaction during construction does not densify under traffic. The asphalt binder in the mixture oxidizes, hardens, and becomes brittle. Pavements undergo thermal contraction and expansion and the stiff asphalt binder/fine aggregate combination becomes brittle and breaks, forming shrinkage or block cracks. Esthetically pleasing fine-graded mixtures used for parking facilities are susceptible to shrinkage and block cracking under certain environmental conditions. Selecting the proper asphalt binder for low volume parking applications will usually reduce the occurrence of block cracking. The proper compaction of the asphalt pavement will also reduce the occurrence of block cracking. Block cracking can be confused with alligator cracking. Alligator cracking is fatigue related and forms much smaller shapes with sharp angles.

Block cracking is rated in the same manner as other types of cracks. The severity of block cracking can be rated in the following manner:

- *Low severity*: A non-filled crack width is less than 6 mm or a filled or repaired crack of any width is present.
- *Medium severity*: A non-filled crack width is 6–12 mm or a non-filled crack of any width up to 12 mm is surrounded by light (hairline) and random cracking or a filled crack of any width is surrounded by light and random cracking. The cracks are moderately spalled.
- *High severity*: Any crack, filled or non-filled, surrounded by medium or high severity random cracking or a non-filled crack width over 12 mm or a crack of any width where the pavement immediately adjacent to the crack is broken or raveled. The blocks are well defined and the cracks are severely spalled.

Block cracking is measured in square meters of surface area that exhibits cracking. Block cracking of low severity can be ignored, until it becomes a rating of medium severity, or an application of a pavement sealer or surface treatment can be applied. Medium severity block cracks are sealed with crack filler followed by a surface treatment or pavement sealer. High severity block cracking requires the pavement to be reconstructed or be covered with an overlay.

Alligator cracking

Alligator cracks are interconnecting cracks that form small pieces ranging in size from 25 to 150 mm. The cracks have the same appearance as the pattern on the skin of an alligator (Plate 8.23). Alligator cracking only occurs from loading or fatigue. The cracks are caused by the failure of the pavement due to traffic loading or inadequate base or subgrade support. Tensile stresses and strains develop at the bottom of the pavement structure. These tensile stresses can exceed the tensile strength of the asphalt mixture, which results in a crack at the bottom of the pavement structure. Repeated loading of the crack causes it to

Plate 8.23 Alligator cracking.

Plate 8.24 Subgrade failure.

then propagate up toward the pavement surface. When the cracks are caused by traffic loading, they only occur in areas of repeated loading, such as wheel paths. Alligator cracking due to inadequate base or subgrade support will have a more random appearance throughout the pavement that receives loading (Plate 8.24). Lack of drainage can cause the subgrade to "wash out" or disintegrate, leading to excessive deflection of the pavement structure, causing alligator cracking. Of all the distresses, alligator cracking is the most commonly occurring distress on asphalt pavements serving parking facilities. The difficulty in providing proper drainage throughout the large spans of pavement in parking lots is the typical cause of the subgrade and base saturation and their corresponding failures.

Alligator cracking is considered a significant major structural distress. Alligator cracking is a symptom of insufficient structural strength in the pavement, weak subgrade, or overloading of the pavement. Insufficient structural strength is either due to the thickness not being adequate for the load being carried or an asphalt mixture deficiency. The actual cracking is caused by excessive deflection of the wearing course over an unstable subgrade or the base courses of the pavement structure (The Asphalt Institute 1983). Unstable base or subgrade support is due to them being saturated with water or being constructed of substandard materials such as clay, poor quality aggregates, etc. When the alligator cracking occurs over spans of asphalt pavement, such as in parking lots, the cause is usually due to poor subgrade support and poor drainage. If alligator cracking occurs in areas that typically do not see vehicle traffic or loading, the cause is a base or subgrade failure, usually do to improper drainage. Alligator crack begins with a crack at the bottom of the pavement structure. The tensile strength of the compacted asphalt mixture is exceeded and a crack at the bottom of the asphalt pavement structure occurs. Repeated loads or flexing of the asphalt pavement structure induces additional tensile stresses, causing additional cracking to occur and the cracks to propagate up toward the pavement surface (Figure 8.2).

Low severity alligator cracking will show up as several low severity longitudinal cracks in the wheelpath or loading area of the pavement. The crack can be distinguished from the typical longitudinal crack in that it has a random appearance and is not caused by base or subgrade movement. There is usually more than one, placed close together, parallel to each other in the wheelpath. Alligator cracking will progress into pavement disintegration, usually in the form of potholes (Plate 8.25). A permanent correction to alligator cracking requires the determination of what caused it, whether induced by loading or water. Subgrade drainage or repair and an increase in pavement thickness or strength are required for a

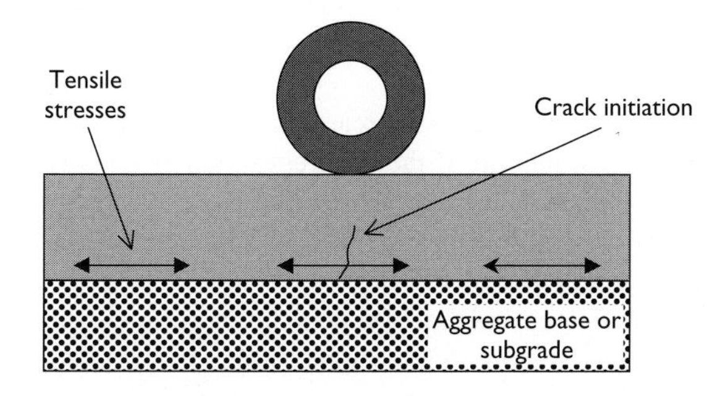

Figure 8.2 Initiation of alligator or fatigue cracking.

Plate 8.25 Alligator cracking beginning to form a pothole.

permanent solution to alligator cracking. The severity of alligator cracking can be rated in the following manner:

- *Low severity*: Fine, longitudinal cracks running parallel to each other with none or only a few interconnecting cracks. The cracks are not spalled. Initially there may only be a single crack in the wheelpath or pavement loading area.
- *Medium severity*: Further development of light alligator cracks into a pattern or network of cracks (Plate 8.26). The cracks may also be slightly spalled.
- *High severity*: The pattern of cracks has progressed so that the individual pieces are well defined and the cracks are spalled at the edges. Some of the pieces may move under traffic or loading. Pieces may begin to disintegrate, forming potholes (Plate 8.27). Pumping of the pavement may also exist.

Alligator cracking is measured in square meter of surface area. There are usually various degrees of severity within the same pavement section. If the different levels of severity can be easily distinguished from each other, they should be measured and recorded separately. If they cannot be easily distinguished from each other, the entire area should be rated at the highest severity present (Local Roads Research Board 1991).

Alligator cracking is a symptom of a deficiency in the pavement structure. A permanent solution would require removing the damaged section, repairing the subgrade if needed, and the installation of a full depth patch. If overloading of the pavement caused the alligator cracking, the permanent solution requires the pavement to receive an overlay or be

Plate 8.26 Medium to severe alligator cracking.

Plate 8.27 Severe alligator cracking.

reconstructed to proper thickness for the applicable loading. Applying a pavement sealer or seal coat to low severity alligator cracking will reduce water intrusion into the base and subgrade. If drainage is the cause of the cracking, improving the pavement drainage should be considered. If the cracking is due to insufficient pavement thickness or overloading, low severity alligator cracking will progress further even with improving the drainage or the application of a pavement sealer. Medium severity and high severity cracking is corrected through the installation of full depth patch overlay, or pavement reconstruction. If overloading is the cause of the alligator cracking, the overlay or reconstructed pavement should be designed with enough thickness to compensate for the additional loading.

Potholes

Potholes are small bowl shaped holes or depressions in the pavement surface (Plate 8.28). The diameter of the pothole is usually 1 m or less. The pothole will have vertical sides and sharp edges near the top of the hole. The pothole will grow in size and depth if water is present in the pothole (Plate 8.29). Potholes can also occur as irregular shaped depressions in the pavement. Potholes are caused by weaknesses in the pavement surface resulting from failure of the base or subgrade, poor drainage or structurally deficient pavement structure. A structurally deficient pavement is due to inadequate pavement thickness or an asphalt mixture that does not have enough residual asphalt binder to glue the aggregate pieces together.

Potholes can be produced when traffic erodes small pieces of an already weakened pavement surface. Potholes can also occur when the subgrade or base of the pavement fails and causes excessive deflection of the pavement surface. High severity alligator cracking will continue to become a pothole or a series of potholes. A pothole is caused by a structural deficiency in the pavement or by some external damage such as tree root ruptures (Plate 8.30). Severely spalled cracks can eventually cause potholes next to the crack (Plate 8.31).

Plate 8.28 Pothole.

Plate 8.29 Water induced pothole.

Plate 8.30 Rupture and pothole.

Plate 8.31 Transverse crack and pothole.

The level of severity of potholes depends on the size and depth of the hole. The frequency or number of holes in the pavement structure is also considered in rating the damage caused by potholes. The severity of potholes can be rated in the following manner:

- *Low severity*: A pothole of less than 25 mm in depth and a diameter of less than 450 mm. Isolated frequency in the pavement structure.
- *Medium severity*: A pothole of less than 25 mm in depth and a diameter greater than 450 mm or a pothole of 25–50 mm in depth and a diameter of less than 450 mm. A random number of potholes, but enough to be noticed frequently in the pavement. For example, over a 500 m^2 of pavement surface, there may be three or four holes.
- *High severity*: A pothole with a depth of greater than 50 mm or a pothole with a depth of 25–50 mm and a diameter greater than 450 mm. A frequent number of potholes in the pavement structure.

Potholes are measured by counting the number of those that are low, medium, and high severity and recording them separately. Potholes generally do not instantly appear but are created through another distress, such as alligator cracking. Wet and freezing weather tend to accelerate the appearance of potholes in the pavement. Pothole repair is completed by two methods:

- Filling the pothole with a patching or pothole filling mixture. The pothole is cleaned out and then filled with the patching mixture. The patching mixture is usually a cutback asphalt mixed with a dense graded aggregate. This is considered a temporary repair and is done until a more permanent solution is provided. Pavements in climates with freezing conditions will have their potholes filled during cold weather with a stockpile patching mixture by a maintenance crew. When the weather improves, the temporary filled pothole may be replaced with a full depth patch.

- Removing the pothole and the immediate surrounding distressed pavement and replacing with a full depth patch. The repair is made by cutting out the pothole all the way to solid pavement on both the sides and bottom. This cutout is then filled with a new base and wearing course. This method is considered a permanent repair because the damaged pavement is removed and is then reconstructed through the use of a full depth patch.

Chapter 9 provides the proper guidelines for temporary filling potholes and permanent patches. Potholes may also be filled with a pothole patching mixture and then the entire pavement receives an asphalt overlay.

Pavement condition rating

The surface and distresses in an asphalt pavement can be evaluated, rated, and given a score or index. A typical rating scale is from 0 to 10 with 0 indicating no defect or distress and 10 indicating a very severe or poor condition. Over the course of time, the condition of the asphalt pavement will progress up the scale. Utilizing the distress descriptions of low, medium, and high severity, a numerical value can be assigned to the distress or condition being evaluated. There are many rating systems available to the engineer or maintenance professional. ASTM D 6433, *Standard Practice for Roads and Parking Lots Pavement Index Surveys*, is one example. The following is a simple and common method that can be used on parking facilities, local roads, and streets.

- zero: *new construction*
- one: *no maintenance required*
- two, three, four: *low severity*
- five, six, seven: *medium severity*
- eight, nine, ten: *high severity*.

Ratings of 2–4 typically require a maintenance activity such as pavement sealers, crack filling, and seal coats. Ratings of 5–7 require crack filling, a substantial seal coat, or possibly an overlay. Ratings greater than seven will require either an overlay or reconstruction. The specific type of maintenance depends on the distress being evaluated. Maintenance guidelines can be tailored to the specific type or end use of the pavement being evaluated. At each score or rating, a specific maintenance activity can be suggested. For example, the owners of a shopping mall require that when raveling in the parking lot becomes a rating of three, the entire parking lot is treated with a single application of a pavement sealer. Future expectations of the use of the pavement should also be considered when determining what maintenance activity should be used. For example, the owners of the shopping mall plan on tearing down the mall and converting it to another use in 2 years. The parking lot is condition surveyed and a rating of 7–8 is applied to the alligator cracking present. It would not be economical to reconstruct the parking lot, when two coats of a pavement sealer would provide a temporary solution.

Rating a pavement will involve evaluating its condition over some length or square area. The condition of the pavement will not be consistent from 1 km to the next. The condition rating should represent the majority of the pavement. Averaging of the distress ratings will provide the best representation of the condition of the majority of the pavement. Isolated areas such as a single pothole should be addressed separately. The purpose of the condition rating is to provide a comparison of the various pavement segments. The pavement segments compete for the limited maintenance funding and personnel available to repair all the distresses. A pavement segment may consist of 5 km of motorway, one block of a city street, or one parking lot area of a shopping center. The actual size of the segments can be determined by the engineer or maintenance professional and will differ depending on the type of roadway or pavement system being evaluated. Regardless of the specific rating given to a pavement section, good engineering judgment (or just good common sense) should be used to determine which segment is to be repaired first. For example, a single pothole appears on the approach apron of a fast food restaurant, with five more potholes appearing in the rear of the parking lot. From both an esthetic and safety standpoint, repairing of the single pothole should take precedent over the five potholes in the rear of the parking lot.

Simple rating systems have been developed for the agency or facility owner that does not have access or funding for a significant engineering staff to evaluate pavements with specialized equipment such as a falling weight deflectometer (FWD), or even a Benkelman Beam. Both these devices are used by highway agencies to determine the structural condition of the pavement by measuring the deflection of the pavement under a static or slow moving load. A simple condition rating system involves using the previously given scale of 0–10 and the descriptive severity levels of each of the distresses. Table 8.1 coordinates the rating score with the distress and required maintenance activity.

It is important to consider that not all the distresses will occur in all the pavements. Some pavements may only have one type of distress (TIC 1989).

Individual forms can be created for the collection of the distress ratings. One form should be used for each pavement length or section being evaluated. The individual sections, which are similar in construction, service, and condition should be grouped or divided together. Rural pavement sections are usually 1–2 km long. Urban pavement sections are usually between one to four blocks. Parking lots can be divided into 2,500–5,000 m^2 sections for large facilities or they can be evaluated in a single evaluation for small facilities. Drainage conditions should be rated, so the evaluation is best completed after a significant rain shower. Any specific condition or unusual observation should also be noted. The frequency of the evaluation depends on the needs of the pavement facility owner, but an annual evaluation is usually adequate. The guidelines for when to take appropriate action will vary, depending on expected service life of the pavement and the amount of surface defects the pavement users or owners will tolerate. For esthetic reasons, a parking facility may receive a pavement sealer or a seal coat when its condition rating becomes two or three. Maintenance activities on a rural highway or motorway may not begin until its condition rating is four or greater.

Most cracking in a pavement initiates at the bottom of the pavement structure and propagates toward the pavement surface. Any increase in the pavement structure either at the point of initial construction or through the use of overlays will either reduce or eliminate the initiation of the crack or it will lengthen the time before the crack appears on the pavement surface. Many formal pavement management systems for highly traveled public roads include the planned use of an overlay at a predetermined later date during the life of the pavement.

Table 8.1 Simple condition rating system

Visible distress	*Maintenance activity*	*Surface condition rating*
None	None, new construction	0
None	None, almost new	1
• No longitudinal cracks, except reflection cracks • Occasional transverse cracks, widely spaced, 15 m or more • All cracks are <6 mm wide	Pavement sealer	2
• Very slight raveling • Surface shows traffic wear • Some longitudinal cracks occurring • Transverse cracks spaced 3 m or more apart • No existing patches	Crack filling	3
• Slight raveling • Longitudinal cracks • Transverse cracks spaced less than 3 m apart • First sign of block cracking • Slight to moderate flushing • Slight to moderate polishing • Occasional patch in good condition	Crack filling, surface defects filled	4
• Moderate to severe raveling • Longitudinal and transverse cracks show spalling • Block cracking <50% of surface • Extensive to severe flushing • Extensive polishing	Crack filling followed by pavement sealer or sealcoat	5
• Severe raveling • Extensive longitudinal and transverse cracks with spalling • Longitudinal cracking in the wheelpath • Block cracking >50% of surface • Patching in medium condition • Slight rutting or shoving (<12 mm deep)	Patching, overlay or surface treatment, potholes filled	6
• Severe longitudinal and transverse cracks with severe spalling • Severe block cracking • Alligator cracking <25% of surface • Patches in medium to poor condition • Moderate rutting or shoving (12–25 mm deep)	Overlay or multiple surface treatments	7
• Occasional potholes • Alligator cracking >25% of surface area	Patching and overlay	8
• Extensive potholes • Patches in poor condition • Rutting >25 mm deep	Base repair and pavement reconstruction	9
• Severe distresses with pavement disintegration	Total pavement reconstruction	10

Traffic growth requires the strength of the pavement structure to be increased at a later date. The overlay increases the strength of the pavement structure and puts a greater distance between the initiation points of cracks and the pavement surface. Thickness of the pavement also puts distance between the load and the bottom of the pavement structure. The greater the distance, the lower the tensile stresses will be at the bottom of the pavement. Economic restrictions and geometric considerations limit the pavement thickness from becoming excessive and to what is actually required for the projected loading.

Chapter 9

Pavement maintenance

Maintenance is an essential practice in providing for the long-term performance and the esthetic appearance of an asphalt pavement. The purpose of pavement maintenance is to correct deficiencies caused by distresses and to protect the pavement from further damage. Various degrees or levels of maintenance can be applied to all pavements, regardless of the end use. A motorway or expressway may require patching or an overlay, while a parking lot may receive an annual application of pavement sealer mainly for esthetic purposes. Pavement maintenance is either preventative or corrective. Preventative maintenance is the procedure performed to protect the pavement and decrease the rate of deterioration of the pavement quality. Corrective maintenance is the procedure performed to correct a specific pavement failure or area of distress. Some procedures will address both functions (Roberts *et al.* 1991). The sealing of cracks for the most part is considered a preventative maintenance measure. Patching is considered a corrective maintenance measure.

A condition rating of the pavement will help determine what pavement maintenance technique is necessary. How much maintenance funding is available and the expected service life of the pavement will generally dictate the frequency and level of maintenance it will receive. The expectations of the pavement owner also determine the maintenance level and frequency. For example, the owners of a hospital parking lot may require an annual application of a pavement sealer to the parking lot. The owners' expectations are that the parking lot has a new, clean, esthetically pleasing appearance at all times. The pavement sealer provides a jet-black appearance and an excellent contrast for the parking stall markings. This "crisp and clean" appearance coincides with what appearance the owners of the hospital want to portray to the general public. A highway agency that owns an interstate highway or motorway is usually more concerned with pavement ride quality, structural integrity, and service life than the cosmetic appearance of the pavement. The sealer used for a parking lot would not last or perform under the heavy traffic of a motorway. In that regard the motorway may receive an asphalt overlay at select intervals to update the ride quality and structural strength of the motorway.

Pavement maintenance can be described by two different categories:

- *Preventative maintenance*: Activities that prevent or reduce further damage to the pavement.
- *Structural maintenance*: Activities that repair or improve the structural integrity of the pavement.

Preventative maintenance activities include:

- Pavement sealers or sealcoats
- Crack filling and sealing
- Surface treatments.

Structural maintenance activities include:

- Pothole filling
- Patching
- Overlays
- Reconstruction.

The type of maintenance technique or procedure used depends on the type and severity of the pavement distress being repaired. Several maintenance techniques may need to be performed on the distressed pavement in order to provide satisfactory results. For example, with medium severity levels of transverse or longitudinal cracking, a pavement sealer or some type of a surface treatment, such as a chip seal or a slurry seal usually follows crack filling or sealing. Alligator cracking can be temporarily addressed with an application of a pavement sealer, but a more permanent solution would require removal of distressed pavement section, followed by an installation of a full depth patch. Some maintenance activities are required when changes are made to utilities or drainage structures located underneath the pavement. These utilities are accessed through a cut made in the pavement and a full depth patch must be installed to replace the removed piece of pavement. These patches need to be installed properly, since they have the potential for additional access points for water to enter the pavement structure. The patches can also affect the ride characteristics and the esthetic appearance of the pavement.

Crack sealing and filling

The most common and widely used maintenance activity for pavements, regardless of use, is crack sealing or filling. Crack sealing and filling is an inexpensive maintenance procedure that will significantly delay further deterioration of the pavement. Expressways, motorways, and small parking lots all benefit and undergo the application of crack sealing or filling. Cracks less than 3 mm wide are too narrow to be sealed or filled. A pavement sealer or surface treatment is adequate to treat these narrow cracks. Cracks that are 3–25 mm can be sealed or filled with an application of a crack sealant or filling material. Cracks that are greater than 25 mm wide are generally too wide to be sealed or filled and should be repaired through the use of a patching mixture or they should be cut out and replaced with a full depth patch.

Crack sealing and crack filling is actually two separate procedures:

- Crack sealing is the installation of a specially formulated crack sealing material either above or into working cracks using unique configurations to prevent the intrusion of water into the crack.
- Crack filling is the placement of crack filling material into non-working cracks to substantially reduce the intrusion of water into the crack. The significant differences are

that crack sealing is applied to working cracks and crack filling is applied to non-working cracks. Crack sealing involves placing sealing material in or on top of the crack. Crack filling involves placing filling material in the crack (Figure 9.1).

A working crack is a crack that has horizontal or vertical movements of 2.5 mm or more. A non-working crack is a crack that has no horizontal or vertical movements or the movements are less than 2.5 mm. Horizontal movement describes the crack expanding or contracting. Vertical movement describes deflections or fault movements that the crack may make. Whether a crack is working or not (moving or not), can usually be determined by the type of crack. Most transverse and reflective cracks are working cracks. Some longitudinal and diagonal cracks may also meet the 2.5 mm movement criterion. Materials that are placed in working cracks must adhere to the crack's sidewalls and be able to flex as the crack expands and contracts. The crack sealant must remain adhered to the crack walls after the crack expands and contracts. Non-working cracks include diagonal cracks, most longitudinal cracks, some block cracks, and some alligator cracks. Because of the relatively close spacing between non-working cracks, little movement occurs. Minimal movement permits the use of less specialized and less expensive crack filling materials. Table 9.1 provides details on deciding whether to seal or fill a crack (SHRP 1994). The filling of alligator cracking, especially extensive alligator cracking, is usually not cost effective since the pavement is failing due to fatigue. The intensity or the amount of alligator cracking within an area usually requires the alligator cracking to be treated with a surface treatment such as

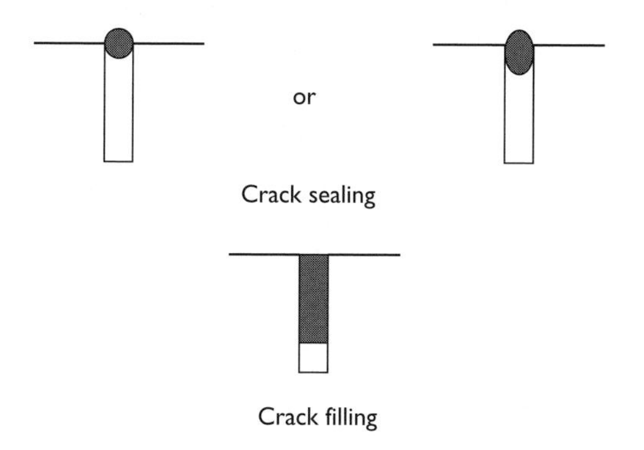

Figure 9.1 Crack sealing and filling differences.

Table 9.1 Sealing or filling cracks

Crack characteristics	*Crack repair procedure*	
	Sealing	*Filling*
Crack width, mm	4–19	4–25
Annual horizontal movement (mm)	⩾2.5	⩽2.5
Crack type	Transverse Reflective Longitudinal	Longitudinal Widely spaced block

a seal coat. If the alligator cracking is severe enough, removal and replacement with a full depth patch is usually warranted.

Crack sealing and filling are procedures that reduce or eliminate water penetration into the pavement structure. They are preventative maintenance procedures and should be applied as soon as cracking of any significance has developed. Repairing the crack reduces the occurrence of it becoming larger and more extensive. Crack sealing should be done as soon as possible after the working crack has developed. Since a working crack can expand and contract with temperature, the crack should be sealed during moderate ambient temperatures, in the range of 7–20 °C. At this temperature range, the working crack is partially expanded, allowing a sufficient amount of crack sealer to be placed in the crack. The crack is also at the middle width of its working range, allowing the correct amount of crack sealer into the crack, so that the sealer will not have to undergo excessive expansion or contraction due to both temperature and horizontal or vertical movements.

Crack filling can be completed at a much wider range of ambient temperatures. The properties of the filling material generally limits how cold of a temperature crack filling can occur at. At colder temperatures the non-working crack is at its greatest width, allowing more crack filler to be placed into the crack. In crack filling, the purpose is to fill the crack and not bridge or cover, as when crack sealing.

Crack sealing and filling materials

Cracks sealing or filling materials are usually formulated specifically for such use. There are many proprietary formulations available and in select cases, generic materials such as asphalt emulsions can also be used. The term *sealant* can be used to describe both crack sealers and crack fillers or just crack sealers. Crack sealers and fillers are either applied hot or applied cold. Hot or cold applied crack materials are also known as *hot or cold pour* sealant. They also are either thermoplastic or thermosetting. Thermoplastic materials become soft when heated and hard when cooled. Thermoset materials are permanently hard, even when heated. Table 9.2 provides typical crack sealing and filling materials. All sealant should provide at least the following:

- Flexibility
- Thermal stability
- Adhesivity
- Durability
- Resistance to flow.

Table 9.2 Crack sealant and filler types

Thermoset	*Thermoplastic*	
	Hot applied	*Cold applied*
Chemically cured silicone sealant	Asphalt binder (cement) Polymer modified asphalt binder Asphalt rubber Fiberized asphalt binder Mineral filled asphalt binder	Asphalt emulsion Cutback asphalt Proprietary or formulated asphalt emulsion sealer Proprietary or formulated cutback asphalt sealer

The requirements for a crack sealer (sealant) or filler is that it must be easily applied and readily flowable in order to facilitate its application and take up expansions and contractions of the crack and the pavement structure. It must also be flexible and be capable of absorbing many cycles of movement without any permanent distortion. Once in the crack, the sealant or crack filler should exhibit thixotropy and not flow under its own weight. The properties of the material must be maintained during exposure to aggressive climatic conditions, such as snow, ice, rain, and temperature extremes. High performance sealers are resistant to the effects of water, UV radiation, and weather without hardening or stiffening in service, while maintaining strong adhesion to the crack (Morgan and Mulder 1995).

Asphalt based crack sealers and fillers will for the most part, provide adequate strength and adhesive properties to perform as a crack sealer or filler. During ambient and high temperatures the adhesive strength of an asphalt based material is usually greater than its cohesive strength. This property makes the material extremely difficult to remove from the crack. The stresses and strains of the movement of cracks are countered by the flexibility and cohesive strength of the asphalt based crack sealer or filler. At low temperatures, asphalt based materials can become stiff and brittle. Cracks can widen or open at low temperatures and expose the crack repair material to tensile strains. During these cases, the asphalt based material's cohesive strength is greater than its adhesive strength resulting in a loss of bond to the crack (Morgan and Mulder 1995). Many polymer or rubber modified asphalt sealers and fillers can perform satisfactorily in a wider range of temperatures than those materials that are not modified. The polymer or rubber improves the flexibility of the asphalt binder. The degree or amount of flexibility depends on the type and amount of rubber or polymer. In the case of rubber, the method of introduction (mixed or melted) also affects the flexibility of the crack sealer. With regard to crack sealers or fillers, rubber modified, especially low modulus rubberized asphalt sealers, tend to be more flexible sealers than an asphalt sealer that is modified with a styrene butadiene styrene (SBS) or other type of polymer (SHRP 1994). The enhanced flexibility of the modified asphalt sealant enables it to follow both rapid and slow crack movements, even at temperatures down to −20 °C. High temperature performance is also enhanced with no flowing, or sagging of the sealant (Morgan and Mulger 1995). Unmodified asphalt based sealants are limited for the most part to use as crack fillers for non-working cracks. Fiberized or mineral filled crack sealants also perform better as crack fillers than as crack sealers, since they are not elastic enough to be used in working cracks.

There are many proprietary sealers and fillers available to engineer or maintenance professional. Some are one or two component systems that chemically react to bond in the crack. A self-leveling crack sealant is a sealant that flows through the crack, requiring no tools to spread the sealer into the crack. Table 9.3 illustrates the various types of crack fillers and sealers and some suggested specifications.

Many government agencies, including the United States Federal Aviation Administration and the United States Army Corps of Engineers have developed their own specifications for crack sealers and fillers. Adopting a specification that is prevalent to the local geography or that is widely used is recommended when specifying a material for crack sealing or filling. Table 9.4 provides the recommended material based on the property desired (SHRP 1994). Tables 9.3 and 9.4 are not all-inclusive, as materials are continually being improved and added.

Any of the rubberized or the newer r polymer modified or chemically reactive sealants will generally provide satisfactory results in most applications. The chemically reactive sealants are typically the most expensive. Any of the rubberized sealants will also provide

Table 9.3 Crack sealers and fillers

Material type	*Product type*	*Application*	*Typical crack width* (mm)	*Typical specification*
Asphalt cement	AC-10, AC-20, AC-30, AC-40, PG52-xx to PG76-xx	Filling	≤6	ASTM D3381 ASTM D5078 ASTM D6373
Polymer modified asphalt cement	AC-20, AC-30, AC-40, PG64-xx to PG82-xx	Sealing Filling	≤25	ASTM D3381 ASTM D3406 ASTM D5078 ASTM D6373
Asphalt emulsion	CRS-2, RS-2, CMS-2, MS-2, HFMS-1	Filling	≤6	ASTM D977 ASTM D2397
Polymer modified asphalt emulsion	CRS –2P	Sealing Filling	≤6	ASTM D2397
Mineral filled asphalt binder		Filling	6–25	ASTM D5078 ASTM D6690
Fiberized asphalt binder	Proprietary products	Filling	6–25	ASTM D5078 ASTM D6690
Asphalt rubber	Asphalt rubber Proprietary products	Sealing Filling	6–25	ASTM D3406 ASTM D5078 ASTM D6690
Rubberized asphalt sealant	Proprietary products	Sealing	6–25	ASTM D1190 ASTM D3406 ASTM D6690
Low modulus rubberized asphalt sealant	Proprietary products	Sealing	6–25	ASTM D1190 ASTM D3406 ASTM D6690
Chemically reactive	Proprietary products	Sealing	≤25	ASTM D5893

satisfactory results for sealing cracks in parking lots. Non-tracking formulated sealers are especially recommended for parking lots, local roads, and driveways. Tracking is vehicles' or pedestrians' picking of the top of the sealer or filler and "tracking" or smearing on the pavement.

Cracking filling is not as a demanding application, since these cracks are non-working. Any of the materials listed for crack filling will perform satisfactorily. The introduction of fiber or mineral matter into a crack sealant provides bulking and viscous properties to the crack filler. The life span of a crack sealer or filler can be up to 5 years, with crack fillers generally not lasting as long as crack sealers, especially the formulated or rubberized sealers.

Crack preparation

Prior to the application of the crack sealer or filler, the crack needs to be prepared to receive the sealant or filler. Proper preparation of the crack provides for better adhesion of the material to the crack. At the very minimum, the crack should be cleaned and dried prior to sealing or filling. A complete crack preparation program would consist of the following:

- Routing or sawing of the crack
- Cleaning and drying
- Application of the sealer or filler
- Finishing and shaping of the sealer or filler
- Blotting of the sealer or filler.

Table 9.4 Various attributes of crack sealers and fillers

Attribute	*Sealant or filler*
Easement of placing	Asphalt cement Polymer modified asphalt cement Asphalt emulsion Polymer modified asphalt emulsion Mineral filled asphalt binder Fiberized asphalt binder Asphalt rubber Rubberized asphalt sealant Low modulus asphalt sealant
Short cure time	Asphalt cement Polymer modified asphalt cement Mineral filled asphalt binder Fiberized asphalt binder Asphalt rubber Rubberized asphalt sealant Low modulus asphalt sealant Chemically reactive
Adhesivity	Asphalt cement Polymer modified asphalt cement Asphalt emulsion Polymer modified asphalt emulsion Mineral filled asphalt binder Fiberized asphalt binder Asphalt rubber Rubberized asphalt sealant Low modulus asphalt sealant Chemically reactive
Resistance to softening when cured	Polymer modified asphalt cement Polymer modified asphalt emulsion Asphalt rubber Rubberized asphalt sealant Low modulus asphalt sealant Chemically reactive
Flexibility	Polymer modified asphalt cement Polymer modified asphalt emulsion Asphalt rubber Rubberized asphalt sealant Low modulus asphalt sealant Chemically reactive
Durability or weather resistance	Rubberized asphalt sealant Low modulus asphalt sealant Chemically reactive
Resistance to abrasion and tracking	Asphalt rubber Rubberized asphalt sealant

Crack cutting

Cracks are cut through the use of a pavement saw or router. Crack routing is the preferred terminology when the cracks are cut with a router. The purpose of cutting or routing the cracks is to prevent the wearing away of loose crack edges and to provide a reservoir for the crack sealant. Cracks are typically cut when they are 6 mm wide or greater. Cutting the crack removes some of the loose or deleterious material from the sides of the crack. Cutting the crack also widens the crack enough to permit an adequate amount of sealer into the crack. The crack should only be slightly widened and not deepened. Cutting the crack creates a reservoir for the crack sealant. Non-working cracks or cracks that will only be filled usually do not require cutting. Routing is the preferred method of cutting cracks. A pavement router is a machine that is equipped with a vertical spinning carbide bit that cuts the crack in the same manner as a wood working router cuts a groove into a piece of wood. The vertical spindle router is one type of pavement router that is very maneuverable and provides the cleanest cuts. The rotary impact router is more productive than a vertical spindle router, but can leave spalls and fractures in the crack (SHRP 1994). Cutting or routing of cracks is an additional step in crack sealing operation that adds extra costs and time to the operation. Only cracks that need to be widened to hold the crack sealant or have an extensive amount of fragile material at the edges should be cut.

Crack cleaning and drying

Adequate cleaning and drying of the crack is the most important procedure in the crack preparation program. A crack sealant or filler will adhere the best to a clean and dry crack. If the crack is dirty the sealant will adhere to the dirt particles and not to the sides of the crack. If the crack is wet, the sealant will not be able to displace all of the water and adhere to the sides of the crack. There are four methods to cleaning and drying a crack:

- Air blasting
- Hot air blasting
- Sand blasting
- Wire brushing.

Air blasting is completed through the use of a blower, similar to a leaf blower, or through the use of compressed air. Blowers move a large quantity of air at a relatively low pressure to clean the pavement surface and subsequent cracks. The low pressure does not usually dislodge all of the deleterious materials from the crack. Air blasting by compressed air of at least 700 KPa provides a more thorough cleaning and drying of the crack than what can be provided by an air blower. The compressed air is provided by a mobile compressor and applied through the use of a wand into the crack. The air blows out any loose material and any dirt and water. Two passes of the air blasting wand through the crack are usually necessary to completely remove any dirt and excessive moisture.

Hot air blasting incorporates the use of a mobile air compressor and heat lance. The compressed air flows through the heat lance into the crack. The heat lance operates at temperatures of 1,300 °C or greater, creating superheated air that effectively dries out the crack. The compressed air removes any loose material and dirt. The heat temporarily softens the sides of the crack, and if crack sealing or filling occurs immediately after the hot air blasting, the crack sealant or filler will have improved bonding to the sides of the crack. This advantage has limited use and is secondary to the crack drying ability of the hot air lance. Two passes of the hot air lance through the crack are usually required. The high temperature of the lance can damage the adjacent pavement, if care is not used.

Sand blasting consists of a mobile air compressor and a sandblasting wand. The sand blasting effectively removes any deleterious material and dirt from the crack. Sand blasting also leaves a smooth surface on the sides of the crack. Sand blasting is usually followed by air blasting, to remove any leftover sand blasting media and moisture from the crack.

Wire brushing of the crack is effective in removing deleterious material and dirt, but does not remove moisture very well from the crack. Power wire brushes are available to mechanically remove any dirt from the crack channel. A typical power wire brush consists of a pavement saw with a wire brush installed inplace of the saw blade. Wire brushing should only be done on dry cracks with very little deleterious or loose material in the crack (SHRP 1994). Wire brushing can be followed by air blasting to remove any remaining dirt and moisture from the crack.

Sealer or filler application

Prior to the application of the crack sealer or filler, several decisions should be made.

1 Determine if the crack is a working or non-working crack.
2 Determine if the crack should be sealed or filled.
3 Determine the type of sealant or filler to be used.
4 Determine the desired configuration of the sealer or filler.
5 Determine if a backer rod will be placed in the crack.
6 Determine the amount of sealant or filler to be used.

Working cracks should be sealed and non-working cracks can be sealed or filled. Most rubberized crack sealers will meet the requirements of both crack sealing or filling along with several other choices listed in Table 9.3.

Crack sealers and fillers can be placed into the crack in various configurations. The configurations can be grouped into four types:

- Flush fill
- Reservoir
- Overband
- Combination.

The flush fill configuration is completed by placing the sealant into an uncut crack, with the excessive sealant struck or planned off. The reservoir configuration consists of placing the sealant into a cut crack and filling the crack until it is slightly below or level with the pavement surface. The overband configuration consists of placing the sealant into and over an uncut crack. The sealant left over the crack is then shaped into a band using a special crack sealing squeegee. If the sealant is not shaped into a band and is just left as a "hump" the configuration is known as a *capped* configuration. The application of the squeegee into a band creates a *Band-Aid* configuration. The Band-Aid is usually 75–125 mm wide. The thickness of the Band-Aid should be about 3 mm thick. A combination configuration consists of the sealant placed into and over a cut crack. A squeegee is then used to shape the sealant into a band that is centered over the crack reservoir. Table 9.5 provides which configuration to use for cut or non-cut cracks. Crack sealing usually requires the cracks to be cut, while when crack filling, cutting of the crack is usually not necessary. Table 9.6 provides the various crack configurations in terms of crack sealing or filling.

Table 9.5 Cut or uncut crack sealing and filling configurations

Uncut	*Cut*
Flush fill	Reservoir
Overband	Combination

Table 9.6 Crack sealing and filling configurations

Crack sealing	*Crack filling*
Reservoir	Flush fill
Combination	Overband

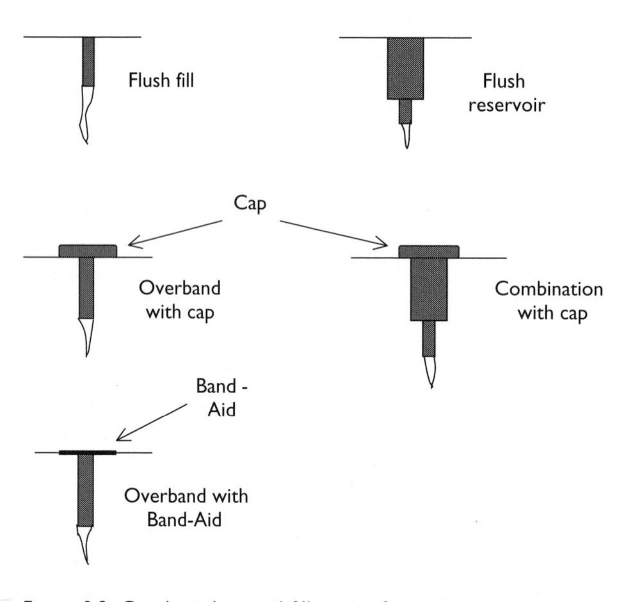

Figure 9.2 Crack sealing and filling configurations.

Figure 9.2 illustrates the four configuration types, flush fill, reservoir, overband, and combination.

On large or deep cut cracks the installation of a bond breaker – a foam backer rod may be placed at the bottom of the crack prior to the application of the sealant. The purpose is to prevent the sealant from forming, running all the way down into the deep portion of the crack. The backer rod adds some integrity to the sealant and reduces the potential for the sealant to shrink or collapse into the crack. Straight working cracks are usually the only cracks that can accept the use of a backer rod. The crack depth is at least 25 mm to allow for the installation of the backer rod. The diameter of the backer rod should also be 25 percent wider than the width of the crack and about half the depth of the crack. For example a 12 mm wide crack that is 35 mm deep should use a backer rod that is 15 mm in diameter. The backer rod is then squeezed into the crack and then filled out in the crack. A small wheel tool, similar to an insect screen repair tool, is used to force the backer rod into the crack.

The desired surface finishing characteristics of the sealant will also determine the type of crack sealant configuration. The most desirable attribute for esthetic and ride quality reasons is a crack sealant that is flush with the pavement surface. The condition of the crack however will also dictate whether to use a flush fill configuration or to use an overband or some combination. Overband and combination configurations will not be as esthetically pleasing in appearance as a flush filled crack. Esthetic criterion is considerably more important in parking facilities than would be on a rural highway. Overband configurations will perform better than flush fill configurations on cracks that have a considerable amount of edge disintegration. The sealant in the cap or Band-Aid provides better coverage and sealing characteristics on the deteriorated edge than a configuration that is just flush with surface. The cap or Band-Aid provides additional coverage to the crack. In certain instances, the use of a Band-Aid or cap may eliminate the need for routing of the crack, since the configuration can seal over any rough or loss edges. The advantage of the cap configuration is that one less step, the smoothing out of the crack sealant, is needed. The disadvantage of the cap configuration is that it leaves a slight bump in the pavement surface. The Band-Aid configuration smoothes out the bump, but the wide band of sealant (75–125 mm) is not very esthetically pleasing.

Crack sealers are typically applied into the crack by the use of a pour pot or through an applicator wand or distributor. The pour pots are either hand held or equipped with small wheels. Pour pots are labor intensive and usually used for small crack sealing or filling operations. Hot applied sealants, especially rubberized sealants are applied through the use of a melter-applicator. The melter is equipped with a kettle that preheats the hot applied sealants to about 200 °C. The sealant is then pumped into a wand type applicator and applied into the crack. Some applicators are equipped with an attachment that allows the sealant to be applied and finished in one step. Cold applied sealants, obviously do not require the use of a melter or kettle and are just applied into the crack through the use of a pour pot or an applicator that is attached to a distributor.

The crack sealant should be applied into the crack with the applicator wand in the crack channel. When using a pour pot the funnel or discharge portion should be as close to the crack as possible. The crack should be filled up from the bottom of the crack channel. The application should be in one continuous motion. The crack should be filled up flush to the surface for flush fill configurations, or slightly overfilled for cap configurations. For overband configurations either enough material should be left to be spread into a band that is 75–125 mm wide; or a second application may be needed.

Shaping and finishing of the crack sealant should immediately follow the application of the sealant. Two persons may be required, one to operate the applicator wand and the other to finish or shape with the squeegee. The squeegee should closely follow the applicator wand. The squeegee is in the shape of a "U" or "V," so that the sealant can be concentrated over the crack. For flush fill configurations the strike off of the squeegee will be flat. A pre-cut mold can be made to create the Band-Aid configuration (SHRP 1994).

One aspect of any crack filling or sealing application is determining the required amount of material to complete the operation. The required information for determining the sealant quantities is:

- Average crack width and depth
- Total length of cracks to be filled
- The density of the sealant
- Wastage factor, usually 15 percent.

The average crack width and depth provides the average cross-section area of the crack. Multiplying this value by the length of the crack provides the volume of the crack to be filled. Multiplying this value by 1.15 provides the required amount of sealant in terms of volume. The volume can then be converted to weight by using the density of the sealant. Example 9.1 illustrates these steps.

Example 9.1

Cracks in a shopping center parking lot were measured after cutting and cleaning. The average crack width is 13 mm. The average crack depth is 25 mm. The length of all the cracks to be filled is equivalent to 326 m. The specific gravity of the crack sealant is 1.12.

Solution

The average cross section of the crack is: $13\,\text{mm} \times 25\,\text{mm} = 325\,\text{mm}^2$ or $0.000325\,\text{m}^2$.
The total volume of the cracks to be filled is: $326\,\text{m} \times 0.325\,\text{m}^2 = 0.106\,\text{m}^3$.
The total required amount of sealant, including wastage is: $0.106\,\text{m}^3 \times 1.15 = 0.122\,\text{m}^3$.
$0.122\,\text{m}^3$ of sealant is equivalent to 122 l.
The weight of sealant required is: $122\,\text{l} \times 1.12\,\text{g/cm}^3 = 136.6\,\text{kg}$.

When the cracks vary substantially in depth or width, the project can be broken into sections and the total crack volume can be estimated by grouping similar sized cracks together and determining the volume in each similar section and then adding the groups together.

Blotting of the crack sealant after application reduces the pickup of the sealant by traffic and pedestrians. Blotting of the crack should be done immediately after the finishing or shaping of the sealant. Blotting can consist of applying a coating of sand on top of the crack, or by spreading a light layer of sand throughout the crack sealing project. This will also provide some abrasive texture for improved skid resistance. Blotting the crack with paper is also effective.

Pothole filling and patching

The patching of a pavement is a permanent solution to a pavement distress, usually a high severity distress. Pothole filling is a somewhat temporary measure, which can also be considered patching. Filling of the pothole will not correct the deficiency in the pavement. The purpose of pothole filling is to temporarily eliminate the pothole as a road hazard and nuisance. Potholes are the result of a rapid disintegration portion of the pavement that was not repaired in time. The purpose of patching is to permanently repair the portion of the pavement that is defective due to:

1 Pavement distress, such as alligator cracking, severe transverse cracking, severe block cracking, etc. (Plate 9.1).
2 Repair a cut in the pavement due to a utility cut or repair (Plate 9.2).

Plate 9.1 Pavement distress patch.

Plate 9.2 Utility cut patch.

Plate 9.3 Vibrating plate compactors.

Potholes can occur on any type of asphalt pavement, including local roads, parking lots, and freeways. Pothole filling can be part of a routine maintenance program or it may be performed as an emergency repair. Potholes can be a hazard to vehicles and pedestrians (Plate 9.3). Potholes generally surface or become more apparent under harsh weather conditions, such as freezing or very wet weather. Freezing and the thawing and wet weather tend to accelerate the progress of a pavement distress into a pothole. In northern climates, the spring season generates the most potholes. Emergency pothole filling usually occurs during inclement weather. Alligator cracking is the most common generator of potholes. Repairing the underlying symptom of the distress that caused the pothole will result in the repair becoming permanent.

Pothole filling is a very simple process. There are four recognized procedures for pothole filling:

1 Throw and go
2 Throw and roll
3 Semi permanent
4 Injection.

"Throw and go" is the least permanent practice in pothole filling. It is also the most common practice. Throw and go is simply placing a patching mixture or material into a pothole and proceeding or *going* onto the next pothole. The patching mixture is put into the pothole usually without cleaning or drying the pothole out. The term *throw* comes from the maintenance crew throwing the patching mixture from the back of a dump truck into the pothole.

The mixture used is a cold patch or stockpile mixture. The compaction of the mixture comes from traffic. Throw and go is a high production method of filling potholes. It is also usually done as a pothole filling program during the cool weather months.

"Throw and roll" uses the same procedure as throw and go, but with one additional step. After putting the patching mixture into the pothole, the dump truck rolls over the patch one or two times. The tires of the dump truck compact the patching mixture into the pothole. The somewhat compacted patch will remain in the pothole for a longer period of time than the throw and go patch. The traffic is also less likely to force the mixture out of the pothole. The compacted mixture is dense and less susceptible to moisture damage, than the uncompacted mixture in the throw and go pothole. The mixture used is a cold patch or stockpile mixture. The goal of both the throw and go and throw and roll methods is to eliminate the pothole as a safety or esthetic issue and not to repair the distress permanently.

The semi-permanent pothole filling method provides a pothole repair that is almost as permanent as a full depth patch. A formal maintenance program for pothole repairs usually uses the semi-permanent filling method. Throw and go and throw and roll are commonly done during emergency or inclement weather repairs. The semi-permanent pothole filling method consists of the following steps:

1 Remove excessive or loose material from the pothole.
2 Remove excessive water from the pothole.
3 Cut or square the sides of the pothole until vertical sides exist. This may involve sawing part of the pothole out. The pothole becomes shaped like a square or a rectangular box (Figure 9.3).
4 Place the patching mixture into the hole.
5 Compact the patching mixture with a vibrating plate compactor (Plate 9.4) or a small roller.

The significant portion of this pothole filling method is the cutting out of some of the damaged or loose pavement surrounding the pothole. The repair is somewhat structurally sound and semi-permanent. The cleaning and drying of the pothole allows the patching mixture to adhere better to the sides and bottom of the pothole. In many cases the repair mixture is a dense graded or fine graded hot mix asphalt (HMA) mixture.

The spray injection procedure for filling potholes consists of the following steps:

1 With compressed air, blow out the loose material and water from the pothole.
2 Spray a tack coat of asphalt binder or an asphalt emulsion on the sides and bottom of the pothole.
3 With a specially designed injection device, blow the patching mixture into the pothole.
4 Place a blotting cover of sand or small aggregate on the filled pothole (SHRP 1994).

Figure 9.3 Semi-permanent pothole repair.

Plate 9.4 Pothole hazard.

The blowing or injection of the patching mixture actually compacts the mixture into the pothole. The compactive effort is at least as effective as the rolling dump truck tire and almost as effective as the vibrating plate compactor. The mixture used is a cold patch or stockpile mixture. The advantage of the spray injection procedure is that it eliminates the compaction step of the semi-permanent pothole repair procedure.

Pothole filling materials

Pothole filling or patching mixtures must possess workability characteristics over long periods of time. It is not unusual to expect a stockpile of pothole filling to remain workable for up to 12 months. The mixtures are called "cold mixtures," "stockpile mixtures," "depot stock," or "cold patch mixtures," since no heat is needed to store, place, or compact the mixture. The mixtures are either generically produced by a local mixing plant or they are proprietary or "branded" mixtures. Proprietary pothole filling mixtures are even available at hardware stores in small buckets. Long workability periods allow maintenance crews to fill potholes over extended periods of time and throughout a season. Pothole filling mixtures are usually produced in a single production at a mixing plant prior to inclement weather setting in. The mixture can be stockpiled at the mixing plant or at the customer's location, such as a public works maintenance yard. Stockpiling the mixture allows pothole filling or patching crews to haul only the material needed for a particular maintenance activity. Homeowners and other small applications of patching mixtures benefit from the availability of the mixture in buckets or bags.

Since pothole filling or patching mixtures require an extended shelf life, HMA generally cannot meet the extended workability characteristics desired. HMA is usually not available year around, especially in very cold climate. However, HMA mixtures do have application in full depth patching programs and generally will last longer than some of the highly workable pothole filling mixtures. Either asphalt emulsion or cutback asphalt provides the long-term workability characteristics of a pothole filling mixture. The addition of solvents, such as kerosene, even in the asphalt emulsions provides the extended shelf life of cold pothole patching mixtures. Soft asphalt binders are also selected to increase the workability characteristics. Asphalt binders that have a penetration value of 200 mm or greater are typical. Whether the pothole filling or patching mixture is used immediately or stockpiled for later use determines the type of asphalt emulsion or cutback used. Table 9.7 provides some recommended asphalt binders for pothole patching mixtures.

The use of an antistripping agent or adhesion promoter is recommended for cold pothole or patching mixtures. The additive is added to the asphalt binder in a dosage range of 1–3 percent by weight of the asphalt binder. The additive helps in coating of the damp aggregate, provides additional adhesion of the mixture to damp potholes, and increases the mixtures resistance to water damage. The antistripping agents or adhesion promoters are proprietary chemicals that are usually fatty amine or amidoamine based. These agents can also increase the workability of the pothole filling mixture by lowering the surface tension of the asphalt binder and lowering the surface contact angle, all causing easier spreading of the asphalt binder on the aggregate.

The aggregate used for cold stockpile patching mixtures should be a high-quality crushed stone with a limited amount of fine material. The quality of the aggregate can be specified by using the values in Table 9.8. A typical gradation for a stockpile patching mixture is given in Table 9.9.

The asphalt binder content of the cold patching mixture is determined by a trial and error method. For the gradation listed in Table 9.9 the residue asphalt binder content should be

Table 9.7 Recommended asphalt binders for cold patching mixtures

Immediate use	*Stockpile use*	
Asphalt emulsion	*Asphalt emulsion*	*Cutback asphalt*
CMS –2	HFMS-2s	SC-250
CMS –2h		SC-800
HFMS-2s		MC-250
		MC-800

Table 9.8 Aggregate quality requirements for cold patching mixtures

Test parameter	*Requirement*
Los Angeles Abrasion, 500 revolutions (%)	$\leqslant$40
Crushed faces (%)	$\geqslant$65
Sand equivalent test (%)	$\geqslant$35

Table 9.9 Typical cold patching mixture gradation

Sieve size	*Total passing (%)*
9.5 mm	100
4.75 mm	20–60
2.36 mm	5–30
1.18 mm	0–10
300 μm	0–6
75 μm	0–2

3.5–4.0 percent. The residue content is the asphalt binder that is left over after the water or solvent has evaporated from the asphalt emulsion or cutback. For example, if the desired residue asphalt binder content is 3.8 percent and the asphalt emulsion has 63 percent solids or asphalt binder in it, the required amount of total emulsion to add is 6.0 percent (3.8 divided by 0.63). Various mixtures are mixed in the laboratory at several different asphalt binder (emulsion or cutback) contents. Any mixture that appears "soupy" has too high an asphalt binder content and should be eliminated. The mixtures are allowed to cure or dry and then visually rated. The asphalt binder content selected should be the lowest content that maintains at least a 90 percent coating on the aggregate mixture. Some agencies have developed additional tests to determine the workability, cohesive strength, and stripping resistance of the cold patching mixture. Workability is difficult to specify since acceptable workability is difficult to define. ASTM D6704, *Standard Test Method for Determining the Workability of Asphalt Cold Patching Material,* is a test method that attempts to apply numerical test values to workability.

Pothole patching is a temporary solution to a pavement distress. Semi-permanent patches can last 2–3 years or more, while the throw and go patches rarely make it through a season. The primary focus of pothole filling or patching is to immediately eliminate a safety or traffic hazard. Eventually a full depth patch will need to repair the distressed portion of the asphalt pavement.

Full depth patching

Most high severity disintegration distresses need to be repaired either by total reconstruction or the use of a full depth patch. Medium and high severity alligator cracking is one common distress that can be repaired through the use of a full depth or partial depth patch. Full depth patches are also used to repaired utility cuts in the pavement (Plate 9.5). The semi-permanent method of pothole filling is the closest method to full depth patching. Full depth patching is simply removing the distress portion of the pavement, usually all the way down to the subgrade or base course and replacing it with usually dense graded HMA. A high quality cold patching mixture can also be used. The asphalt mixture is properly laid and compacted in the patch. The full depth patch is structurally sound and becomes an integral part of the asphalt pavement.

When a patch is properly constructed, it will provide several years of service. The basic concepts used in full or partial depth patching applies regardless of why the patch is being installed. The same techniques are used for utility cut patches as used for repairing alligator cracking. The same concepts are applicable regardless of the patch being used on

Plate 9.5 Utility cut patching.

a parking lot or on a freeway or motorway. The procedure for installing a full depth patch consist of the following.

1 Excavate the distressed portion of the pavement to sound pavement at the edges and bottom. If necessary excavate all the way to the subgrade. Create vertical sides for the patch. In case of utility cuts, the purpose of the cut will determine how much to excavate (Plate 9.6).
2 Tack coat the bottom and sides of the excavation with a tack coat (Plate 9.7). An asphalt emulsion or an asphalt cutback will be adequate.
3 Dump and spread (Plate 9.8) the HMA or patching mixture into the excavated area (Plate 9.9) If the excavation is deep, the mixture may need to be placed and compacted in two separate lifts.
4 Compact the mixture with a vibrating plate compactor (Plate 9.10) or a tamping foot compactor (Plate 9.12). Two or three passes are usually adequate. For larger patches or patches that are long and narrow, such as utility cuts, two or three passes of a small roller (Plate 9.11) will provide good compaction for the patch.
5 Place the final lift or layer slightly above the adjacent pavement and compact until flush with the adjacent pavement.
6 Seal the newly formed edges or joints with a crack sealer or an asphalt emulsion.
7 Open to traffic after cooling, or if using a cold patching mixture, allow to cure before opening to traffic.

Plate 9.6 Patch excavation.

Plate 9.7 Tack coat application.

Plate 9.8 Dumping the mixture.

Plate 9.9 Spreading the mixture.

Plate 9.10 Vibrating plate compactor.

Plate 9.11 Small roller.

Plate 9.12 Tamping foot compactor.

When excavating out for the patch, the use of a pavement saw will make straight vertical edges. The cutout should be square or rectangular and should be 100–200 mm into the sound portion of the asphalt pavement. After excavating the pavement the existing subgrade or base course should be inspected for defects. If the subgrade is contaminated or saturated with water, it should also be removed. Compressed air and shovels or brooms can assist in blowing and cleaning out the excavation. Compressed air can also help remove any standing water in the excavation or the use of an infrared or surface heater can be used. Gas torches or lances will also work, but must be used carefully so that the surrounding pavement is not damaged.

Prior to dumping or spreading an asphalt mixture into the excavation, a tack coat should be applied. A thin coat is applied to the sides and bottom of the excavation. A typical application rate is 0.5 l/m^2 of area. A tack coat for patching consists of a rapid setting asphalt emulsion or cutback asphalt. Slow setting emulsions will also work, but additional time will be needed for the emulsion to break prior to the spreading of the patching mixture. There has been some research on the fact that waiting for the emulsion to break prior to the spreading of HMA is unnecessary. It is believed that the heat from the HMA will break the emulsion and drive out any remaining water. The tack coat should be sprayed to just "paint" the tack coat on the excavation. Excessive application of the tack coat should be avoided.

The asphalt mixture is usually spread into the patch by hand, using rakes and lutes. A lute is similar to a rake, but has smaller teeth. Excavations or holes that are deeper than 100 mm should be placed in two or more layers or lifts. Each lift should be compacted separately. The excavation or hole should be slightly overfilled with the last lift to allow for compaction to bring the patch flush with the existing pavement surface. The amount of overfilling will depend on the amount of compactive effort, which is usually 6 mm of extra thickness for every 25 mm of compacted thickness. If the patch is small, compact the mixture with a vibrating plate or tamping foot compactor. If the excavation is large enough, the use of a small roller will also provide adequate compaction. Since the sides of the patch confine the mixture, adequate compaction can be achieved with two or three passes of the roller or compactor. On very large patches, compaction can be specified using the same methods for specifying and determining the required amount of compaction or density for mainline asphalt pavements. In these cases, however, reconstruction of pavement should be considered before installing very large patches.

Patching materials

Hot mix asphalt is usually the best patching mixture with regard to performance. The same dense graded mixtures used for the mainline pavement will usually work for patching. However, patching is usually relatively small and requires a significant amount of handwork. A fine graded or smaller stone size mixture is preferred for the typical small patch or utility cut. Table 9.10 provides two typical gradations that can be used for HMA patching mixtures. The finer graded mixtures are better suited for parking lots, since the texture will blend in with the existing asphalt pavement and be esthetically more pleasing.

The asphalt binder selected for a hot patching mixture can be the same grade that would be selected for the mainline pavement. Some manufacturers of HMA patching mixtures will

Table 9.10 Typical hot mix asphalt patching gradations

Sieve size	*Total passing (%)*	
	12.5 (mm)	*9.5* (mm)
12.5 mm	100	
9.5 mm	90–100	100
4.75 mm	60–80	80–100
2.36 mm	35–65	65–100
1.18 mm		40–80
600 μm	20–65	
300 μm	10–25	10–40
75 μm	2–10	2–10

offer the mixture with a softer grade asphalt binder in order to increase the workability of the mixture. However, for heavy-duty pavements, such as an expressway or motorway, the asphalt binder used should be the appropriate choice for the climate and traffic. HMA mixtures are designed for relatively immediate use and are not stockpiled. The mixtures may be stored in a heated or insulated silo or in the "hot box" of a maintenance or patching vehicle. The mixture design of HMA mixtures for patching should follow the same method that is used for the mixture design for the mainline pavements. However, for all practical purposes, patching mixtures do not vary from project to project. The typical mixture design is an established "recipe" design that does not vary. If the patching mixture were to be stockpiled or have any extended use, the use of an asphalt emulsion or cutback would be selected instead of an asphalt binder. Table 9.7 provides some recommended asphalt emulsions and cutbacks for cold patching mixtures. The design of cold patching mixtures can use the same method as that used for cold stockpile or pothole filling mixtures. Cold patching, cold stockpile, and pothole filling mixtures are usually just one mixture.

The significant differences between pothole filling and patching are that patching requires the excavation of the pavement while pothole filling does not. Full depth patching is considered permanent because:

1 The distressed portion of the pavement is removed.
2 New construction techniques are applied to the installation of a high quality patching mixture.
3 The mixture is adequately compacted.

After patching has been completed on a distressed pavement a surface treatment may be installed that covers up the patch. The pavement becomes more uniform in appearance and the surface treatment provides additional protection to the patch. Also at about the time in the pavement's life that patching is considered, the surface treatment is needed to extend the pavement's life.

Surface treatments

Surface treatments is a broad category, encompassing several types of asphalt and coal tar sealers or asphalt aggregate combinations. What is common is that they are 25 mm or less in thickness, can be placed on any type of asphalt pavement surface. Pavement sealers describes more accurately the application of an asphalt emulsion or coal tar to a pavement surface. An asphalt surface treatment consists of a thin layer of asphalt concrete formed by the application of an asphalt emulsion, cutback or asphalt binder plus aggregate to protect or restore an existing pavement surface. An asphalt treatment that consists of a sprayed on coat of an asphalt binder, emulsion or cutback that is followed by aggregate that is spread on top is known as a *chip seal*. A chip seal is also known as *surface dressing, armor coat*, or *seal coat*. The term "seal coat" can also be used interchangeably with the term "chip seal" or even the broader term "surface treatment." Applications of a pavement sealer has also been called seal coat. The interchanging of the terminology can be confusing. For the purposes of this text, the term "seal coat" will refer to the application of a pavement sealer. The engineer or maintenance professional responsible for the maintenance of parking lots will usually encounter the term "seal coat" as referring to the application of a pavement sealer.

The surface treatment will perform one or more of the following functions:

1 Provide a weather resistant surface.
2 Provide a fuel or oil resistant surface.
3 Provide an esthetically pleasing coating to the pavement surface.
4 Fill or seal hairline or cracks under 3 mm width.
5 Fill distortions or rutting.
6 Provide a skid resistant surface.

One function that a surface treatment will not provide is structural strength. The lack of any significant aggregate interlock and thickness in a surface treatment results in no structural strength and is not considered when determining the overall required thickness for an asphalt pavement. Surface treatments can be subgrouped into three classes:

1 Pavement sealers
2 Chip seals
3 Slurry seals.

The difference between the three is that pavement sealers contain essentially no aggregate (except possibly small amounts of clay or sand) and chip seals and slurry seals contain a significant portion of aggregate. In some instances a thin layer of an asphalt mixture, usually less than 25 mm thick, is used as a surface treatment. The mixture is usually dense graded HMA with a maximum aggregate size of 9.5 mm or less. *Thin surfacing* or *scratch coat* are terms that can be used to describe the thin application of an asphalt mixture.

Pavement sealers

Pavement sealers are used to restore or rejuvenate an oxidized asphalt pavement surface. They are also used to fill hairline cracks that are less than 3 mm wide. Some sealers provide an improved or "new" appearance to an aged asphalt pavement and can protect the asphalt pavement from fuel or oil damage. The most common pavement sealers are:

1 Fog seals
2 Asphalt emulsion seal coat
3 Coal tar seal coat.

Pavement sealers can be both preventative and corrective maintenance. The most common use of pavement sealers is for preventative maintenance and to improve the appearance of an asphalt parking lot. Sealers can cover up old parking stripes allowing new stripes to be painted on the parking lot.

Fog seals

A fog seal is a very light application of an asphalt emulsion. A fog seal is used to renew old or oxidized pavement surfaces that have become brittle and to seal hairline cracks. It also coats aggregate particles at the surface. A fog seal can prolong the pavement life and possibly delay major maintenance if applied in time (The Asphalt Institute 1998a).

The asphalt emulsion is sprayed, usually through an asphalt distributor truck (Plate 9.13, Plate 9.14), on the pavement surface. The application rate is in the range of

Plate 9.13 Sideview of a distributor truck.

Plate 9.14 The back of a distributor truck.

0.4–0.7 l/m^2. The asphalt emulsion is usually diluted one to one with water to provide an asphalt binder content of around 30 percent. The asphalt emulsion used is a slow setting emulsion, either anionic or cationic. SS-1, SS-1h, CSS-1, and CSS-1h type asphalt emulsions can be used for fog seals. Over application of the fog seal should be avoided. Over application is led to tracking of the residue asphalt emulsion by vehicles and pedestrians and has the potential for skid resistance problems. If excess application of the fog seal is encountered, a light application of the sand should correct the problem. Fog seals do not last very long with an annual application usually considered adequate. Fog seals are also used as a dust palliative on gravel and dirt roads.

The asphalt distributor or sprayer is the most common and essential piece of equipment for surface treatments, including pavement sealers and chip seals. The distributor applies tack coats, prime coats, road oils, fog seals, and any other liquid material to the pavement surface. The distributor's function is to uniformly apply an asphalt binder, emulsion, or cutback at a specified rate to the pavement surface. The distributor also needs to maintain this rate regardless of the distributor's travel speed. The distributor consists of a truck or trailer mounted insulated tank and metering pump. The tank usually is equipped with a heating system to maintain the temperature of the asphalt material, if needed. Rapid set emulsions are usually sprayed at elevated temperatures, but under 100 °C. Table 9.11 provides typical spray temperatures for various emulsions. At the back of the tank is a spray bar (Plate 9.15) equipped with nozzles (Plate 9.16). The asphalt material is sprayed through the spray bar and nozzles onto the pavement surface. The asphalt material flows through either slots or swirls that are cut into the nozzles. Slotted nozzles are the most common. The spray may be equipped with extensions to spray widths up to 10 m. If needed, the individual nozzles may be shut off when variations in the spray width are required. The asphalt material is continually recirculated through the spray bar to prevent solidification of the asphalt material in the spray bar. The tank should also be equipped with a thermometer (Plate 9.17). The distributor is usually equipped with a hand spray wand for application in small areas (The Asphalt Institute 1982).

The spray bar height can be changed in order to change the amount of material coverage on the pavement surface. The higher the height of the spray bar, the more coverage and overlap each nozzle will provide (Figure 9.4). Double coverage by the nozzles provides the

Table 9.11 Spray temperatures for pavement sealers and surface treatments

Material type	*Emulsion type*	*Spray temperature* (°C)
Asphalt binder	AC-2.5, AC-5	140–200
Asphalt binder	asphalt penetrations >120 dmm	130–180
Asphalt cutback	70 grade	50–105
Asphalt cutback	250 grade	75–130
Asphalt cutback	800 grade	95–150
Asphalt cutback	3,000 grade	110–175
Slow set emulsion	SS-1, SS-1h, CSS-1, CSS-1h	25–40
Medium set emulsion	MS-1, MS-2, MS-2h, HFMS-1, HFMS-2, HFMS-2h, HFMS-2s, CMS-2	30–50
Rapid set emulsion	RS-1, RS-2, HFRS-2, CRS-1, CRS-2, CRS-2P	50–90
Coal tar sealers	≥40% solids	10–35

Plate 9.15 Distributor spray bar.

Plate 9.16 Spray nozzles.

Plate 9.17 Tank thermometer.

best results. The spray bar height is constantly maintained even as the load on the distributor becomes lighter. Normally, as the distributor became lighter on the truck springs and shocks, the spray bar height would slightly increase. Most modern distributors can compensate for the loading by automatically adjusting the spray bar height.

In order to prevent the spray pattern from interfering with each other, the nozzles are offset from the spray bar axis at an angle of 15–30° (Figure 9.5).

The total flow rate of the asphalt material onto the pavement surface is controlled by three variables:

1 Proportioning flow valve
2 Asphalt pump output measured by a tachometer and pressure gauge
3 Bitumeter equipped with an odometer.

A Bitumeter is a device that measures the asphalt distributor's travel rate. It is equipped with a rubber tire wheel mounted on a retractable frame that records the distributor's progress in meters per minute and total meters traveled. The device then computes the liters per square meter sprayed on the pavement surface and reports it on a display in the cab.

The required amount of asphalt material to be sprayed for a given area can be determined by equation 9.1.

$$\text{Total liters} = W * L * R \tag{9.1}$$

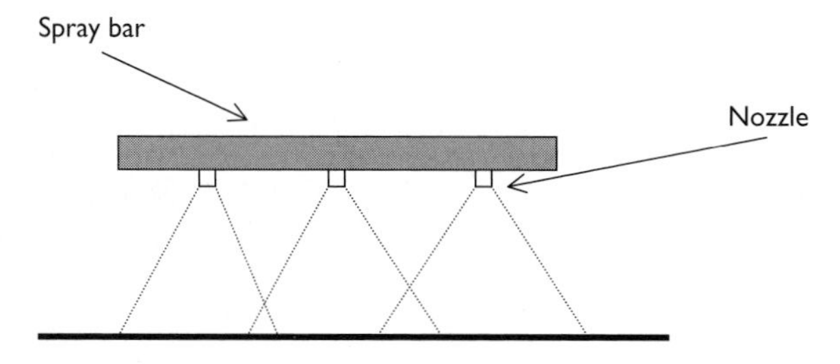

Figure 9.4 Spray nozzle coverage and overlap.

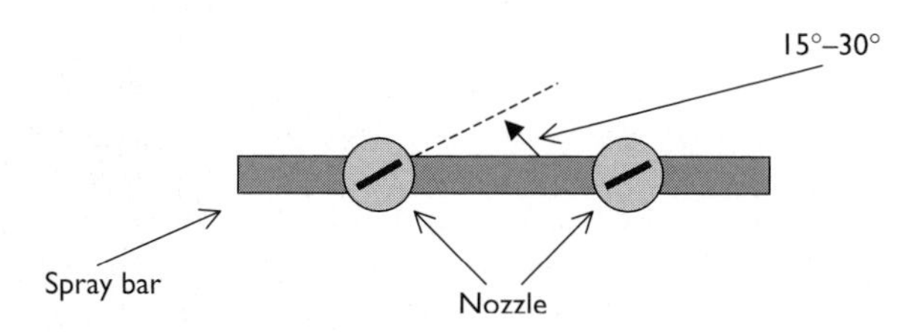

Figure 9.5 Spray nozzle angle.

where W is the pavement or area width or spray bar width (m); L, the pavement or area length (m); R, the application rate (l/m^2).

Example 9.2

Determine the amount of liters of diluted CSS-1 emulsion to be sprayed as a fog seal on a parking lot 500 × by 1,350 m at an application rate of 0.5 l/m^2.

Solution

Total liters required = (500 m)(1,350 m)(0.5 l/m^2) = <u>337,500 l</u> of diluted emulsion.

If the emulsion is diluted one to one; 168,750 l of CSS-1 will be required to spray the fog seal.

Example 9.3

Determine how much of a pavement a distributor with a 2 m spray bar and 10,000 l tank spraying a fog seal at 0.6 l/m^2 can spray in one pass.

Solution

$$\text{Using equation 9.1 : length} = \frac{10{,}000\ \text{l}}{(2)(0.6)} = \underline{8{,}333\ \text{m}}$$

Coal tar and asphalt sealers

A very common surface treatment for parking lots and driveways is an application of a coal tar or asphalt based sealer. It is usually the only type of surface treatment a parking lot will receive. The application of these types of sealers is known as *sealcoating*. The sealer is also known as a *sealcoat.* The term's sealcoat and pavement sealer can be used interchangeably and is so throughout this text. Usually the primary purpose of applying these sealers is for cosmetic purposes. Sealcoats do not add structural strength or integrity to an asphalt pavement. They are used to improve the overall appearance of the asphalt pavement and to protect it from potential damage from the environment. Sealcoats can provide an excellent background contrast to the white or yellow paints used to stripe or mark parking stalls (Plates 9.18 and 9.19). The sealer or sealcoat can:

- Fill hairline cracks
- Bind together slightly raveled surfaces
- Slow oxidation and water penetration
- Resist motor oil and fuel spills
- Provide a rich dark black color to the pavement surface (Plate 9.20).

Sealers that are formulated with refined coal tar pitch provide additional protection to the pavement surface since they are also highly resistant to damage by motor oil or gasoline. Aircraft refueling areas, gasoline stations, and most parking lots receive the occasional motor oil drip or fuel spillage from a vehicle. Fuel resistant sealers protect the asphalt pavement from damage. Asphalt based sealers that are formulated with acrylic or other polymers have also proven to be resistant to damage by motor oil and fuel. Fuel resistant sealers are also known as aliphatic solvent resistant sealers. Aliphatic solvents are petroleum distillates such as gasoline, kerosene, diesel fuel, and motor oils.

Plate 9.18 Parking stalls.

Plate 9.19 Contrast of pavement markings and a sealcoat.

Plate 9.20 Dark black color.

A pavement sealer or sealcoat consists of a base material such as refined coal tar or asphalt cement and water. Crude coal tar is a by-product from the heating of coal to form coke. Coke is carbon that is used in the manufacture of steel and other foundry processes. Crude coal tar is processed to form refined coal tar, pitches, creosote, and chemical oils. Refined coal tar is used in applications such as cosmetics, shampoos, and pavement sealers (PCTC 1997). Gilsonite and acrylic polymers can also be used as base materials for the pavement sealer. An emulsifier is used to disperse the base material into the water. The emulsifier also acts as an adhesion promoter, or an amine based adhesion promoter is added. Clay is added to thicken the sealer. In most cases fine sand or slag is also added at the project site to provide slip resistance properties to the sealcoat.

Sealer materials and manufacture

The most common sealer or sealcoat material in use today is a refined coal tar sealer. Asphalt and acrylic polymer sealers are becoming more common. Refined coal tar offers resistance to solvents and provides a deep or "jet" black appearance to the pavement. Carbon black is sometimes added to asphalt and polymer based sealers to assist in obtaining a deep black color. Coal tar and asphalt sealers are often formulated with various types of polymers including neoprene, acrylics, and latex to improve the longevity of the sealcoat.

ASTM D490, *Standard Specification for Road Tar*, is often used to grade or specify the coal tar used in the manufacture of refined coal tar based sealers. The specification covers 14 grades of tar through the following designations, RT-1, RT-2, RT-3, RT-4, RT-5, RT-6, RT-7, RT-8, RT-9, RT-10, RT-11, RT-12, RTCB-5, and RTCB-6. RT stands for road tar and RTCB stands for road tar cut back (road tar and petroleum distillate). Refined coal tar sealers are made with coal tar that meets an RT-11 or RT-12 grade. Table 9.12 provides the ASTM D490 specification requirements for RT-11 and RT-12. The amount of coal tar in a pavement sealer formulation is typically 30 percent. Additional materials, including clay, bring the total solids in the sealer formulation to 40–50 percent.

Asphalt based sealers are formulated using slow set asphalt emulsions. The asphalt emulsion is diluted with water to final asphalt content of 30–40 percent. The asphalt emulsion is typically an anionic slow set such as an SS-1 or SS-1h emulsion, but pavement sealers are also successfully formulated with cationic slow set emulsion such as a CSS-1 or CSS-1h emulsions.

A standard pavement sealer consists of refined coal tar or an asphalt emulsion, an emulsifier, clay, and water. The clay is added to create a thixotropic slurry. A fine graded sand or slag is added to the sealer, usually later, at the time of application. The amount of sand or slag is 0.5–0.8 kg/l of sealer. Table 9.13 provides typical formulations for both a coal tar and an asphalt emulsion based sealer or sealcoat.

The manufacture of sealcoats requires the ability to disperse water and clay into either coal tar or an asphalt emulsion. A high shear mixer or a colloid mill is required to produce stable sealer formulations. The colloid mill is the same as one would use to produce asphalt emulsions and is operated in a continuous operation. The high shear mixer is used in the batch manufacture of sealers. The high shear mixer produces a fine particle dispersion, similar to a colloid mill.

The batch process consists of a mixing tank, high shear mixer, and a storage tank. The mixer first disperses the clay into 60–80 °C water. The clay used is usually a refined ceramic ball clay. The mixer produces a clay slurry. The surfactants or emulsifiers are added to either the clay slurry or to the coal tar. If additional materials are formulated into sealer,

Table 9.12 RT-11 and RT-12 specifications

Test parameter	*Specification*	
Road tar grade	*RT-11*	*RT-12*
Specific gravity, 25 °C	⩾1.16	⩾1.16
Float test, ASTM D139, 32 °C	100–150 s	150–220 s
Total distillate, to 170 °C, mass	⩽1.0%	⩽1.0%
Total distillate, to 270 °C, mass	⩽10.0%	⩽10.0%
Total distillate, to 300 °C, mass	⩽20.0%	⩽20.0%
Ring and ball softening point of distillate residue	40–70 °C	40–70 °C
Total bitumen, ASTM D4, mass	⩾75%	⩾75%

Source: American Society of Testing Materials D490 (ASTM 2002).

Table 9.13 Typical sealcoat or sealer formulations

Ingredient	*Coal tar based (%)*	*Asphalt emulsion based (%)*
Refined coal tar, RT-12	30	
Asphalt emulsion, SS-1h		52
Emulsifier/adhesion promoter	0.4	0.2
Ball clay	14	14
Latex	1.5	2.6
Water	54.1	32.8
Total	100	100

such as polymers or fibers, they are added to the clay slurry. The refined coal tar is heated to 90–110 °C and then mixed into the slurry using the shear mixer at high speed. If an asphalt emulsion is being added, it is added at 60 °C–80 °C and mixed into the slurry. Care should be taken to prevent the asphalt emulsion from foaming or by entraining air into the asphalt emulsion. The additional required amount of water is then mixed into the formulation.

The continuous process requires the use of a colloid mill. The clay and other fillers are predispersed into 60–80 °C water. The refined tar or asphalt binder is pumped into the colloid mill in the same method as used to manufacture an asphalt emulsion. When producing an asphalt based sealer using the continuous process, an asphalt cement or binder is used instead of an asphalt emulsion. The asphalt emulsion is produced *in situ* during the manufacture of the asphalt based sealer. The emulsifier and other chemical surfactants can be added either to the clay slurry or the coal tar or asphalt binder (Cleaver 2001).

Sealer additives

Various types of additives can be added to the sealer, usually during the manufacturing process or post added at the pavement site. These additives include:

- Surfactants
- Latex
- Polyvinyl acetate

- Acrylic polymer
- Rubber
- Chloroprene polymer
- Acrylonitrile Butadiene rubber
- Sand or slag
- Coloring agents.

Surfactants or surface active agents, are usually amines that are used as dispersants and emulsifiers during the manufacturing process of the pavement sealer. They can be post added to the sealer to improve the coating ability of the sealer, the spreading characteristics, and the adhesion of the sealer to the pavement. Improved adhesion of the sealer will increase its longevity.

Latex is used to thicken the sealer, assist in the dispersing or suspending of the sand or slag, quicken the drying or curing time of the sealer, and produce a uniform application color.

Polyvinyl acetate is used to thicken the sealer, assist in dispersing or suspending sand, and to quicken the drying time of the sealer.

Acrylic polymer provides the same functions as polyvinyl acetate and in addition provides resistance to ultraviolet light and aliphatic solvent damage.

Rubber includes several types such as natural latex or rubber, styrene butadiene rubber (SBR), and butyl rubber. Provides the same functions in the sealer as acrylic polymer with the additional benefit of increased resistance to water and chemicals such as acid.

Chloroprene polymer or Neoprene is used to increase the sealer's resistance to certain chemicals including mild acids, greases, and oils.

Acrylonitrile Butadiene rubber provides the same benefits to a sealer as rubber.

Sand or slag is usually post added to the sealer. It is a fine graded aggregate that increases the skid resistance of the sealer. It creates a thin mastic that provides wear resistance for the sealer and fills hairline or minor cracks in the pavement surface.

Coloring agents such as carbon black are used to deepen or darken the sealer. Carbon black is a powder that will provide a permanent black color to the pavement. Other coloring agents are used in specialty sealers such as those for tennis courts, running tracks, and other athletic applications. Colored sealers are often used on footpaths or asphalt sidewalks.

The various types of polymers are added at a rate of 3–5 percent by weight of the refined coal tar or asphalt emulsion in the sealer (Dubey 2002). For example, if 30 percent by weight of total sealer is coal tar, 0.9–1.5 percent by weight of the sealer is the polymer.

Fine aggregate is added at a typical rate of 0.5–0.8 kg/l of sealer. There are instances that up to 2 kg of fine aggregates per liter of sealer have been used. The fine aggregate is washed and is either a silica sand or slag. Table 9.14 provides a typical gradation for the fine aggregate to be used in sealcoating. The fine aggregate is slowly added into the sealer prior to application. During the addition of the fine aggregate, the sealer is slowly agitated through a stirrer, such as a paint mixer or similar mixer. The addition of the fine aggregate displaces a volumetric portion of the sealer. Approximately one liter of sealer is displaced by every 2.5 kg of fine aggregate. The displaced sealer is replaced by additional water. For example, if 0.6 kg of silica sand is added to the sealer, an additional 240 cc of water also needs to be added to the sealer.

Table 9.14 Fine aggregate gradation for sealcoating

Sieve size	*Total passing (%)*
2.36 mm	100
1.18 mm	95–100
600 μm	60–95
300 μm	10–40
150 μm	0–10
75 μm	0–2

Sealer specifications

Users of pavement sealers are concerned with the performance of the finished sealer and not necessarily with all of its individual components. Specification of the individual components used in the manufacture of the sealer is usually limited to the base material (coal tar or asphalt emulsion) and the added sand or slag. ASTM D3320, *Standard Specification for Emulsified Coal Tar Pitch (Mineral Colloid Type),* the Federal Aviation Administration (FAA) P-625 *Coal Tar Pitch Emulsion Seal Coat* specification and Federal Specification R-P-355e, *Coal Tar Emulsion Coating for Bituminous Pavements*, are commonly referenced specifications in the United States. The FAA engineering brief #46 provides guidance for the application of the FAA P-625 specification. Some specifications may have been discontinued by their sponsor or developer, but are still in practice by others who have adopted the specification. Federal Specification R-P-355e is one example. Federal Specification R-P-355e can be replaced by the new ASTM D5727, *Standard Specification for Emulsified Refined Coal Tar (Mineral Colloid Type).* In the United States, parking lot maintenance professionals who specify one of the listed specifications will find readily available products that provide excellent performance. The performance of a coal tar pitch sealer can also be specified by ASTM D4866, *Standard Performance Specification for Coal Tar Pitch Emulsion Pavement Sealer Formulations Containing Mineral Aggregate and Optional Polymeric Admixtures.* Reputable sealcoat manufacturers can also provide guidance for sealcoating projects. A similar process can be used in Europe and other countries.

When developing or adopting a specification, there should be purpose or reason for each portion of the specification. The ability to check for specification compliance should also be considered before adopting any type of specification. Stand-alone specifications that are readily accepted in the industry will meet the needs of the customer or user and provide for readily available, quality products. Specifications that needlessly restrict the producer of seal coats will generally cost the customer with no additional benefit. Table 9.15 provides some generic specifications for a refined coal tar and an asphalt emulsion sealer.

Sealer or sealcoat application

Sealcoating is a maintenance activity that protects the pavement from possible damage and beautifies the pavement. Sealcoating will not repair a structurally deficient pavement or correct pavement distresses that are severe. Any necessary structural repairs, patching, and crack sealing/filling should be done prior to sealcoating. Only low severity alligator cracking should be treated with sealcoating. Medium to high severity alligator cracking needs to be corrected prior to sealcoating.

Table 9.15 Typical sealcoat or sealer specification

	Requirement	
	Refined coal tar emulsion, ASTM D5727	*Asphalt emulsion*
Refined coal tar	ASTM D490 RT-12	
Asphalt emulsion		ASTM D977 SS-1h
Non volatile, solids (%)	$\geq$47	$\geq$45
Water (%)	$\leq$53	$\leq$55
Non volatile ash (%)	30–40	30–40
Latex (if applicable) solids (%)	$\geq$40	$\geq$40
Drying time	$\leq$8 h	$\leq$8 h

A new asphalt pavement should not be sealcoated until it is at least 60 days old. The light oils and aromatic or volatile compounds in the asphalt binder need to cure prior to the application of a sealcoat. If the volatile compounds are trapped under the sealcoat, the surface may soften and the sealcoat may track or flake and peel off. If residual oils are remaining on the pavement surface, the sealer will not adequately bond to it. Spraying water onto the pavement surface and making a couple of observations can test the pavement surface. If the water beads on the surface, there are still residual oils on the pavement. If the water flows in sheets, the residual oils are lost and the sealcoating operation can take place.

A new asphalt pavement can benefit from sealcoating, since the sealer acts as a barrier to protect the damaging effects of weather, petroleum distillates, and chemicals. The sealer coats the pavement surface, providing the barrier effect (Dubey 2002). The pavement surface must be clean prior to the sealer application. Power brooming or pressure washing will clean the surface. Oil spots from vehicles are sometimes primed or treated with a proprietary oil spot primer prior to sealcoating. Parking blocks, shopping cart corrals, and other parking accessories should be removed if possible. If they cannot be removed, they should be masked off with Kraft or butcher paper.

The amount of sealer required for the project needs to be determined. Sealcoats are typically applied in two coats at an application of 0.4–0.6 l/m^2 of pavement surface. If the pavement is oxidized or slightly raveled, the first coat should be at 0.6 l, with the second coat lighter, at 0.4 l/m^2. The amount of sand required to be blended also needs to be determined. Some bulk sealer manufacturers provide the sealer already dispersed with the sand. The sealer usually must be diluted with water in order to adequately spray and apply the correct film thickness. A film thickness that is too thick will "bake" rather than cure, leading to tracking problems. A film thickness that is too thin will have reduced life expectancy. Sealers with non-volatile solids greater than 30 percent need to be diluted with water. The dilution rates will vary by manufacturer, but usually in the range of 15–20 percent by weight. The manufacturer of the sealer will provide the proper dilution rates. Most sealers are shipped in concentrate form and will require some dilution.

Example 9.4

Determine the total amount of ready to use sealer, sand, and dilution water for a two coat seal coat project on a parking lot 500 $\times$ 1,350 m. The application rate is 0.5 l/m^2. Sand of about 0.6 kg per liter of sealer is required and the manufacturer recommends a dilution rate of 20 l for every 100 l of sealer (20 percent).

Solution

The area of the parking lot is: (500 m)(1,350 m) = 675,000 m^2
The amount of sealer material applied is: (2 coats)(675,000 m^2)(0.5 l/m^2) = 675,000 l
The amount of sand required is: (675,000 l)(0.6 kg/l) = 405,000 kg or 202.5 tonnes
The amount of additional water due to sand is: (405,000 kg)/(2.5 kg/l) = 162,000 kg or 162,000 l
The amount of dilution water required is: (675,000 l) (20%) = 135,000 l
The amount of sealer to purchase is: (675,000) − (162,000) − (135,000) = 378,000 l

The air and pavement temperature should be at least 10 °C during sealer application and for at least 8 h afterward. Significant rainfall during the sealcoating application can cause it to wash out prior to its drying. If the ambient air temperature exceeds 30 °C, a light mist of water should be applied to the pavement prior to the application of the sealer. The pavement sealer or sealcoat is applied through the use of a squeegee, brush, and spray wand or a distributor. The area being sealcoated is "cut in," similar to cutting in during painting. The sealer is applied along the edge of the pavement or parking lot. The edge is coated first, then the sealer is applied perpendicular to the pavement edge. The sealer is applied in parallel strips over the entire pavement surface. The sealer is then spread uniformly across the surface with a squeegee (Plate 9.21) or brush. Areas near curbs and drainage apertures will need to be hand brushed. The first coat needs to dry completely prior to the application of

Plate 9.21 Sealcoat squeegee.

Plate 9.22 Sealcoat blocked off.

the second coat. The sealer is dry when it is not sticky to the touch. The first coat should also be clean prior to the application of the second coat. Loose dirt can be blown or swept off. The second coat should be applied at right angles to the first coat. After the second coat is applied, the area should be blocked off (Plate 9.22) for at least 24 h to ensure a thorough drying or curing of both coats (Camillo 2001). A third coat is sometimes applied to parking lot drive lanes. After curing, the pavement surface and the parking stalls can then be striped. The application techniques are the same for both refined coal tar emulsion sealers and asphalt emulsion sealers.

Sand or fine aggregates should be used in almost all sealcoat applications. The primary purpose of incorporating fine aggregate into the sealer is to provide for pedestrian slip resistance. The fine aggregate will offer a limited amount of low speed skid resistance. Table 9.16 provides some recommended guidelines on application coverage and fine aggregate requirements. The sealer manufacturer should specify the maximum amount of fine aggregate that can be incorporated into the sealer. The amount of fine aggregate added to the sealer is typically in the range of 0.5–0.8 kg per liter. Some specifications may permit up to 2.0 kg, but that is generally considered excessive and can turn the sealer into mastic.

The condition of the existing pavement also influences the amount of sealer applied to the surface. Excessively oxidized or rough surfaces will require a greater amount of sealer than a new or dense pavement surface. The application rates provided in Table 9.16 assume that the condition of the pavement surface is smooth and almost new. Table 9.17 provides the increased application rates based on surface conditions. These values would be added to the values listed in Table 9.16. The application rates provided are typical and sound judgment should be used to increase or decrease these rates based on local conditions or past experience.

Table 9.16 Typical sealcoat application rates

Pavement use	*Number of coats*	*Application rate* (l/m^2)	*Sand rate* (kg/l)
Commercial parking area	Two sand coats	0.5	0.6
Parking lot drive lanes	One prime coat (no sand) Two sand coats	0.6	0.8
Residential drives	One sand coat	0.4	0.5
Refueling areas	Three sand coats	0.5	0.7

Table 9.17 Surface condition application rate addition

Pavement surface condition	*Sealer application increase* (l/m^2)
Smooth or dense surface	+0
Medium or average texture, slightly oxidized	+0.1
Some raveling, slightly oxidized	+0.3
Rough texture or severe oxidation	+0.4
Excessively rough texture	+0.5

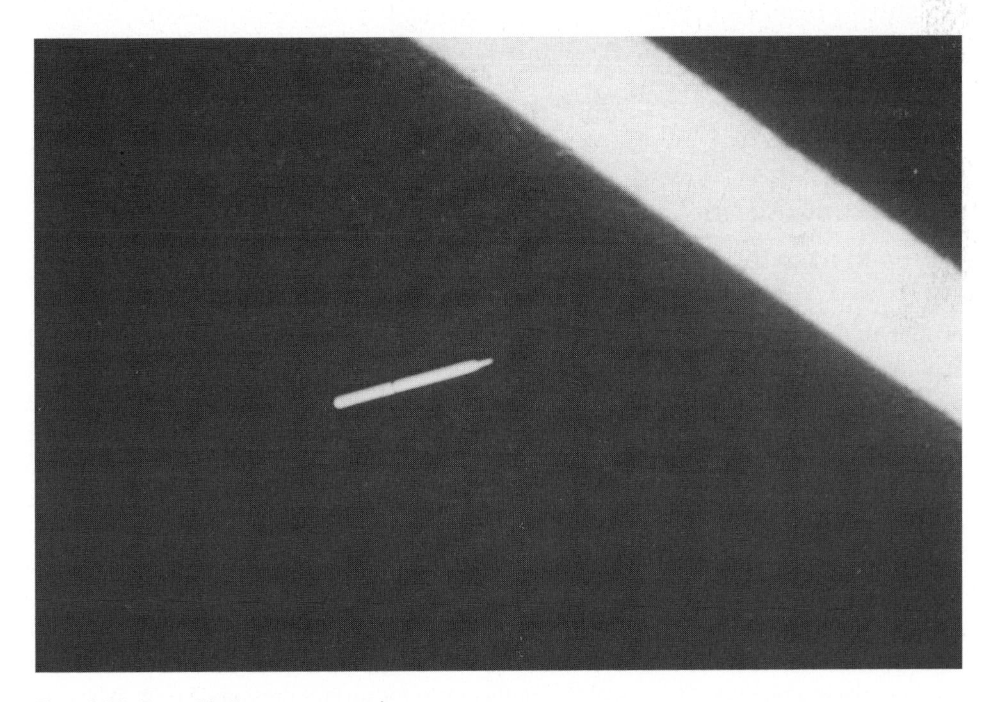

Plate 9.23 Deep black pavement sealer.

The frequency of sealcoating varies, mostly due to the expectations of the owners of the facility. Traffic and water eventually wears away the majority of the sealcoat, requiring another sealcoat to be applied to the pavement. Sunlight will also weather the sealcoat. Many parking lot owners will sealcoat once every year for esthetic reasons (Plate 9.23). The additional cost

Plate 9.24 Sealcoat on top of a crack.

of applying new pavement markings to the parking lot every year must also be considered. With regard to the preventative maintenance benefits of sealcoating, an application of the sealer every two to four years is usually considered adequate. Pavement maintenance such as crack sealing and patching is still required in addition to sealcoating. The sealcoat will not cure or fill large cracks (Plate 9.24).

Chip seals

A chip seal (Plate 9.25) is a surface treatment and another form of maintenance that can be applied to an asphalt pavement. Chip sealing is a form of sealcoating, with the exception that a thin layer of coarse aggregate is spread onto a previously sprayed application of an asphalt binder, emulsion, or cutback. The term *chip* refers to a nearly single size, and usually crushed coarse aggregate. "Chip seal" is another term for surface dressing. Sealcoat has also been used to describe a chip seal, but for the purposes of this text, sealcoat only refers to an application of a pavement sealer, usually containing a small amount of fine aggregate, to the pavement surface. Chip seals also do not add any structural strength to the pavement and is considered either a maintenance activity or a method to upgrade a gravel or dirt road. Chip seals are typically not applied to parking lots since they do not provide much esthetic value and can be damaged by the considerable amount of steering movements that occur in

Plate 9.25 Chip seal or surface dressing.

a parking lot. If the chip seal consists of large pieces of aggregate, shopping carts and pedestrians may encounter difficulty with the aggressive texture. The exposed aggregate of a chip seal does not provide the deep black appearance of a pavement sealer, nor does it provide for the contrast against pavement markings.

A chip seal protects the pavement from water damage and preserves its structural integrity. The application of the chip seal will also fill hairline and minor cracks. Cracks greater than 3 mm wide should be filled or sealed prior to the application of the chip seal. Structural deficiencies in the pavement will also need to be repaired or patched. The chip seal also provides a macro texture (Plate 9.26) to the pavement, providing skid resistance properties. A chip seal will provide a uniform appearance to a pavement that is crack filled and patched.

A chip seal is a single surface treatment that involves spraying an asphalt material, followed immediately by a thin or one stone thick aggregate cover which then set into the asphalt material by a roller. A single surface treatment can be designated by the size of the aggregate used to cover the asphalt material. A sand seal is a surface treatment that uses a fine aggregate as the cover aggregate. A chip seal is a surface treatment that uses a coarse aggregate as the cover aggregate. Multiple surface treatments involve repeating the process two or three times or combining a chip seal followed by a sand seal. The maximum size of the cover aggregate for each successive application is reduced by 50 percent (Roberts *et al.* 1991).

The most common use of chip seals is on rural or low volume roads. It is a relatively inexpensive way to upgrade a gravel or dirt road to an all weather road. Chip seals are also inexpensive in terms of cost per kilometer of roadway maintenance. Rural or county or township road maintenance departments can apply chip seals with only a small amount of

Plate 9.26 Chip seal texture.

equipment. An asphalt distributor, aggregate spreader, and a pneumatic tired roller are all that is required to construct a chip seal. Chip seals have also been applied on high volume or traffic roadways. The traffic volume of the roadway dictates the application rate of the asphalt material and the aggregate.

There are two opposite forces at work on a chip seal. The roller and subsequent traffic create a force causing chip embedment into the pavement. An opposite force resists embedment. This resistant force is the underlying pavement and the size of the chip seal aggregate (Nicholls 1998).

Chip seal materials

Asphalt emulsions are the most common types of asphalt materials used for chip seals. The emulsions have replaced for the most part, the cutback asphalt used for chip seals. Air quality concerns are restricting or eliminating the use of cutback asphalt. Soft asphalt binders can also be used for chip seals. Table 9.18 provides a list of acceptable asphalt materials for chip seals.

The specifier or designer of the chip seal should select materials that are locally available and preferably has experience with. Table 9.11 provides spray temperature guidelines for the various asphalt materials that can be used for chip seals. Rapid setting emulsions either cationic or anionic are the materials of choice for chip seals. Cationic emulsions are thought to provide better adhesion to the aggregate in the chip seals, but can perform poorly with dirty or dusty aggregates. The cationic emulsions tend to break prematurely on the dust or dirt and do not have time to bond properly to the aggregate. Anionic emulsion can be used

Table 9.18 Asphalt materials for chip seals

Asphalt material class	*Material type*
Asphalt binders	Penetration ⩾120 dmm
Cutback asphalt	RC-250, RC-800, RC-3000 MC-800, MC-3000
Anionic emulsion	HFE-90, HFE-150, HFE 300 RS-1, RS-2 HFRS-2
Cationic emulsion	CRS-1, CRS-1P, CRS-2, CRS-2P

with dusty or dirty aggregates and the high float versions tend to provide better chip retention or adhesion than the anionic emulsions that are not high float versions. One to two liters per square meter of pavement surface is a typical binder application rate.

The use of adhesion promoters or agents is especially common in chip seals. The adhesion agent is a surfactant that is usually an amidoamine, a tallow amine, or a combination of both. The adhesion agent is added to the asphalt binder, cutback asphalt, or asphalt emulsion at a rate of 0.2–1.0 percent by weight of the asphalt material. The adhesion agent displaces the surface water that may be present on the chip seal aggregate, allowing for better adhesion of the asphalt material and chip retention. The adhesion agent also promotes a better bond between the chip seal and the pavement below. The adhesion agent may be added directly to a cationic emulsion or introduced into the asphalt binder prior to emulsification. When adding an adhesion agent to an anionic emulsion, it is best to add it to the asphalt binder prior to emulsification. Most adhesion agents are cationic in nature and could cause the anionic emulsion to break prematurely, if added in excessive quantities directly to the emulsion. Adhesion agents in an aqueous solution can also be added directly to the chip seal aggregate prior to spreading. The "doped" solution is sprayed on the aggregate prior to discharge from the spreader.

The aggregate for a chip seal can also be precoated with an asphalt binder prior to spreading. The aggregate is typically precoated at an HMA mixing plant. The precoated aggregate is more expensive, but can be used for dusty aggregates and to improve adhesion to the pavement.

Chip seal aggregates are a mostly single size, crushed coarse aggregates. Chip seals are usually a one stone layer and are 25 mm thick or less, and the top size of the aggregate should reflect the desired thickness of the chip seal. The amount of the traffic, and if the chip seal is part of a multiple surface treatment, also influences the size of the aggregate. The most desirable shape of the chip seal aggregate is cubical. The initial rolling and the traffic orient the aggregate in the chip seal. The roller and traffic will orient the aggregate particle to lay on its flat or elongated side. The more elongated or flaky the aggregate is, the more susceptible the chip seal will be to either bleeding in the wheel paths or excessive aggregate loss in the non-wheel path areas. Due to traffic orientation of the aggregate, the chip seal will be thinner in the wheel path. Traffic does not have a pronounced effect on cubical aggregate orientation.

A flakiness index, or a flat and elongated particles (ASTM D4791) specification, can limit the amount of aggregates that are very thin or elongated. The average least dimension (ALD) of the stone is even a better measurement of the stone shape than the flakiness index or flat and elongated particle measurements. The ALD takes into account both flakiness and the gradation of the aggregate. It is a reduction of the median particle size after taking into account for flat particles. It allows the texture depth, space available for the asphalt material, and the aggregate spread rate to be predicted. The interlock between the aggregate particles can also be predicted (Whiteoak 1991). The aggregate must also be durable and provide some skid resistance. The abrasion resistance of the aggregate can be measured by test like the Los Angles abrasion test and the skid resistance can be predicted through the polished stone value (PSV) test. A balance of a high PSV and a low Los Angeles abrasion value is the most desirable for chip seal aggregates. Table 9.19 provides some quality requirements for chip seal aggregates. The PSV requirements may vary based on traffic levels. Table 9.20 provides some minimum PSVs for various traffic levels.

Table 9.19 Typical aggregate quality requirements for chip seals

Test parameter	*Requirement*
Los Angeles Abrasion, 500 revolutions (%)	⩽40
Crushed faces (%)	⩾65
Flat and Elongated Particles, ASTM D4791 (%)	⩽10
Polished stone value, ASTM D3319	⩾40

Table 9.20 Typical PSV criteria based on traffic

Design index	*Daily ESAL*	*Minimum PSV*
DI-1	⩽5	40
DI-2	6–20	45
DI-3	21–75	50
DI-4	76–250	55
DI-5	251–900	60
DI-6	901–3,000	65

Table 9.21 Typical gradations of chip seal aggregates

Sieve size	*Nominal top size* (mm)			
	25	19	12.5	9.5
	Total passing (%)			
25.0 mm	100	100	100	100
19.0 mm	90–100	100	100	100
12.5 mm	20–55	90–100	100	100
9.5 mm	0–15	40–70	85–100	100
4.75 mm	0–5	0–15	10–30	85–100
2.36 mm		0–5	0–10	10–40
1.18 mm			0–5	0–10
300 μm				0–5

The grading of the aggregate should be as close to single size as possible. A one size aggregate has a lot of room between the particles for filling with an asphalt material. The availability of local aggregates will have a significant influence on what aggregate gradation to specify. The aggregate should also be crushed and washed. Dirty or dusty aggregate can cause adhesion problems for the chip seal, leading to chip loss and potential vehicle damage. When using an asphalt emulsion the aggregate can contain up to 3 percent moisture. Table 9.21 provides some typical aggregate gradations for chip seals.

Chip seal design

Chip seal design involves determining the asphalt material spray rate, the cover aggregate spread rate, and the top size or gradation of the cover aggregate. The influencing variables are the traffic level and the existing pavement condition. The spray rates include the total asphalt material including the asphalt binder and water in an asphalt emulsion. The most common design concept is based on work done by Norman McLeod. The concept is based on the principle that the asphalt binder or residue asphalt binder should generally fill 60–70 percent of the aggregate void space after the particles have been fully oriented. The greater the amount of aggregate or spread rate, the greater the amount of asphalt binder that will be required. The application of a given cover aggregate should be only one stone thick when possible. The amount of aggregate should remain constant regardless of the asphalt material type or the pavement condition (Alaska DOT 2001). Traffic and pavement type or use determines the maximum top size of the aggregate.

Various empirical methods have been developed in determining the amount of aggregate to spread for a chip seal. The desired thickness of the chip seal also influences the amount of aggregate to spread. If the chip seal is part of a multiple chip seal surface, treatment will also influence the aggregate spread rate. The median particle size and the ALD of the aggrgate must be determined. The median particle size is the theoretical sieve size through which 50 percent of the material passes.

The flakiness index, FI, is a measure of the percent by weight of flat particles. It is determined by testing a small sample of aggregate particles for their ability to fit through a slotted plate. If the aggregate particles can fit through the slotted plate, they are considered to be flat. If they cannot fit, they are considered to be cubical. The lower the FI, the more cubical the aggregate is (Alaska DOT 2001).

The ALD can actually be measured by measuring the dimensions of each aggregate particle based on a 200 piece sample size. Alternatively the ALD can also be calculated using the median particle size and the FI of the aggregate. ALD is a measurement of the ability of the aggregate in the chip seal to be able to pack together. A higher ALD value means greater packing together of the aggregate, higher asphalt binder, demand and a greater macro texture or texture depth. The space between the aggregate provides room for the asphalt binder volume and the design texture depth after the aggregate is rolled or embedded into the pavement. The space that is available is 15–40 percent of the ALD volume with 30 percent being typical (Whiteoak 1991).

The loose weight or unit weight of the aggregate also needs to be known. ASTM C29, *Standard Test Method for Bulk Density (Unit Weight) and Voids in Aggregate*, can be used to determine the unit weight of the aggregate. A container of known volume is filled with the aggregate in three layers and rodded 25 times in each layer. The container is weighed and the unit weight of the aggregate, W, is calculated. The unit weight of the aggregate and the specific gravity of the aggregate are used to determine the voids in the aggregate. The voids in the loose aggregate, V, can be determined by equation 9.2.

$$V = 1 - \frac{W}{1000G_{sb}} \quad (9.2)$$

where V is the voids in the loose aggregate, in percent as a decimal; W, the loose unit weight of the aggregate, kg/m^3; G_{sb}, the bulk specific gravity of the aggregate.

The median particle size, M, can be calculated as the weighted average of the mean size of the largest 20 percent, the second largest 60 percent, and the smallest 20 percent particles. The median particle size, M, can be calculated using equation 9.3.

$$M = 0.1(b + a) + 0.3(c + b) + 0.1(d + c) \quad (9.3)$$

where M is the median particle size, mm; a, the 100 percent passing aggregate size, mm; b, the 80 percent passing aggregate size, mm; c, the 20 percent passing aggregate size, mm; d, the 0 percent passing aggregate size, mm.

The ALD can physically be measured base on an aggregate sample or it can be calculated using the value determined from the median particle size and the FI. If the FI is zero, meaning the aggregate is cubical, the mean particle size is equal to the average least dimension, ALD. When there is a significant amount of flat or elongated particles or the FI is high the ALD value can be significantly smaller than the median particle size. Equation 9.4 can be used to calculate ALD using the median particle size and FI.

$$ALD = \frac{M}{1.139285 + (0.011506)(FI)} \quad (9.4)$$

where ALD is the average least dimension, mm; M, the median particle size, mm; FI, the flakiness index, %.

Knowing the selected aggregate median particle size or the ALD and the loose unit weight (kg/m^3), the aggregate spread rate can now be determined. Equation 9.5 can be used to determine the aggregate spread rate.

$$S = 0.001\,(M)(W)(1 + E) \quad (9.5)$$

where S is the aggregate spread rate, kg/m^2; M, the median particle size or ALD; W, the aggregate unit weight, kg/m^3; E, the wastage factor, for loose aggregate, usually 5–10 percent (0.05 – 0.1).

Typical values can also be used for the ALD or median particle size values. Table 9.22 provides some typical values. The midpoint of the values is normally used. Calculating the actual ALD or median particle size on the tested gradation of the selected is preferred and not difficult.

Example 9.5

Determine the spread rate for a chip seal using a 19 mm top size chip seal aggregate. The loose unit weight of the aggregate is 1,377 kg/m^3 and the wastage is 5 percent. The aggregate is cubical in shape or has a FI of zero.

Table 9.22 Typical ALD values

Nominal top size (mm)	*Typical ALD* (mm)
25	16–20
19 or 20	11–15
12.5	7–9
9.5 or 10	5–6.5

Table 9.23 Typical chip seal thickness based on traffic

Design index	*Daily ESAL*	*Minimum chip seal thickness* (mm)
DI-1	≤5	6
DI-2	6–20	6
DI-3	21–75	9
DI-4	76–250	12
DI-5	251–900	19
DI-6	901–3,000	25[a]

Note
a Multiple surface treatments are recommended.

Solution

Use the typical gradation for a 19 mm top size listed in Table 9.21. Pick the values that are close to the criteria for calculating the median particle size, M:

$$M = 0.1(12.5 + 19) + 0.3(12.5 + 4.5) + 0.1(2.36 + 4.75) = 9.04$$

This is slightly smaller than the typical ALD value in Table 9.22, but is satisfactory for use.

$$S = 0.001(9.04)(1{,}137)(1 + 0.05) = \underline{10.8\,\text{kg/m}^2}$$

When using this method to determine the aggregate spread rate, the maximum top size of the aggregate is first picked based on traffic or the desired chip seal thickness. Usually the heavier the traffic, the thicker the chip seal should be. Table 9.23 provides typical chip seal thickness values related to traffic. The aggregate spread rate of the typical single size aggregate leaves the thickness of the chip seal about one stone thick. The top size of the aggregate has the greatest influence on the thickness of the chip seal. Table 9.24 provides guidance on selecting the proper aggregate size, based on traffic. Typical spread rates are also included in Table 9.24.

In a double or triple surface treatment or chip seal, the largest size of the stone in the first course determines the total chip seal thickness. The subsequent courses or chip seals fill the voids of the first cover aggregate. For each succeeding course the top size of the cover stone

Table 9.24 Typical single chip seal aggregate and spread rates based on traffic

Design index	*Nominal top size* (mm)	*Typical spread rates* (kg/m²)
DI-1	9.5	8–11
DI-2	9.5	8–11
DI-3	12.5	11–14
DI-4	19	14–16
DI-5	25	22–27
DI-6[1]	31.5, 37.5, 40[1]	32–40

Note
1 Multiple surface treatments are recommended.

Table 9.25 Chip seal spray adjustment factors

Pavement condition or texture	*Adjustment factor* (l/m²)
New surface	−0.05
Flushed surface	−0.2
Heavily flushed surface	−0.27
Smooth or non porous	0
Slightly porous, slightly oxidized	+0.14
Slightly pocked, porous, oxidized	+0.27
Heavily pocked, porous, oxidized	+0.40

should be approximately 50 percent of the size of the previously laid chip seal. For example, if a traffic index of DI-5 requires the top size of the first chip seal to be 25 mm, the second chip seal should have a top size of 12.5 mm. If a third chip seal is placed as part of a multiple surface treatment, the third chip seal should have a maximum top size of 6.3 mm or 5 mm (United States, 4.75 mm). No allowance is made for wastage in the additional surface treatments (The Asphalt Institute 1998a).

The final part of the chip seal design process is determining the amount of asphalt binder to spray. The aggregate in the chip seal will align themselves under traffic with about 20 percent voids between the particles. Enough asphalt binder or emulsion should be sprayed to fill up 60–75 percent of the voids with the residual asphalt binder. It is important to point out that when using an asphalt emulsion, initially almost 100 percent of the aggregate voids will be filled, but after the emulsion water displacement and curing 60–75 percent of voids should be left filled. The condition of the existing surface also determines the amount of asphalt material to spray. A highly oxidized and rugged texture will require more asphalt material than a smooth, dense new surface. During the chip seal design process, the amount of asphalt material required to partially fill the voids is first determined and then adjusted with factors based on the existing pavement surface. During the construction stage these factors may need to be adjusted based on varying conditions of the existing pavement. Table 9.25 provides adjustment rates based on the existing pavement surface. These factors are added or subtracted from the asphalt binder amount that is based on the aggregate gradation. These factors are also residue asphalt binder amounts. When using an asphalt

Table 9.26 Asphalt spray quantity traffic factors

Traffic design index	*Traffic factor (T)*
DI-1	0.85
DI-2	0.75
DI-3	0.70
DI-4	0.65
DI-5	0.60
DI-6	0.55

emulsion the spray quantity is increased to compensate for the water in the emulsion. For example, if using a CRS-2 asphalt emulsion containing 66.5 percent asphalt binder and a residue asphalt binder adjustment of 0.2 l/m^2 is required the new adjustment becomes 0.3 l/m^2 ($\{0.2/0.665\} = 0.3$).

The asphalt spray rates, A, are based on the assumption that the asphalt binder should fill up 60–70 percent of the chip seal void space after the particles have become fully oriented. The total void space (asphalt binder and air) is assumed at 20 percent. Equation 9.6 can be used to determine the asphalt spray rates.

$$A = \frac{(0.4\ MTV + PS)}{R} \tag{9.6}$$

where B is the asphalt material or binder spray rate, l/m^2; M, the median particle size or ALD; T, the traffic factors (Table 9.26); V, the voids in the loose aggregate, in percent as a decimal; PS, the pavement surface adjustment factors (Table 9.25); R, the residue asphalt content, decimal form, that is, 65% or 0.65.

Example 9.6

Using the median particle size determined in Example 9.5, a traffic index of DI-3, determine the spray quantity of a CRS-2 asphalt emulsion containing 66 percent asphalt binder on a slightly porous surface. The voids in the aggregate are 46 percent.

Solution

As previously determined, $M = 9.04$, from Table 9.26, $T = 0.70$, and from Table 9.25 $V = +0.14\ l/m^2$. R is 66 % or 0.66.

$$A = \frac{[(0.4)(9.04)(0.70)(0.46) + 0.14]}{0.66} = 1.98\ l/m^2 \text{ of CRS}-2 \text{ emulsion}$$

The quantity of asphalt sprayed can also be provided in ranges based on the maximum top size of the aggregate. The spray quantity still needs to be adjusted based on the existing pavement conditions using the factors in Table 9.25. Table 9.27 provides some typical emulsion spray rates based on traffic and aggregate size.

Table 9.27 Typical single chip seal asphalt emulsion spray rates based on traffic

Design index	*Nominal top size* (mm)	*Typical spray rates, asphalt emulsion* (l/m^2)
DI-1	9.5	0.7–0.9
DI-2	9.5	0.7–0.9
DI-3	12.5	0.9–1.6
DI-4	19	1.5–2.0
DI-5	25	1.8–2.3
DI-6[1]	31.5, 37.5, 40[1]	2.1–3.0

Note
1 Multiple surface treatments are recommended.

Multiple chip seal design

There are various types of multiple surface treatments, with the most common being a double chip seal. In isolated cases a triple chip seal may be used. The following procedure can be used in the design of multiple chip seals:

1 Complete a chip seal design for each layer of cover aggregate, as if it were the only layer.
2 The maximum top size of each succeeding layer should not be more than about 50 percent the maximum top size of the proceeding layer.
3 Make no allowance for wastage.
4 Except for the first chip seal, make no correction for the underlying surface texture.
5 Add together the amounts of asphalt material (asphalt binder, cutback, or emulsion) determined for each layer of cover aggregate to obtain a total asphalt material requirement.
6 For double chip seals, apply 40 percent of the total aggregate requirement for the first layer of cover aggregate, and the remaining 60 percent for the second layer of aggregate. For triple chip seals, the total asphalt material is summed from all three layers, then apportioned for the first, second, and third cover aggregate layers in portions of 30, 40, and 30 percent respectively (Alaska DOT 2001).

Chip seal construction

Prior to any surface treatment, the structurally deficient portions of the pavement should be repaired. Patches should be installed where required and significant cracking should be filled or sealed. Patches and crack repairs should have enough time to cure prior to the application of the chip seal. The chip seal will not correct any structural weaknesses or drainage problems in the existing pavement. The construction of a chip seal consists of the following steps:

1 Make improvements to the pavement's drainage, if needed.
2 Patch potholes.
3 Seal or fill significant cracks.
4 Make structural related repairs, including patching.

5 Clean the pavement surface.
6 Spray the asphalt material at the specified rate and temperature.
7 Immediately spread the cover aggregate at the specified rate behind the spray application. An asphalt emulsion will still be brown in color.
8 Roll the cover aggregate to seat them into the asphalt material.
9 After curing or setting, sweep any loose stones (The Asphalt Institute 1982).

Four pieces of equipment are used in constructing chip seals:

1 Power broom
2 Asphalt distributor
3 Aggregate spreader
4 Pneumatic roller.

The first step in the chip seal process is the removal of all dirt, dust, and clay from the existing pavement surface. Any standing water needs to be drained or removed. Sweeping with a power broom (Plate 9.27) will remove any significant dirt or debris from the pavement surface.

When the chip seal is going to be placed directly on a granular base or a dirt or gravel road, the existing surface should be shaped into the desired cross-section and compacted. A prime coat should then be applied. The prime coat should be completed sufficiently ahead of the chip seal operation, so that the prime coat has penetrated the base as completely as possible and has properly cured or set. Prime coats are usually applied at a rate of 2.0 l/m^2 for a very tight and impervious base up to a rate of 7.0 l/m^2 for a very porous or loose base.

Plate 9.27 Power broom.

Plate 9.28 Asphalt distributor.

Medium curing and rapid curing cutback asphalt were the asphalt material of choice for prime coats. The MC-30 and RC-70 were very commonly used prime coat materials. Due to air and ground water quality concerns, asphalt emulsions have replaced the use of cutback asphalt as prime coats. Slow setting penetrating emulsion prime (PEP) or asphalt emulsion prime (AEP) has been specially formulated for use as prime coats. A normal slow setting emulsion such as an SS-1 or CSS-1 can be used if they are bladed or mixed into the top 50–75 mm of the aggregate or gravel base. The asphalt emulsions are diluted 1 : 1 with water for typical porous applications. If the base is very tight or has a significant amount of fines, the asphalt emulsion may be diluted with up to 10 parts of water in order to facilitate penetration (The Asphalt Institute 1998a).

The asphalt distributor (Plate 9.28) applies the asphalt material for the chip seal and has been previously discussed in the section on fog seals. The aggregate spreader is responsible for spreading the correct amount of aggregate in a uniform manner. The aggregate spreader ranges from a vane type or tailgate type attached to a truck tailgate to a self propelled type (Plate 9.29). Tailgate spreaders are one of two types:

- Series of vanes attached to a steel plate to provide coverage across the width of the dump bed.
- Truck mounted hopper equipped with a feed roller that is activated by small wheels driven by the truck tires.

In either case the truck backs up as it discharges the aggregate onto the asphalt film. Backing up the truck prevents the tires from picking up the asphalt film.

Plate 9.29 Aggregate spreader.

Mechanical spreaders contain hoppers and a distribution system (Plate 9.30) to ensure a uniform spread rate and to completely cover the entire lane width. Mechanical spreaders are either truck attached or self-propelled (Plate 9.29). In both types the aggregate is dumped into a receiving hopper for spreading. Both the truck attached and the self-propelled spreader contains an auger and a spread roll in the hopper that ensures a positive uniform feed rate of material.

The aggregate (chipping) spreader spreads the aggregate on top of the asphalt material film, directly behind the asphalt distributor. The aggregate should be spread immediately or within 2 min of the application of the asphalt material. The aggregate is discharged a few meters behind the asphalt distributor in a veil in front of the spreader's tires. The hopper of the self-propelled spreader is at the rear of the spreader, with the dump trucks backing to the hopper and discharging the aggregate. This allows for continuous operation with starting and stopping kept to a minimum. The aggregate spreader speed should not exceed 90 m/min, when spreading aggregate.

As with the asphalt distributor, the aggregate spreader must also be calibrated to ensure that the proper amount of aggregate is being spread. The amount of aggregate that is discharged from the spreader is controlled by gate openings (Plate 9.30) and feed rate. The aggregate spread rate can be confirmed by the use of a series of rubber mats laid side by side, widthwise across the pavement. The aggregate spreader proceeds across the rubber mats with the control gates open, allowing aggregate to drop onto the rubber mats. The aggregate collected on each of the mats is removed and weighed. The transverse aggregate distribution is calculated and the aggregate spread rate can then be adjusted, if needed.

Plate 9.30 Aggregate distribution gates.

The conformance procedure is outlined in the ASTM D5624, *Standard Test Method for Determining the Transverse Aggregate Spread Rate for Surface Treatment Applications.*

The roller is the final piece of equipment used in the construction of a chip seal. The roller orients the aggregate and seats the aggregate into the sprayed asphalt material. The aggregate orientation is usually on the flattest section of the particle. The roller operates continuously covering the entire width of the pavement. Usually only one pass is needed, since the function of the roller is not to obtain density but only to seat the aggregate.

The pneumatic roller (Plate 9.31) is the most common roller used in the construction of chip seals. For single chip seals, the pneumatic roller provides the best results. A steel wheel roller used on a single chip seal will have a tendency to bridge some of the aggregate and not properly seat them. The pneumatic roller forces the aggregate firmly into the asphalt binder film without crushing the aggregate. The tire presses into the small depressions that are between the aggregate particles and knead the particles into the asphalt binder (The Asphalt Institute 1998a). Pneumatic rollers provide uniform pressure over the rough surfaces, while steel wheel rollers will only contact the high spots. Steel wheel rollers are often used on the second or third course of a multiple chip seal in order to force the smaller aggregate particles between the larger particles underneath.

When using an asphalt emulsion as the binder, the aggregate needs to be rolled or embedded before the asphalt emulsion breaks or cures. The asphalt emulsion should break after one or two passes of the roller. In order to have complete coverage and embed the aggregate quick enough, more than one roller is often used. To properly seat and orient the aggregate, the roller speed should not exceed 10 km/h. The outer edges of the pavement are rolled followed by the center of the pavement. Care must be taken not to make sudden starts,

Plate 9.31 Pneumatic roller used for chip seals.

stops, or turns on a newly laid chip seal. The roller should weigh at least 10 tonnes, have a minimum of nine tires and have the tire pressure inflated to 400–600 kPa (Alaska DOT 2001).

Transverse joints are constructed at the beginning and the end of the project and whenever there are delays. Transverse joints are constructed by starting and stopping the asphalt material spray and the aggregate on large Kraft or butcher paper. The paper should be positioned across the lane to be treated and positioned so that the forward edge of the paper is at the desired location. The distributor starts spraying on the paper at the correct application rate and speed so that when it reaches the exposed surface, the spray will be at the correct application rate and coverage. The paper is removed, exposing the seam.

Longitudinal joints are constructed when placing chip seals adjacent to each other. The edge of the aggregate spread should coincide with the edge of the full thickness of applied asphalt material. This allows a width that can be overlapped when the asphalt material is sprayed in the adjacent lane, because a portion of the spray fan from the end nozzle will be a single film. When the aggregate is then spread full width in the next lane, there will be no aggregate build up at the joint. It is important to prevent aggregate from building up at the joint, causing a bump. The width of the exposed asphalt material will vary, depending on whether the height of the spray bar and the nozzles are set for single, double, or triple coverage, and on the spacing of the nozzles (The Asphalt Institute 1982).

Sweeping is completed to remove any loose or excessive aggregate from the chip seal. The chip seal must have cured enough to hold the embedded aggregate solidly in place. Depending on the asphalt material used, the humidity, and temperature, proper curing can take anywhere from one day to a week. Traffic can be allowed on the chip seal prior to complete curing, but the speed should be kept less than 35 km/h during the first 24 h of curing. This is usually accomplished through the use of a pilot vehicle or the new chip seal is temporarily closed to traffic.

Chip seal troubleshooting

When a chip seal is properly designed and constructed, it will provide several years of satisfactory performance. The useful life of a chip seal varies with traffic, underlying pavement conditions, and the environment, but can range from 2 to 7 years, with 4 to 5 years being typical. When a chip seal fails prematurely, it is usually due to some problem with the construction. Chip seals fails consist mostly of excessive loss of aggregate or chippings. Excess asphalt binder being exposed or flushing is the other common type of chip seal failure. The following precautions will prevent most chip seal failures:

- Do not construct a chip seal during the rain or when rain may occur within 24 h of construction.
- Chip seals should not be constructed when the air temperature is less than 10 °C. The temperature of the pavement surface should be 20 °C or higher.
- Use aggregate that is clean and not saturated with water.
- Spray the asphalt material at the proper temperature.
- Calibrate and confirm the spray rate and the aggregate spread rate.
- Spread the aggregate immediately or within 2 min of the asphalt spray application.
- Roll the aggregate immediately and before the asphalt emulsion breaks (if used).
- Keep excessive traffic and speed off the chip seal until it cures.
- Do not sweep the chip seal until it is fully cured.

The three most common problems or appearances with a chip seal are:

1 Streaked appearance
2 Bleeding or flushing
3 Loss of cover aggregate.

A streaked appearance or streaking is the presence of longitudinal or transverse grooves in the chip seal surface. Longitudinal lines of asphalt material are the most common form of streaking and are mostly cosmetic. Longitudinal streaking is parallel to the road direction and is caused by:

- Spray nozzles improperly set or misaligned, causing uneven application.
- Spray bar not set at the correct height.
- Asphalt material pump speed too low, causing insufficient pressure at the spray bar ends.
- Clogged nozzles.

Transverse streaking is usually caused by a pulsation of the asphalt material pump, due to worn parts, improper pump speed, or poor control of pump speed.

Bleeding is defined as an excess of asphalt material in the wheel path or traffic areas. Flushing (fatting up) is similar to bleeding but is usually more widespread throughout the chip seal. Too much asphalt material for the required amount of aggregate is the typical cause. If the aggregate particle is too small or has a high flakiness index, the particles will lie flat and the layer of aggregate becomes too thin, exposing the asphalt binder film. Water vapor pressure under the chip seal or trapped in the aggregate, will sometimes cause the aggregate particle to become loose, exposing the asphalt film underneath.

The loss of the cover aggregate or chips can be caused by any of the following:

- Insufficient asphalt binder.
- Road surface dirty or very wet.
- Dirty or dusty aggregates.
- Delay of aggregate application, following the spray application.
- Allowing an asphalt emulsion to break prior to the aggregate application.
- A dense graded gradation, as opposed to a single size aggregate.
- Insufficient rolling.
- Construction during cool or wet weather.
- Opening to traffic too soon or traffic too fast (Plummer 1996).

Types of chip seals

There are various types of chip seals that can be used with the single chip seal being the most common and the foundation of the other types of chip seals.

Single chip seal is an application of an asphalt material directly on a road, followed by a single application of aggregate. It is the most common method and is adequate for most roads with light to medium traffic levels.

Double chip seal is a second chip seal put directly on top of a single chip seal. There is a second application of asphalt material followed by the addition of a layer of aggregate that has a maximum top size of about 50 percent of the first chip seal. There is no increase in thickness since the smaller particles fit between the larger aggregate particles. A double chip seal will provide about three times the service life for about one-and-a-half times the cost (The Asphalt Institute 1998a). The packing in of additional smaller aggregate by the single application results in a chip seal with a significant reduction in aggregate loss. A double chip can be used for roadways with higher levels of traffic.

Triple chip seal is similar to a double chip seal with even smaller aggregate for the third aggregate application. A triple chip seal is often used for high speed or high traffic applications.

Sand seal is the same as a single chip seal but with the use of fine aggregate or sand as the aggregate layer. It is used usually for very low volume traffic or where good coarse aggregate is not available and provides a tight water-resistant barrier for the pavement below. A sand seal can also prevent the further loss of material from an old or oxidized pavement surface that is showing signs of abrasion. A sand seal can also provide an excellent foundation or pad (pad coat) for embedding the aggregate in a chip seal. The asphalt material application rate is 0.7–1.25 l/m^2 followed by 5–11 kg/m^2 of sand.

Sandwich seal is a chip seal that is constructed by spreading a single layer of large aggregate (15–20 mm) directly on the road surface, followed by the application of the asphalt material, and lastly followed by a second application of aggregate, but much smaller (5–12 mm). The application of the larger aggregate is used to overcome flushing or bleeding problems in the existing pavement surface and the smaller aggregate locks in the larger aggregate.

Table 9.28 Typical cape seal application rates

Nominal top size of chip seal (mm)	*Chip seal spray rates, asphalt emulsion* (l/m^2)	*Chip seal aggregate spread rates* (kg/m^2)	*Type II slurry seal* (kg/m^2)
19	1.3–1.8	14–16	2.5–4.5
25	1.6–2.2	22–27	3.5–5.5

Cape seal is a single chip seal followed by an application of a slurry seal. The slurry seal fills the voids between the aggregate in the chip seal and provides a uniform black appearance to the chip seal. The slurry seal bonds the chip seal aggregate to prevent loss and in return the chip seal aggregate prevents traffic abrasion and erosion to the chip seal. The asphalt binder content in the chip seal should be slightly reduced from typical chip seals, because of the asphalt binder content in the slurry seal. The chip seal should cure fully prior to the application of the slurry seal. A Cape seal is a highly durable surface treatment and can be used on many different applications, including high volume. Table 9.28 provides some typical application rates for a Cape seal (The Asphalt Institute 1998a).

Slurry seal

A slurry seal is a surface treatment that can be used for both preventative and corrective maintenance. A slurry seal differs from a chip seal in that the aggregate is dense graded and premixed with an asphalt emulsion before it is laid. Chip seals are usually not permitted on aircraft runways or taxiways due to the possibility of loose aggregate ingestion by a jet engine. Loose aggregates are not a common problem with slurry seals and is often used on runways. Slurry seal does not need to be compacted like a chip seal, but it can be if desired. Slurry seal was initially developed in Germany during the 1930s and became popular during the 1960s with improvements in the lay down machines. Like other surface treatments, a slurry seal will not increase the structural strength of the pavement or correct major deficiencies. The same repairs that would be done prior to a chip seal should also be done prior to the application of the slurry seal. A slurry seal will fill minor cracks, slow or stop raveling, improve skid resistance, and provide protection from water damage. The slurry seal will also reduce oxidation and increase the service life of pavements that are already in relatively good condition. The benefits of a slurry seal are:

- Fast application and quick return of traffic.
- No loose cover aggregate.
- Dark black appearance and esthetically pleasing.
- Excellent surface texture and skid resistant.
- Ability to correct minor surface irregularities.
- Very minimum loss of curb heights, manholes, and other drainage structures.
- No need for manhole or drainage structure adjustments.
- An excellent surface treatment for urban applications.
- An excellent surface treatment for aircraft runways (The Asphalt Institute 1998a).

A slurry seal is a mixture of dense graded aggregate, an asphalt emulsion, mineral filler, and water. In some cases additional additives are added to the slurry to modify the curing

Figure 9.6 Slurry seal unit.

characteristics of the slurry seal. The slurry seal is applied in one step through a special slurry seal applicator in a thickness of 3–10 mm. Since the thickness is thin, the maximum top size of the aggregate is usually no greater than 9–10 mm and can be as small as 4.75 or 5 mm. The slurry seal applicator or machine mixes the slurry seal as it is being laid. The machine is a self-contained and continuous flow-mixing unit (Figure 9.6).

Slurry seal materials

The basic materials used in a slurry seal are a dense graded aggregate and a slow setting emulsion. When a slurry seal contains polymer modified asphalt emulsion and all crushed aggregate, it becomes known as microsurfacing, micro asphalt, or micro seal. The significant advantages of microsurfacing is that it is highly stable and can be applied in lifts up to 50 mm.

Only asphalt emulsions are used in the construction of a slurry seal. The most common asphalt emulsion used for slurry seals is a CSS-1 or CSS-1h. Slow setting anionic emulsions can also be used (Table 9.29). Cationic emulsions are usually preferred, since they tend to break faster and provide better adhesion on negatively charged aggregates. Most aggregates, especially calcareous types such as limestone, are negatively charged. A slightly faster breaking version of the slow set emulsion is the quick set (QS) or cationic quick set (CQS). Most slurry seals today use the quick set version. The emulsifier provides the quick set properties. The CQS version is by far the most common with the CQS emulsifier being an amidoamine added at a dosage rate of 1.5–2.5 percent by weight of the total asphalt emulsion. Microsurfacing requires a polymer modified cationic quick set asphalt emulsion. The QS type emulsions meet the requirements of standard slow set emulsions except that the cement-mixing test is waived. The asphalt emulsion should meet the relevant specification such as or equivalent to ASTM D977 for anionic emulsions and ASTM D2397 for cationic emulsions.

Aggregates, excluding the mineral filler, constitute about 82–90 percent by weight of the slurry seal. Similar to HMA, the aggregate gradation contributes to a significant portion of the slurry seal performance. The reactivity of the aggregate can also affect the setting characteristics of the slurry seal. Highly charged aggregates can set very fast when combined with positively charged asphalt emulsion. In this regard, once an aggregate source is selected for the slurry seal, it should not be altered to another source, without a mixture design verification. For the best results, 100 percent crushed aggregate should be used. The aggregate should be clean, well graded, and uniform (USDOT 1994). The quality of the aggregate used should meet the requirements in Table 9.30 or similar requirements.

Table 9.29 Slurry seal emulsions

Anionic	*Cationic*
SS-1, SS-1h, QS-1h	CSS-1, CSS-1h, CSS-1P, CQS-1h, CQS-1P

Table 9.30 Slurry seal aggregate quality requirements

Test	*Requirement*	
	Slurry seal	*Microsurfacing*
Sand equivalent value, ASTM D2419	⩾45%	⩾60%
Sodium sulfate soundness, ASTM C 88	⩽15%	⩽15%
Los Angeles abrasion, ASTM C 131	⩽35%	⩽30%

Table 9.31 Slurry seal gradation types and application rates

Sieve size	*Slurry seal type*		
	I	*II*	*III*
	Total passing (%)		
9.5 mm	100	100	100
4.75 mm	100	90–100	70–90
2.36 mm	90–100	65–90	45–70
1.18 mm	65–90	45–70	28–50
600 μm	40–65	30–50	19–34
300 μm	25–42	18–30	12–25
150 μm	15–30	10–21	7–18
75 μm	10–20	5–15	5–15
Residual asphalt binder content, % weight of aggregate	10–16	7.5–13.5	6.5–12
Application rate (kg/m^2)	3.5–5.5	5.5–9.0	8.0–14.0

The International Slurry Seal Association (ISSA) recommends three slurry seal grades. The slurry seal grades are known as Type I, Type II, and Type III, with the aggregate gradation distinguishing each type. All of the gradations are dense graded. Type I is the finest and is used for crack filling and as a thin slurry seal course. The maximum aggregate top size is 4.75–5 mm. Type I provides good crack penetration and sealing properties. Type I slurry seal is often used as a crack filling technique prior to a chip seal or overlay. Type I slurry seal will also perform well in light traffic applications and parking lots.

Type II slurry seal is the most commonly used slurry seal and can be used for most traffic applications. Type II slurry seal can protect the pavement from oxidation, water, and excessive tire wear. The Type II slurry seal will also increase skid resistance and can correct surface raveling (The Asphalt Institute 1998a).

Type III slurry seal is the coarsest gradation and is used for heavy traffic. The Type III gradation is also used for microsurfacing. Table 9.31 provides the gradations and other guidelines for the three types of slurry seal.

The three gradation types are the gradation of the slurry seal mixture, including the mineral filler if needed.

Mineral filler is added to alter the consistency and cohesive characteristics. The mineral filler can also alter the break characteristics and improve the segregation resistance of the slurry seal mixture. The required amount of mineral filler is determined at the mixture design stage and by how much is needed to meet the required gradation. Portland cement or hydrated lime is often used as mineral filler since both can also alter the break characteristics of the slurry seal. Normally up to 2 percent of Portland cement or 1 percent of hydrated lime by weight of dry aggregate can be used. The Portland cement type is Type I or Type II A. Mineral fillers such as limestone dust may also be used and for the most part will be nonreactive and do not significantly chemically alter the setting characteristics of the slurry seal.

Field control additives are used to increase or decrease the break time of the slurry seal. It is important that the slurry seal does not break prior to being spread or take too long to break. The ambient temperature and the pavement temperature have a significant influence on the breaking characteristics of the asphalt emulsion in the slurry seal. Retarding or "slowing" down the break of the slurry seal is the most common need, mostly due to construction during hot weather. If cool conditions develop, an alternative additive can be introduced and the break can be accelerated. Portland cement or hydrated lime can be used to accelerate or speed up the break of the slurry seal. Addition rates will vary, up to 3 percent. Aluminum sulfate or aluminum chloride can be used to retard or slow down the break of the slurry seal, typically at dosages of 1 percent or less. Field control additives are stored on the slurry seal unit, allowing the operator to vary the dosage rate depending on actual field conditions.

Water is sprayed on the aggregate during the mixing process to alter the consistency of the slurry seal. The amount of mixing water introduced will vary, depending on the moisture content of the aggregate and the ambient temperature. Increasing the water content of the slurry seal mixture will make it more workable, up to a point that it becomes soupy. Too much water will cause the mixture to become segregated and will severely retard its breaking time. Additional water carried on the slurry seal unit can be adjusted as varying field conditions are encountered. More water is used during hot temperatures than at cooler temperatures. The total moisture content (mixing water and aggregate moisture content) should be between a range of 4–12 percent by weight of aggregate (USDOT 1994). Water is often lightly sprayed by the slurry seal unit on the pavement surface, immediately prior to the application of the slurry seal. Just enough water is sprayed to lightly dampen or fog the pavement surface. Water spray bars are located on the slurry seal unit immediately in front of the spreader box. Excessive water should be avoided and will cause puddles to form on the pavement surface.

Slurry seal design

The slurry seal design is used to determine the amount of asphalt emulsion, mixing water, and mineral filler to add to the dense graded aggregate. The gradation of the aggregate is predetermined by the type of slurry seal selected (Table 9.31). The application determines which slurry seal type to use, with Type II for most applications and Type III for microsurfacing.

The asphalt emulsion content, mixing water, and mineral filler all have significant effects on the consistency and cohesive characteristics of the slurry seal. Tests that can measure

Table 9.32 ISSA recommended slurry seal tests and performance guidelines

Description	*ISSA test method*	*Requirement*
Slurry seal mix time	TB-113	Controllable at the highest expected construction temperature for 180 s
Slurry seal consistency	TB-106	2.5 cm
Wet cohesion	TB-139	⩾12 kg-cm at 30 min set time ⩾20 kg-cm at 60 min set time
Cured cohesion	TB-139	⩾24 kg-cm at 60 °C
Wet track abrasion loss	TB-100	⩽ 540 g/m^2 after a 60 min soak time ⩽ 805 g/m^2 after a 6 day soak time
Wet stripping	TB-114	⩾90% retained coating
Loaded wheel tester and sand adhesion	TB-109	⩽540 g/m^2 at 1,000 cycles

those properties are used to determine the optimum proportions of each of the ingredients. The ISSA has developed several laboratory test methods to determine the proper slurry seal mixture design and predict its performance. Many of these test methods have been adopted by various highway agencies. Table 9.32 provides the recommended ISSA tests and guidelines for the design of slurry seals. Not all of the ISSA tests need to be used for proportioning the materials in the slurry seal design. The mixing time, wet cohesion, and slurry seal consistency tests are the main proportioning tests.

The slurry seal mixing time test is the main trial mixture procedure for slurry seal design. The test consists of trial hand mixing for 30 s and 5 min, proportions of aggregate, mineral filler, mixing water, and the asphalt emulsion. Trial 100–200 g samples of the proposed aggregate is dried to a moisture content of less than 1 percent. The mineral filler, usually hydrated lime or Portland cement, is added at various amounts of 0, 1, and 2 percent. The mixing water and asphalt emulsion are then added and mixed for 30 s. At the end of the 30 s mixing time, half of the mixture is removed and spread and cast on a piece of paper. The other half of the mixture is mixed for 5 min or until the mixture stiffens and the asphalt emulsion breaks. The spread mixture is periodically depressed with an index finger until the slurry becomes firm and is no longer appreciably displaced by the finger. The amount of time this takes is the "set time." Set time can actually be defined as the lapsed time after casting when a slurry seal mixture cannot be remixed into a homogenous mixture. It is also set when no lateral displacement is possible and none of the asphalt binder can be tracked from its surface. The amount of mixing water and asphalt emulsion to add to the trial mixture will vary with the percent asphalt emulsion and the emulsion asphalt binder content. Type II formulations will typically require a total liquid (mixing water and emulsion) content of 28–32 percent. Each 1 percent addition of mineral filler will also require an additional 1.0–1.5 percent of mixing water (ISSA 1990). The amount of mixing water and asphalt emulsion is the one that provides a controllable mixing time for 180 s at the highest expected construction temperature. A mixing time of 60 s at 21 °C is the very minimum.

The slurry seal consistency test is a test that uses a small funnel or cone that can provide a numerical value to consistency. The cone test is used to determine the amount of water required in forming a stable and workable mixture. The test uses the same cone used for the fine aggregate absorption test as described in ASTM C-128, *Standard Test Method for*

Specific Gravity and Absorption of Fine Aggregate. The cone is 75 mm in height, a 40 mm diameter at the top, and a 90 mm diameter at the bottom. A flow scale is produced with seven concentric circles in 1 cm increasing radii on a piece of paper. The cone is centered on the scale. The same trial mixture proportions are produced as for the mix time test. The sample size is 400 grams of the slurry seal mixture. The mixture is mixed for 30 s and then placed in the cone. The cone is immediately removed and the outflow of the slurry mixture is measured at four points, 90 ° apart. The optimum fluid content provides a radial flow of 2.5 cm. A graph is developed for the various water and asphalt emulsion combinations versus radial flow rate, allowing the optimum mixing water and asphalt emulsion content to be determined in conjunction with the other tests.

The wet cohesion and the cured cohesion test provide numerical strength values to measure the setting and curing characteristics of the slurry seal. A cohesion tester provides the strength value. The cohesion tester puts forces on the slurry seal to simulate power steering movements and rotating tires. The cohesion tester is a device that can apply various amounts of pressure to a cast slurry seal mixture or pad through a rubber foot placed on top of the casting. A twisting force or torque is applied through a torque wrench and the resisting force by the slurry seal mixture to the rubber pad is measured and reported in kg-cm. A slurry seal mixture is prepared at the previously determined optimum asphalt emulsion, mixing water, and mineral filler content. The mixture is cast in pad and tested at a 30 and 60 min cure time. The mixture can also be tested at a fully cured, 60 °C state. The mixture should exceed 12 kg-cm at 30 min of cure time and exceed 20 kg-cm at 60 min of cure time. Set is considered meeting or exceeding 12 kg-cm at 30 min of cure time. Traffic can be started on a slurry seal that has achieved a torque of 20 kg-cm. The mineral filler content and asphalt emulsion content can be adjusted to meet the requirements. The addition of a set accelerator such as Portland cement or hydrated lime will increase the torque values.

A quick set slurry seal is defined as a slurry seal that achieves 12 kg-cm in 30 min. A quick traffic slurry seal is defined as a slurry seal that achieves 20 kg-cm in 60 min. The wet track abrasion is another performance related test. The wet track abrasion simulates the wearing effects of traffic. The wet track abrasion measures the amount of material that is abraded by a rotating rubber hose from a cured slurry seal pad that is under water. The wet track abrasion test has been correlated to the wearing characteristics of field applied slurry seal. The quality of the aggregate, the residue asphalt binder content and the mineral filler content all significantly affect the abrasion resistance of a cured slurry seal. The slurry mixture is prepared at the previously determined optimum asphalt emulsion and mineral filler contents. An 800 gram sample is prepared and cast into a wet track mold. The sample is cured in a 60 °C oven for a minimum of 15 h. The sample is then weighed and placed in a 25 °C water bath for about 1 h or for 6 days. The sample is then placed in a shallow pan and covered with 25 °C water. The sample and pan are placed in the wet tack abrasion machine. The wet track abrasion machine is actually a small mechanical restaurant mixer equipped with a 127 mm long piece of 19 mm diameter rubber hose. The ISSA Technical Bulletin 100 (TB-100) provides specific information on the type of mixer and molds to use. The rubber hose is the same or similar to a heater hose used in an automobile. The hose is mounted in the modified mixer head horizontally. The mixer operates in low speed. After the sample is placed in the wet track abrasion machine, the mixer is turned on and the rubber hose rotates on the surface of the sample for 5–6 h. The sample is removed and weighed again. The amount of material lost is determined and knowing the sample mold area, grams per square meter of sample area can be calculated. The amount lost should be less than 805 grams/m^2 (ISSA 1990).

The wet stripping test is a stripping test on a cured 10 g sample of the slurry seal mixture at the optimum content of the various materials. The sample is placed in a beaker of boiling distilled water for 3 min. After 3 min, cold water is run on the beaker and sample until any free asphalt film flows over the top of the beaker. The water is then decanted from the beaker and the sample is placed on a paper towel to dry. After drying, the sample is visually evaluated for retained asphalt coating on the aggregate. A retained coating of 90 percent or greater is satisfactory. Below 75 percent is considered unsatisfactory.

The loaded wheel tester is another performance test for the cured slurry seal mixture. The loaded wheel compacts and wears the slurry seal through the use of a loaded, reciprocating rubber tire. The test can be used for slurry seal design purposes to establish maximum limits of the asphalt binder (asphalt emulsion residue) and to predict the occurrence of severe rutting or flushing under traffic. The apparatus is similar to the various loaded wheel testers used for HMA, but is much smaller. The apparatus and procedure is described in the ISSA Technical Bulletin, TB-100. A slurry seal mixture is made at the previously determined optimum material contents. The mixture is placed in a specimen mold. The specimen is cured in a 60 °C oven for a minimum of 12 h prior to testing. The test itself is run at 25 °C. The specimen is placed in the loaded wheel tester and the wheel (tire) is run 1,000 cycles over the sample. The sample is removed and any loose material is washed off with water. The sample is again dried at 60 °C to a constant weight. The sample is weighed and the amount of material abraded or lost is determined. Knowing the area of the sample mold, the grams lost per square meter can be calculated. After weighing, the sample is placed again in the loaded wheel tester and this time has 300 g of hot (82 °C) fine Ottawa laboratory standard sand spread on top of the sample. The loaded wheel is immediately placed on the sample and an additional 1,000 cycles is completed. All the loose sand is removed with a vacuum cleaner and the specimen is weighed again. The increase in weight to sand adhesion is determined (ISSA 1990). The weight increase due to the sand is usually just noted or reported.

The loaded wheel tester is useful in determining the maximum amount of asphalt binder that the mixture can contain before rutting or flushing. The loaded wheel tester is not always part of the mixture design process and can be considered optional. However, when completing a microsurfacing mixture design or a design for a high traffic facility, the loaded wheel testing should be included.

In summary, the slurry seal design process involves selecting the optimum asphalt emulsion content, mixing water amount, and mineral filler amount initially by the mixing, setting, and consistency tests. The performance of the selection is verified by the cohesion test, the wet track abrasion test, the wet stripping test, and possibly the loaded wheel tester. Modifications to the asphalt emulsion or residue content and the mineral filler content are carried out after evaluating the results of the performance tests. Many slurry seal producers will evaluate several raw materials to develop a formulation that is economical, versatile, and provides high performance. Once such a design is selected, the proportions and raw materials will not vary. In that regard, the slurry seal design process will generally produce recipe type formulations.

Slurry seal construction

Slurry seal and microsurfacing construction is a one-step process utilizing the slurry seal machine, which is a self-contained and continuous flow-mixing unit. The slurry seal machine proportions, mixes, and spreads the mixture all in one step. The machine must be capable of storage of the raw materials and adequate replenishment needs to be considered. Generally the aggregate is stockpiled close to the project site and tankers deliver the asphalt emulsion. The remaining materials are replenished as needed. Once an application rate is chosen, the amount of raw materials needed are estimated by the slurry seal design and ordered. If the application rate is not specified, the midpoint of the rates given in Table 9.31 should suffice. For example, an application rate of 7.25 kg/m^2 of a Type II slurry seal should perform adequately for most applications. The condition of the existing pavement surface has a significant effect on the slurry spread rate. If the pavement surface is smooth, a lower rate can be used, if the pavement surface pocked, raveled, or full of voids, the higher application rate should be used. The square meter area of the project is calculated and the amount of slurry seal and raw materials required for the project can be determined.

Example 9.7

Determine the raw materials required for a Type II slurry seal spread at an application rate of 7.25 kg/m^2. The road is a 6 m wide residential street that is 2 km long. The slurry seal has a CQS-1 asphalt emulsion content of 11 percent and a hydrated lime content of 1.5 percent. 2.5 percent mixing water is also required.

Solution

First determine the amount of slurry seal required (add 5 percent for wastage):

$$(6\,\text{m})(2{,}000\,\text{m})(7.25\,\text{kg/m}^2)(1.05) = 91{,}350\,\text{kg or } \underline{91.35\text{ tonnes}}$$

All of the optimum contents are by weight of aggregate. Convert the contents to percent by weight of mixture:

CQS emulsion: $(11)/(100 + 11) = 0.0991$ or 9.91%
Hydrated lime: $(1.5)/(100 + 1.5) = 0.0147$ or 1.47%
Mixing water: $(2.5)/(100 + 2.5) = 0.0243$ or 2.43%
Aggregate: $100 - (9.91 + 1.47 + 2.43) = 86.19\%$

Quantities:

CQS emulsion: (91.35 tonnes) (9.91%) = $\underline{9.05\text{ tonnes}}$
Aggregate: (91.35 tonnes) (86.19%) = $\underline{78.73\text{ tonnes}}$
Hydrated lime: (91.35 tonnes) (1.47%) = $\underline{1.34\text{ tonnes}}$
Mixing water: (91.35 tonnes) (2.43%) = $\underline{2.22\text{ tonnes or } 2{,}220\text{ l}}$

Slurry seal and microsurfacing is made by the slurry seal machine or mixer. The machine is self-propelled and equipped with a pugmill type mixer. Figure 9.7 provides a schematic view of a slurry seal unit. The individual raw materials are stored separately on the machine. Most slurry seal machines have capacities of 6 tonnes or more. The aggregate and mineral filler are stored in hoppers. The aggregate is the largest hopper on the machine and occupies the most space. A conveyor belt is under the aggregate hopper for metering and moving the aggregate to the pugmill or mixing chamber (Figure 9.8). The mineral filler is also

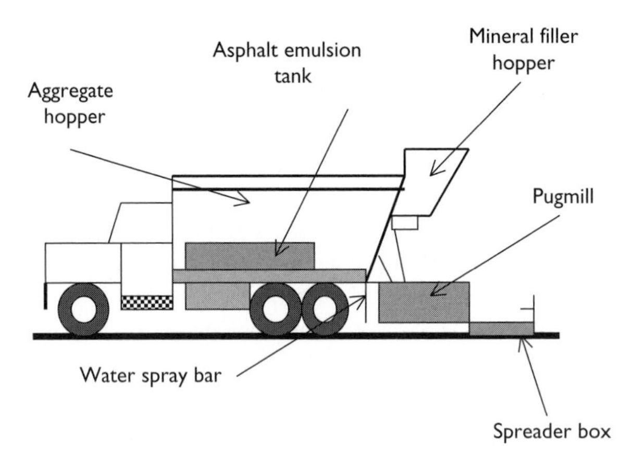

Figure 9.7 Slurry seal unit schematic.

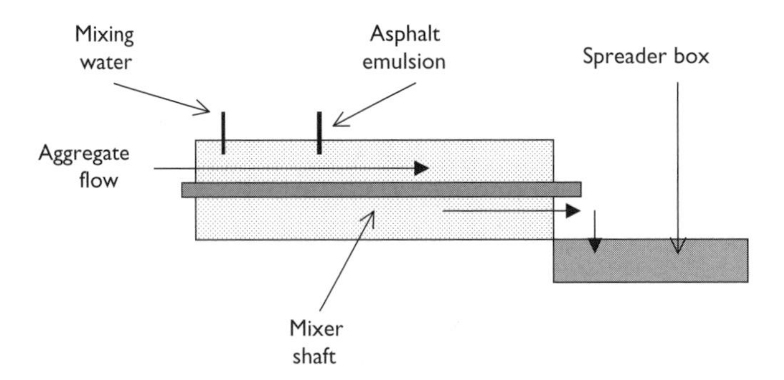

Figure 9.8 Slurry seal pugmill mixer.

metered into the pugmill. The asphalt emulsion and the mixing water are stored in tanks on the machine and metered into the pugmill by pumps and flow meters. The mixing water is added first at the beginning of the pugmill to wet the aggregate, followed by the asphalt emulsion being sprayed on the aggregate. The field control additive, which may also be the mineral filler (Portland cement or hydrated lime), is also stored on the machine. There is an automated system for sequencing and proportioning all the raw materials to ensure a constant flow of slurry and a consistent slurry mixture. The pugmill mixer may be equipped with a single or double shaft mixing blades.

Slurry seal and microsurfacing is applied to the existing pavement surface through a spreader box that is attached to the slurry seal machine. The mixed slurry is introduced into the spreader box, which then spreads the slurry as the slurry seal machine travels down the road. The spreader box may be equipped with augers to spread the mixture, especially stiff microsurfacing mixtures, throughout the box, providing uniform coverage. The spreader box has a flexible squeegee at the edge of the box. The squeegee is in contact with the pavement surface to prevent loss of the slurry. The spreader box is also adjustable in width to allow varying widths of slurry seal to be spread. The spreader box is also adjustable to provide a uniform spread over varying crowns and grades. It is equipped with a strike off at the trailing edge of the box to provide for a uniform and smooth surface. In front of the spreader box is a water spray bar that provides for a fogging or damping of the pavement surface immediately prior to slurry sealing. The spray bar should provide water coverage of 0.1–0.3 l/m^2. The travel speed during slurry sealing should be between 20–50 m/min. Slurry seals or microsurfacing should not be placed when the ambient or the pavement temperature is less than 10 °C. Slurry seals should not be applied when there is a chance of freezing within 24 h of application. High humidity and rain will lengthen the time it takes for the slurry seal to set.

During the application or spreading of the slurry seal, there should be no lumping, balling, or unmixed aggregate visible in the spreader box. Enough slurry should be carried in all parts of the spreader box, so that complete and uniform coverage can be achieved. On the other hand, overloading of the spreader box must also be avoided. Streaking or tearing of the slurry seal by oversized aggregate should be repaired immediately by a hand squeegee. Excessive buildup of slurry seal at existing longitudinal and transverse joints should be avoided. Overlapping longitudinal joints should be no more than 150 mm. Hand squeegees can be used to tailor and improve the slurry seal at joints, drainage fixtures, curbs, and other similar areas. They are also useful for repairing minor imperfections in the slurry seal.

Compaction of the slurry seal is not needed. Rolling of the slurry seal with a pneumatic roller may be required in areas of very light or no traffic, such as aircraft aprons parking lots. The pneumatic roller embeds the lower portion of the slurry seal into the existing pavement surface improving the bonding of the slurry seal to the existing pavement. One to two passes of a 10–12 tonne roller with tire pressures at 350–450 kPa is adequate. The slurry seal must have already undergone initial set and be cohesive enough so that the roller tires do not pick up any of the slurry. Steel wheel rollers should not be used since they will bridge the aggregate and have the possibility of crushing the larger aggregate pieces and mark the surface (The Asphalt Institute 1998a).

A slurry seal can be opened to traffic after it has set and shows no sign of tracking. No displacement should occur when subjected to a lateral force. A slurry seal that achieves 20 kg-cm of torque is ready to be open to traffic. Under ideal conditions, the quick traffic slurry seal could be open to traffic as early as 1 h after being spread.

Slurry seal troubleshooting

Slurry seals are generally troublefree with most concern involving setting time and surface texture. Change in the slurry seal design and the use of field control additives will solve most setting and curing problems. Special attention should always be placed on the ambient temperature and humidity. Low temperatures and high humidity will cause the slurry seal to take longer to set and cure.

Transverse rippling is a surface texture effect that can be caused by:

- Application too thin
- Low application rate
- Slurry seal machine travel speed too high.

The use of a rubber strike off instead of a steel strike off will also reduce the occurrence of rippling.

Longitudinal streaking is caused by worn strike off plates and drag mops (where used). Tear or drag marks can be caused by:

- Worn or dirty strike off
- Oversized aggregate
- Insufficient application rate or material in spreader box
- Premature breaking or setting of the mixture in the pugmill or spreader box.

Tear or drag marks can be easily avoided by following proper construction techniques, scalping oversized aggregate, proper mixture design, and the use of the correct field control additive based on field conditions (USDOT 1994).

Microsurfacing

Microsurfacing is a form of slurry seal. The construction techniques and precautions that apply for slurry seals also apply for microsurfacing. It is designed using the same methods and uses the same materials except for a few noted exceptions:

- Uses the Type III slurry seal gradation
- Requires all crushed aggregate
- Uses a polymer modified cationic quick set emulsion (CQS-1P).

The addition of the polymer and requiring all crushed aggregate increases the stability and stiffness of the cured slurry seal or microsurfacing. The increased stability and stiffness allows the slurry to be laid in thicknesses upto 50 mm as rut-filling mixtures. Rutting on pavements that are otherwise stable can have the ruts filled to eliminate vehicle hydroplaning in the ruts or channels. Special rut-filling spreader boxes allow deep pavement ruts or channels to be filled with the microsurfacing mixture. Microsurfacing is typically spread in thicknesses of 9–16 mm. These thicknesses require application rates of 11–16 kg/m^2. Microsurfacing can also be spread in two or more lifts. The stronger microsurfacing can undergo heavier and greater traffic loads than a standard slurry seal. Microsurfacing will set fairly fast due to the use of quick set emulsifiers in the asphalt emulsion.

The aggregate quality requirements are higher than for standard slurry seal (Table 9.30). Higher quality aggregates are required for the higher traffic levels that microsurfacing often endure. The asphalt emulsion for microsurfacing is a polymer modified cationic quick set (CQS-1P). The polymer is typically a SBR or latex added at the rate of 3 percent by weight of emulsion.

Microsurfacing is applied using the same construction methods that are used for standard slurry seal. A properly equipped slurry seal machine can mix and apply microsurfacing. Due to the stiffness of the microsurfacing mixture, pugmill should be a twin shaft mixer and the spreader box should be equipped with augers. Special rut-filling spreader boxes should be available when needed. The microsurfacing emulsion will break with 2–4 min of mixing with the aggregate. Microsurfacing can generally accept traffic 1 h after spreading.

Slurry seals and microsurfacing are very esthetically pleasing and can provide excellent performance as a surface treatment. Similar to pavement sealers or sealcoats, slurry seals can be colored by various colors with black, green, and red being the most common. They will last longer than a pavement sealer or sealcoat and should be considered for heavily raveled or oxidized parking lot and residential street surfaces. A slurry seal provides a tight dense surface and provides some slip resistance.

Life spans are similar to chip seals and also depend on traffic. It is not unusual to have microsurfacing treatment last 7–10 years. Microsurfacing is more expensive to produce than standard slurry seals due to the higher quality aggregate and specialized polymer modified asphalt emulsion. Microsurfacing is a Type III slurry seal and is applied at a heavier rate than a Type II standard slurry seal. If the benefits of microsurfacing warrant the additional cost, it should be considered. Microsurfacing is often a proprietary form of thin surfacing, usually having a commercial trade name and marketed by specialized contractors or producers.

Chapter 10

Estimating and specifying asphalt pavements

Estimating an asphalt pavement consists of determining the amount and type of materials needed for a project. An estimate of the amount of material needed for a project will provide the owner of the facility with what the expected cost of the project should be. Estimating the material quantities prior to beginning the project will help ensure that the owner receives what is being paid for. The process and detail of construction estimating is enough of a subject, that it is often covered by its own text. Throughout the chapters in this book are various examples of determining material quantities. The purpose of discussing estimation in this text is to provide the reader with enough information to determine how much asphalt material would be required for a project. Estimating material quantities for drainage structures, lighting, traffic control devices, parking barriers, and other related transportation materials is beyond the scope of this text. That information is often available in separate texts on construction estimating.

Determining the amount of asphalt mixture requires knowing the area of the project to be paved, the compacted specific gravity of the mixture, and the thickness that the mixture will be placed. The boundaries of the project will define the area being paved. Calculating area is a simple length times width calculation. The expected use or traffic level determines the thickness being placed. Preliminary engineering should provide both the area being paved and the required thickness of the pavement. The compacted density or the material weight per square meter will depend on the type of mixture being placed and the compactive effort being placed. The more compactive effort given to a mixture, the denser and thinner the mixture becomes. The pavement for the project should be estimated and specified in terms of compacted thickness. The contractor or asphalt producer will convert the compacted thickness estimates to how much material to ship. The density of the mixture can be determined from the asphalt mixture design, provided by the asphalt mixture producer, or estimated using typical values.

Example 10.1

Determine how much dense graded hot mix asphalt (HMA) to order for a parking lot that is 240 m × 350 m. The compacted thickness of the HMA is 75 mm. The maximum theoretical, G_{mm}, specific gravity of the HMA is 2.453 and the density requirement is 93 percent of G_{mm}.

Solution

First determine the volume of pavement required: (240 m)(350 m)(0.075 m) = 6,300 m^3.
The specific gravity of the compacted mixture is: (93%)(2.453) = 2.281 or 2,281 kg/m^3.
The weight of the asphalt mixture required is: (6,300 m^3)(2,281 kg/m^3) = <u>14,370 tonnes.</u>

Table 10.1 Typical material densities that can be used for estimating

Material	*Density* (kg/m^3)
Water	1,000
Asphalt cement or binder	980–1,070
Asphalt emulsion	980–1,020
Porous asphalt	1,820–2,250
Dense graded HMA	2,150–2,550
Hot rolled asphalt	2,300–2,400
Single size aggregate chippings	1,700–1,900
Granular or aggregate bases	2,000–2,200
SMA	2,300–2,500

Estimating and purchasing by weight is less likely to have discrepancies than purchasing by volume. Volume is temperature dependent and also includes air voids or air volume. If the laboratory density of the asphalt mixture is not yet known or is not available, Table 10.1 can be used as a guideline.

In most cases the producer of the asphalt mixture can provide the density information. It is important to always use the compacted or spread density. The compacted or spread density is the amount that the contractor or producer should be held accountable for. By maintaining consistency, there will less likely be discrepancies between the owner and the contractor.

Specifications

When purchasing an asphalt pavement, the owner expects the pavement to perform its intended function and to last a certain amount of time. These expectations are accepted regardless of whether the asphalt pavement is for a parking lot or for a motorway or expressway. The owners of a parking lot will have different expectations of pavement service and life than the owners (usually a government agency) of an expressway.

The plans and specifications for a project are used to describe to the contractor or builder specifically how a project should be built. The owner of the project expects it to be built as described. The plans provide the dimensions, pavement thickness, layout, and all other aspects of the project. The specifications describe the materials, workmanship, and other general requirements for the project (Roberts *et al.* 1991). A build sheet is similar to specifications and is more specific for the particular project. Owners or agencies that are responsible for many projects will prepare a standard specification that will apply, where applicable to all of their projects. Highway departments generally will have a standard specification book or manual and particular sections will be referenced in the plans for the project. Highway agencies often develop a single specification for a particular process or material such as a specification on thin surfacings or porous asphalt. Architects or consulting engineers may prepare their own specifications to be used on the roadways or parking lots they are designing or overseeing. The architect or designer will create a specification for the entire project, including the materials used on the project. Specifications for smaller projects are often referenced in the standard specifications of a large highway agency such as The Department of Transport or the Federal Highway Administration. Such usage should be viewed with some caution. The highway agency specifications that are referenced must be applicable for the particular project and be achievable. For example, specifying that the HMA base course for a shopping center in Miami meet the Illinois Department of Transportation HMA binder course specifications may not be achievable. The materials available in the Miami area may not meets the Illinois Department of Transportation criterion. This problem happens more often when involving a design firm based in a different geographical area than the project. It also occurs during building parking lots for national franchises or chains. It is also important to reference specifications that will meet the needs of the project owner and perform satisfactory. The particular specification should be realistic and relevant to the application. Referencing a specification that is for HMA for a expressway or motorway, will more than likely specify a mixture that will be too stiff and possibly ravel when placed on a low volume application like a fast food restaurant parking lot.

A specification for a project actually consists of two parts:

1 The preparation of an adequate and relevant specification.
2 Confirmation or compliance that the specification has been met.

The purpose of a specification is to ensure that the product (in this case, an asphalt pavement) is acceptable and will perform to the buyer's expectations. In general, a specification is only as good as its compliance. If there is no practical method of verifying that a particular specification has been met, the specification has very limited value. Likewise, if there is no practical method to *meet* a specification requirement, then it also has limited value. Performance expectations differ, depending on the buyer. The buyers of asphalt pavements are for the most part, either commercial owners or public agencies. Public agencies may require a pavement to last 20 years and provide a smooth and skid resistant surface. A commercial buyer may only need a parking lot for 10 years and will find esthetic value of most importance. The specification should always be tailored to the needs of the buyer and be realistically and economically met by the contractor or builder.

Pavement specifications can be classified by three concepts or methods:

1 Method specification
2 End result or performance specification
3 Quality control/quality assurance specification.

Method specifications

The method specification concept is the most common form of specifying asphalt pavements. The method specification has been in use for the longest and is how most public highway agencies specify just about anything that has do with roadwork. The method specification describes in specific detail almost the entire process of building an asphalt pavement. It describes the materials, equipment, and procedures in order to provide a satisfactory product for the buyer. It must also be in sufficient detail to describe all the variables required to produce an asphalt pavement. The method specification also requires an owner's agent or inspector to be involved with the project at all times in order to ensure compliance with the specification. The inspector is involved and thoroughly familiar with the contractor's operations during the building of the asphalt pavement.

How the method specification is involved in the entire pavement process is illustrated in the following outline:

1 Approved sources of aggregate and asphalt binder.
2 Test specifications on the aggregate and asphalt binder.
3 Approved sources of equipment, including the hot mix plant, paver, and rollers.
4 The buyer or highway agency provides the asphalt mixture design.
5 The agency proportions the asphalt mixing plant.
6 The agency describes how the mixture should be produced in the mixing plant.
7 The agency describes how the mixture should be loaded and hauled to the project site.
8 The agency describes in detail how the mixture should be placed.
9 The agency describes in detail how the mixture should be compacted.
10 The agency tests the mixture for density, smoothness, and mixture design.
11 The agency pays the contractor for the asphalt pavement.

A method specification describes in detail the entire process of producing the asphalt pavement and how it will be acceptable to the buyer (highway agency). A method specification can also be a specification that does not require any laboratory or field testing. A common method specification is a method type specification describing compaction. For example, the specification, would state: *Compact the asphalt mixture with five passes of a 10 tonne static, tandem, steel wheel roller at a speed that shall not exceed 5 km/h.* In this type of specification, it is assumed that five passes of the roller will achieve compaction without any actual confirmation of that the mixture is adequately compacted. The specification is based on past experience that five passes adequately achieved density. If a method specification is not updated or changed with material or equipment changes, it may not provide an adequate performing pavement. A method specification routinely specifies the type of equipment the contractor should use on the project. The disadvantages of a method specification is that:

1 Based on past experience and performance.
2 Does not specify the pavement on actual or accepted performance.
3 Requires an inspector and adequate staffing on behalf of the buyer.
4 Does not have a method to identify deficiencies in the final pavement.

Most highway agencies have modified method specifications to include some form of laboratory and field testing to check for conformance. The compaction specification may include a requirement of the amount of density to be achieved and verified by a density measurement. The pavement may also be required to meet a smoothness requirement that will be confirmed with a piece of equipment such as a profilometer. These types of requirements or specifications that are based on some type of acceptance testing, usually at the end of construction is a form of an end result specification. Testing that is done during the project by the contractor is considered quality control testing that allows the contractor to immediately correct for deficiencies. Testing that is done after the project is completed and is done by the buyer or owner is acceptance testing. The buyer may do quality control type testing as a form of acceptance. Acceptance testing is done to make the decision to accept and pay for the project and award a bonus or penalty for certain milestones that are exceeded or not even met. Acceptance testing may also allow the contractor the opportunity to correct any deficiencies.

Most highway agencies no longer use a "pure" method specification, but require more testing for specification compliance and project acceptance. Many architects and engineers use these types of modified method specifications for the projects that they are overseeing. The decision to use a method specification should be based on the ability of the buyer or owner to confirm specification compliance. This may include the hiring of a consulting engineer or laboratory to monitor the project.

End result performance specifications

An end result performance specification is simply criteria for accepting a finished product, in this case, an asphalt pavement. In the purest sense, the materials, equipment, and methods used to produce the product are irrelevant to the buyer, as the product is acceptable when purchased. In terms of consumer purchases, the process of purchasing an automobile car is an informal performance specification. The consumer enters the automobile dealership with certain expectations that the automobile should meet. The consumer test drives the automobile and then decides to purchase it. The test drive of the automobile is an informal performance specification. The consumer at this point does not really care how the automobile was manufactured or where its raw materials came from, as long the vehicle performs to expectations. If the vehicle passes the test ride, then it meets the performance specification.

An end result performance specification for asphalt pavements is centered on various types of acceptance laboratory and field testing. The properties that the testing measures should relate to the desired performance of the pavement. Tests that are completed during production are not actually end result tests but rather quality control tests. Tests that are for quality control are:

- Aggregate and asphalt binder testing
- Cold feed gradations
- Hot bin gradations (batch mixing plant)
- Mixture testing to verify mixture design (air voids, VMA, etc.)
- Asphalt binder content verification
- Density testing during compaction.

All of these tests are used as quality control checkpoints and also to advise the producer or contractor when and how adjustments to the pavement producing process should be made. The tests either affect the hot mixing plant or the compaction process. These tests are usually known as process control tests.

End result acceptance tests that have been used for acceptance of the project after it has been completed are:

- Density (final by cores or nuclear)
- Pavement thickness
- Moisture sensitivity
- Mixture properties verification (on cores or slabs cut from the pavement)
- Smoothness.

This list is not all-inclusive, since more extensive performance tests are being developed and validated. However, acceptance tests are used by the buyer to accept the pavement from the builder to the extent that it is built as required. They should not be confused with tests that predict the performance of the pavement.

End result tests are only on the finished asphalt pavement. All of these tests directly correlate to the buyer's expectations of performance. A properly compacted pavement of the required thickness should provide the required performance for the vehicle loading. The mixture property and the moisture sensitivity test all correlate to the pavement's susceptibility to distresses. Smoothness is a reflection of the pavement's ride characteristics. The specification will clearly spell out the frequency and amount of acceptance testing. For example, a density core may be cut for every 1,000 tonnes of HMA laid. If the density is found to be deficient, the buyer may still accept the pavement for some reduced payment or require that it be removed and replaced.

For an end result performance specification to be effective, it should clearly identify the following items:

- Types of testing to be performed
- Testing lot sizes
- Number of tests per lot
- How lots will be selected
- Sampling method
- Acceptable test target values
- Acceptable tolerances of the target values
- Action to be taken when the test values are outside the acceptable limits.

A lot is the amount of material that is evaluated for acceptance and pay purposes. Each lot is evaluated as a separate project based on test results from a specified number of random samples taken from the lot. Lot sizes can vary but are usually the amount of mixture that can be placed in one day. A common lot size is anywhere from 100 to 1,000 tonnes, depending on the type of test being completed. Table 10.2 provides some typical testing frequencies. The number of tests per lot will also vary but is usually five. A larger number of tests will provide a better representation of the true test value of the entire lot size. The number of tests is limited by the ability to perform the tests, with five being considered adequate. The lot can be divided into sublots with one random sample taken from each sublot (Roberts *et al.* 1991).

Table 10.2 Typical test frequencies

Test	*Frequency (per 1,000 tonnes of HMA produced)*
Aggregate stockpile gradation	1
Aggregate cold feed gradation	1
Hot bin gradations (batch plants)	1
Asphalt binder content verification	2
Mixture properties verification	1
Compacted density	8
Thickness verification	5
Smoothness	Entire project

If the buyer were not involved in the production of the asphalt pavement, the only tests that the buyer can complete would be the density, thickness, and smoothness requirements. Currently, without extensive forensic testing, these are the only "after the fact" tests that a buyer can perform on a finished pavement to determine if it is acceptable. End result performance specifications will become more common when more extensive or relative acceptance tests are developed. True performance tests for asphalt mixtures are currently being evaluated and are yet to be implemented on a routine basis.

Quality control/quality assurance specifications

The problem with end result performance specifications is that there are not many reliable methods to determine if a pavement was built as designed or required without being involved in the process of building the pavement. A combination method specification and end result performance specification should fulfill those needs. The combination specification is the quality control/quality assurance specification or QC/QA specification. This specification bridges the two by completing both process control testing and acceptance testing.

What is unique to the QC/QA specification is that the producer or contractor will perform the quality control testing while the buyer or agency will perform the quality assurance testing. The specification will require that the dual or "split" samples be taken one for the producer and one for the buyer. The buyer may perform testing on the split sample along with the producer or random sample the split samples and test those accordingly. The purpose of the specification is to move the quality control testing from the buyer to the producer. Method specifications often require the buyer to do the quality control testing, which substitutes for acceptance testing. The key in making a QC/QA specification work is confidence in the abilities of both the buyer and the contractor to perform the testing accurately and consistently. The reproducibility of test results by either party is key to the success of a QC/QA program. The implementation of laboratory certification and technician training is part of any success program. Training both the buyer and the contractor personnel reduces the amount of discrepancies in the test results.

The buyer performs the quality assurance or acceptance testing such as final density, thickness, and smoothness. Density is often checked by a nuclear gage during the pavement construction process and accepted by a core density. However, nothing prevents the nuclear gauge density testing to be used for acceptance or quality assurance. Testing completed during the quality control stage should confirm that the pavement constructed will perform to the buyer's satisfaction.

The quality control testing should be carefully monitored since it is part of the buyer's acceptance plan. The use of quality control charts allows the analysis and evaluation of test results during the progression of the project. A quality control chart is a graph of all the test values plotted in the order of sampling and testing. The charts show the condition of the process at a glance. By showing the process data and variability for successive tests, the charts can illustrate changes in the process (NAPA 1997). Charts illustrate trends of the test data, so that corrective actions can be taken, if needed. The charts can provide the effects of changes in mixture production or process changes. For example, changes in the asphalt binder content can affect compactive efforts. If density becomes significantly harder or easier to achieve than normally, the quality control charts on the mixture production should be checked.

Specification development

Modern specifications are usually a hybrid of the method and performance specifications. The specification should be easy to understand, consistent and attainable. When developing a specification for a project, the designer must be able to answer three questions:

1 What benefit does our purpose achieve by the requirement?
2 Is the requirement reasonable and attainable?
3 What are the consequences if the requirement is not met?

If answers can not be provided for these three questions, the requirement should not be included in the specification.

All paving specifications should include the following parts:

1 Full project name and description
2 Related documents and references
3 General requirements including submittals
4 Quality control or quality assurance
5 Project conditions
6 Materials
7 Construction
8 Acceptance tolerances.

Incidentals to the project such as shopping cart corrals, pavement markings, parking blocks, and traffic control can also be added.

The references in a specification should be current and relative to the project. Specification proliferation often occurs when specifications are copied or referred to without checking on the applicability of the reference or requirement. When writing specifications, all references should be confirmed for suitability.

The specification must describe the test method used to determine compliance. When the test method is described or referenced, the test data can be compared with no confusion. The specification must also list the number of tests to be taken and how they are taken. The number of tests can be taken randomly or representatively. A random sample is a sample that has an equal chance of being selected as any other sample. A representative sample is obtained so that it represents the average of the lot or population being tested. A representative sample is usually obtained by taking a number of small random samples and combining them to obtain a representative sample.

The test or specification requirements should be established as a function of the number of tests. The test results should be near the specification target value and have low variability. Ideally the population test value should equal the specification target value. An average test result, calculated form a large number of tests on random samples will provide a good estimate of the population average. Small numbers of test samples may not provide a good estimate of the population average, because of the higher variability. Smaller sample sizes or infrequent samples increase the chance that a good pavement may be rejected or a poor pavement accepted (Roberts *et al.* 1991).

The specifications must also list a target or accepted value for any test required. Requiring a test with no acceptance criterion is not useful. The target or specification value is the value

Table 10.3 Typical asphalt construction tolerances

Test or function	*Target tolerances*
Aggregate gradation	
Sieve size (mm)	*Percent passing range* (±, %)
≥12.5	8
9.5	7
4.75	7
2.36	6
1.18	6
600 μm	5
300 μm	5
75 μm	3
Dense graded mixture, job mix formula	
Sieve size (mm)	*Percent passing range* (±, %)
≥12.5	6
4.75	5
2.36	5
300 μm	4
75 μm	1.5
Asphalt binder content	0.3
Marshall stability	≥95% of requirement
Construction	
Density	±2%
Base or binder course thickness	±13 mm
Surface course thickness	±6 mm, no minus
Smoothness	±6 mm for every 3 m

that the builder or contractor will attempt to achieve when constructing the asphalt pavement. All specific performance requirements should have tolerances where applicable. The ability of large construction equipment to achieve extremely narrow or specific targets is questionable. It is impossible to produce an asphalt mixture or pavement that will exactly meet the specification target value. The tolerance should be enough to accept normal material variability but not so much that low quality material is accepted. A specification tolerance is a function of both testing and material variability. If test method variability is higher than the variability in the material being tested, the test method should not be used for the specification (Roberts *et al.* 1991). A specification tolerance that is tighter than the variability in testing is not an acceptable tolerance. Table 10.3 provides some common tolerances for various aspects of a paving project that are capable of being met with good construction practices.

A specification should have a clear objective and understanding of what needs to be performed. It should also clearly state the recourses if the objective or specification requirements are not met. Specifications will either be removed and replaced or have penalties or a combination of both. An absolute specification, especially a critical requirement will require that the builder or contractor remove the unacceptable material and replace it. Objectives or specifications that are not as critical will pay for marginal or unacceptable materials at a discount or *penalty* to the contractor. Most specifications will penalize for marginal materials and require removal and replacement for completely unacceptable materials.

An engineering report is prepared prior to the specification. The engineering report should include, but not limited to, geotechnical conditions, traffic analysis, required subgrade improvements, required drainage, pavement thickness design, recommended asphalt mixtures, or the mixture design and the economic alternatives for the various pavement thickness combinations.

The terminology of the specification should be easy to understand by everyone involved. If necessary, a glossary should also be included, especially when using new or unknown processes. Terms and requirements should be easily defined and be able to be measured by both the buyer and the contractor. The word, *shall*, describes what the contractor must do or provide and the word, *will*, denotes what the buyer will provide. Being consistent in the use of words will minimize misinterpretations during the inspection and acceptance of the asphalt pavement (Roberts *et al.* 1991).

Specifications tend to be specific to certain regions. The example specifications are specific to imaginary projects in the United States and may not directly translate to a project, say in Germany. However, the concepts are universal and if the specification is met, it will produce a quality product anywhere. The equivalent test procedures and specifications to the provided ones can be easily determined. The American Society of Testing Material standards are becoming recognized worldwide and there already exists many regional alternative or equivalent methods and specifications to the ASTM ones provided.

Chapter 11

Draft parking lot specification

Sunya Gold Shopping Center
Elmhurst, Illinois
Parking lot Specification

Hot Mix Asphalt Paving Section

GENERAL

Description of work

Furnish and construct a hot mix asphalt pavement for the Sunya Gold Shopping Center according to the materials, workmanship, and other applicable requirements of ASTM D3515 and other applicable requirements of the standard specifications of the Illinois Department of Transportation. The hot mix asphalt pavement shall consist of a base or binder course and a surface course. The hot mix asphalt pavement will also include a granular, unbound aggregate sub-base course, and all required drainage structures as specified.

Related documents

Drawings and general provisions of the Contract, including General and Supplementary Conditions apply to this section. The Standard Specifications for Road and Bridge Construction, January 1, 2002 of the Illinois Department of Transportation and the 2002 Annual Book of ASTM Standards, Section Four, Construction, Volume 4.03 apply to this section.

Submittals

1 *Product data*: For each product specified. Including technical data and tested physical and performance properties where applicable.
2 *Asphalt mixture design*: An approved asphalt mixture design for the Illinois Department of Transportation Class I, Type one, two, and three mixtures or an asphalt mixture design completed specifically for this work, meeting all referenced requirements and signed and sealed by a engineer registered in the State of Illinois.
3 *Shop drawings*: Indicate pavement markings, lane separations, parking blocks, barriers, and defined parking spaces. The parking capacity is 1,300 automobiles. Indicate dedicated handicap spaces with the proper and current markings.

4 *Qualification data*: Firms or persons specified in quality control and quality assurance, with the ability to demonstrate their capabilities and experience. Include lists of completed projects with project names and addresses, and names and addresses of architects and owners.
5 *Material test reports*: Indicate and interpret test results for compliance of materials with the requirements specified.
6 *Material certificates*: Certificates signed by the manufacturer certifying that each shipment of material complies with the requirements.

Quality control

1 *Contractor or builder qualifications*: Engage an experienced contractor or builder who has completed hot mix asphalt paving similar in quantities, materials, design, and the extent to that indicated for this Project and with a record of being successful in service performance.
2 *Producer qualifications*: Engage a firm experienced in the production of hot mix asphalt similar to that indicated for this Product with a record of in service performance. It is permitted that the Producer and the Contractor or Builder be the firm. The firm must be registered to do business with the Illinois Department of Transportation and City of Chicago Department of Public Works.
3 *Testing agency qualifications*: Demonstrate to the Architect's satisfaction, based on the Architects evaluation of criteria substantially conforming to ASTM D3666, *Standard Specification for Minimum Requirements for Agencies Testing and Inspecting Road Materials*, that the independent testing agency has experience and capability to satisfactorily conduct testing indicated without delaying the work.
4 *Regulatory requirements*: Conform to applicable standards and regulations of authorities having jurisdiction for asphalt paving work on public property.
5 *Preconstruction conference*: Conduct a meeting with the Architect at the Project site or other suitable location to review methods and procedures relating to asphalt paving including, but not limited to, the following:

   (a) Review proposed sources of paving materials, including capabilities, and location of the asphalt mixing facility that will manufacture the hot mix asphalt.
   (b) Review the condition of the subgrade, including clearing, grading, and installation of drainage structures.
   (c) Review requirements for protecting paving work, including the restriction of traffic during the installation period and for the remainder of the construction period.
   (d) Review and finalize the construction schedule for paving and related work. Verify the ability of materials, personnel, and equipment required to complete the work without any delay.
   (e) Review inspection, testing and acceptance requirements, governing regulations, and proposed construction procedures.
   (f) Review forecasted weather conditions and procedures during inclement weather.

Delivery, storage, and handling

1 Deliver the pavement sealer and pavement marking paint to the Project site in original packages with seals unbroken and bearing the manufacturer's labels, containing the

brand name and type of material, date of manufacture, and directions for storage. Pavement sealers may be purchased in bulk along with a manufacturer certification that the material conforms to the requirements. Pavement marking paints shall not be delivered in bulk, but in individual containers no larger that 200 l.

2 Store the pavement sealer and pavement marking paint in a clean and dry location and within the temperature range recommended by the manufacturer for storage. Protect the materials from expose to direct sunlight.

Project conditions

1 Do not apply asphalt materials or hot mix asphalt if the subgrade is wet. Do not apply any material when it is raining or snowing.
2 Do not apply asphalt emulsion prime coats and tack coats when the ambient temperature is less than 15 °C.
3 Do not apply hot mix asphalt base or binder course when the ambient temperature is less than 4 °C.
4 Do not apply the hot mix asphalt surface course when the ambient temperature is less than 7 °C.
5 Do not apply the pavement sealer when the ambient temperature is less than 10 °C.
6 Do not apply the pavement marking paint when the ambient air temperature is greater than 35 °C and less than 10 °C.

Aggregates

1 *Coarse aggregate*: Sound angular crushed stone or crushed gravel complying with ASTM D692, *Standard Specification for Coarse Aggregate for Bituminous Paving Mixtures*, or the Illinois Department of Transportation Standard Specification's functional equivalent.
2 *Fine aggregate*: Sound angular natural sand or manufactured sand prepared from stone or gravel or a combination thereof, complying with ASTM D1073, Grading 1, 2, or 4, *Standard Specification for Fine Aggregate for Bituminous Paving Mixtures*, or the Illinois Department of Transportation Standard Specification's functional equivalent.
3 *Granular sub-base aggregate*: Sound angular crushed stone or crushed gravel complying with ASTM D2940, *Standard Specification for Graded Aggregate Material for Highways and Airports*, or the Illinois Department of Transportation Standard Specification's functional equivalent.
4 *Mineral filler*: Stone dust, flyash hydrated lime, or other inert material complying with ASTM D242, *Standard Specification for Mineral Filler for Bituminous Paving Mixtures*, or the Illinois Department of Transportation Standard Specification's functional equivalent.

Asphalt materials

1 The asphalt binder used in the hot mix asphalt shall be a PG64-22 meeting the requirements of ASTM D6373 *Specification for Performance Graded Asphalt Binder*. A 50/70 penetration graded asphalt binder may be substituted for the PG64-22.
2 The asphalt emulsion for the tack coats and prime coats shall be a CSS-1 meeting the requirements of ASTM D2397, *Specification for Cationic Emulsified Asphalt*.

Asphalt pavement

1 *Granular sub-base*: 300 mm of compacted sound angular crushed stone or crushed gravel complying with ASTM D2940.
2 *Hot mix asphalt for parking and automobile drives*: 120 mm of compacted Illinois Department of Transportation Class I Type 2B binder course mixture. 50 mm of compacted Illinois Department of Transportation Class I Type 3C surface course mixture. 120 mm of a 19 mm mixture and 50 mm of a 12.5 mm dense graded mixture meeting the requirements of ASTM D3515, *Standard Specification for Hot-Mixed, Hot-Laid Bituminous Paving Mixtures* may be substituted by permission of the Architect.
3 *Hot mix asphalt for truck unloading areas, fire lanes, and truck drives*: 200 mm of compacted Illinois Department of Transportation Class I Type 1B binder course mixture. 50 mm of compacted Illinois Department of Transportation Class I Type 1D surface course mixture. 200 mm of a 19 mm mixture and 50 mm of a 12.5 mm dense graded mixture meeting the requirements of ASTM D3515, *Standard Specification for Hot-Mixed, Hot-Laid Bituminous Paving Mixtures* may be substituted by permission of the Architect.
4 *Refuse container pads*: 200 mm of Portland Cement Concrete containing a minimum of 180 kg of Portland Cement Type I. The Portland Cement Concrete shall meet the requirements of ASTM C94, *Standard Specification for Ready Mixed Concrete* or the Illinois Department of Transportation Standard Specification's functional equivalent. The sub-base shall consist of 350 mm of compacted sound angular crushed stone or crushed gravel complying with ASTM D2940.
5 *Hot mix asphalt mixture design*: The hot mix asphalt shall be designed using either the Marshall or SHRP SuperPave® mixture design method. The mixture design properties shall meet the requirements of Table 11.1.

Pavement sealer

A refined coal tar sealer meeting the requirements of ASTM D5727, *Emulsified Refined Coal Tar*. No more than 15 percent dilution water by weight of total sealer may be added. The sealer shall contain 0.6 kg/l of fine aggregate meeting the requirements or recommended by the f sealer manufacturer. Two coats of the sealer at an application rate of 0.5 l/m^2 shall be applied.

Table 11.1 Mixture design properties

Test parameter	*Base or binder course*	*Wearing course*
Marshall design method		
Compaction blows	50	50
Air voids	3.0–5.0%	3.0–5.0%
VMA	≥13.0%	≥14.0%
Marshall Stability, kN	≥3.3	≥3.3
Marshall Flow, 0.25 mm	8–18	8–18
SuperPave design method		
Gyrations, N_{design}	50	50
Air voids	3.0–5.0%	3.0–5.0%
VMA	≥13.0%	≥14.0%
VFA	70–80%	70–80%

Pavement marking paint

A commercially available paint formulated specifically for pavement markings consisting of a latex water based emulsion. The color of the paints shall be white and yellow, locations as specified on the plans, and meeting the requirements of ASTM D6628, *Standard Specification for Color of Pavement Marking Materials.*

Construction

1 The subgrade should be prepared and compacted as described in the engineering geotechnical report. The subgrade should be compacted to not less than 95 percent of the standard laboratory density as described in ASTM D698, *Moisture Density Relations of Soils using a 2.5 kg Hammer and a 305 mm Drop*. The subgrade should be proof rolled with a pneumatic roller prior to placing the granular sub-base. Notify the Architect of any observed unsatisfactory subgrade conditions prior to paving.
2 Place and compact the granular sub-base to the required compacted thickness as shown on the plans. The granular sub-base should be compacted to not less than 100 percent of the standard laboratory density as described in ASTM D698.
3 Place an application of a prime coat on the granular sub base at an application rate of $0.8\,l/m^2$.
4 Place the hot mix asphalt base or binder courses and surface courses to the required compacted thickness. Compacted thicknesses of 200 mm or less can be placed as one lift. The placement temperatures shall be between 125 and 175 °C.
5 Begin placing the hot mix asphalt along the centerline of the crown for the drives with crowns and on the high side of one way slopes on parking areas. Place the hot mix asphalt to grades and cross slopes as indicated on the plans. Place the hot mix asphalt in consecutive strips of not less than 3 m wide, except where edge strips or confinement requires less width to be placed.
6 After the first strip has been placed and compacted, place succeeding strips, and extend rolling to overlap the previous strips.
7 Complete the entire hot mix asphalt base or binder course prior to the placement of the surface course. The application of a tack coat at an application rate of $0.3\,l/m^2$ shall be used when there are more than 30 days are between the placement of the base or binder course and the surface course on top. The tack coat shall be placed between the base or binder course and the surface course.
8 Promptly correct surface irregularities in the hot mix asphalt placed behind the paver. Use suitable hand tools to remove excess hot mix asphalt forming high spots. Fill depressions with hot mix asphalt using suitable hand tools to provide a smooth surface.

Joints

Construct construction joints to ensure a continuous bond between the adjoining pavement sections. Construct joints free of depressions with the same texture and smoothness as the mainline or other sections of the hot mix asphalt course. The number of construction joints needed should be kept to a minimum. Construction joints should be constructed in the following manner:

1 Clean contact surfaces and apply a tack coat at an application rate of $0.3\,l/m^2$.
2 Offset longitudinal joints in successive courses to a minimum of 150 mm.

3 Offset transverse joints in successive courses to a minimum of 600 mm.
4 Construct transverse joints as either a butt or taper joint.
5 Construct longitudinal joints as a vertical joint.
6 Compact the joints as soon as possible to a density that is within 2 percent of the specified mainline density requirements.

Compaction

1 Begin compaction as soon as the placed hot mix asphalt will bear the roller weight without excessive displacement. Compact the hot mix asphalt with a vibrating plate compactor in areas that are inaccessible to rollers.
2 Complete compaction before the hot mix asphalt cools to 85 °C.
3 *Initial rolling*: Begin initial rolling immediately after rolling joints and the outside pavement edge. Examine the pavement surface immediately after initial rolling for indicated crown, grade, and smoothness. Repair surfaces by loosening displaced hot mix asphalt, filling in with new hot mix asphalt and compacting to the required elevations.
4 *Intermediate rolling*: Begin intermediate rolling immediately after initial rolling when the hot mix asphalt is still hot enough to achieve the required density. Continue rolling until the hot mix asphalt has been uniformly compacted to the required density. The roller used for initial compaction may also be used for intermediate rolling.
5 *Required density*: The average density shall be a minimum of 94 percent of the reference maximum theoretical density obtained from the hot mix asphalt mixture design. The maximum theoretical density of the hot mix asphalt shall be measured by ASTM D2041, *Standard Test Method for Theoretical Maximum Specific Gravity and Density of Bituminous Paving Mixtures*. The average density is the average of five tests completed by the nuclear density gage across the full paving width. The number of average density tests completed shall be one every 250 tonnes of hot mix asphalt placed.
6 *Finish rolling*: Finish rolling the paved surfaces to remove roller marks.
7 *Edge shaping*: While the surface is being compacted and finished, trim edges of the pavement to the proper alignment as shown on the plans. Bevel the edges while the hot mix asphalt is still hot, with the back of a rake or lute. Compact thoroughly using a hand tamper or a vibrating plate tamper.
8 *Repairs*: Remove paved areas that are defective or contaminated with foreign materials. Remove the paving course from areas affected and replace with hot mix asphalt. Compact by rolling to the specified density, elevation, and smoothness.
9 *Protection*: After final rolling do not permit vehicles on the pavement until it has cooled to 30 °C or until the following day.

Pavement construction tolerances

1 The compacted thickness of the base or binder course shall be within 13 mm of the required thickness.
2 The compacted thickness of the surface course shall be no more than 6 mm thicker than the required thickness. No thickness deficiencies are permitted on the surface course.
3 The final pavement smoothness shall be determined by a 3 m straight edge applied transversally or longitudinally to the paved areas. The smoothness of the base or binder course shall be 6 mm for every 3 m. The smoothness of the surface course shall be 3 mm for every 3m. On the crowned drives, the allowable variance for the crown is 6 mm.

Pavement scaler

1 The application of the pavement sealer should not begin until 30 days after the final placement of the surface course. When ambient temperatures are averaging 25 °C or less, the application of the pavement sealer should not begin until 45 days after the final placement of the surface course.
2 The pavement shall be swept and clean prior to the application of the pavement sealer. The application of the pavement sealer should be before the installation of any parking accessories or incidentals such as parking blocks.
3 The first coat of pavement sealer should dry completely prior to the application of the second coat.

Pavement markings

1 The pavement sealer shall cure 15 days prior to the application of any pavement markings.
2 The layout, colors, and placement of the pavement markings shall be verified with the architect prior to application.
3 The pavement surface shall be clean and dry prior to the application of the pavement markings.
4 Apply paint with mechanical pavement painting equipment to produce pavement markings of the dimensions on the plans with uniform and straight edges. The paint should be applied at the manufacturer's recommended rates to provide a minimum wet film thickness of 0.4 mm. Pavement markings on drives shall receive an application of glass spheres spread by broadcast and uniformly into the wet pavement markings at an application rate of 0.7 kg/l of pavement marking paint.

Quality assurance

1 *Testing agency*: Owner or Architect will engage a qualified independent testing agency to perform field inspections, tests, and test reports. The testing agency will conduct and interpret test results and will state in each report whether the tested work complies or deviates from the specified requirements.
2 Additional testing on deficient or corrected work to determine that the corrected work complies with the specified requirements will be done at the contractor's expense.
3 The acceptance of the completed hot mix asphalt pavement thickness will be determined by the methods described in ASTM D3549, *Standard Test Method for Thickness or Height of Compacted Bituminous Paving Mixture Specimens*.
4 *Surface smoothness*: The finished surface of each hot mix asphalt course will be tested for compliance with smoothness tolerances.
5 *In place density*: Samples of uncompacted paving mixtures will be obtained by the testing agency according to ASTM D979, *Standard Practice for Sampling Bituminous Mixtures*. The reference maximum laboratory density will be determined by averaging the results from three samples of hot mix asphalt delivered daily to the project site and tested according to ASTM D2041. The in-place density for acceptance will be determined by nuclear density test for core from every 500 m^2 of installed asphalt pavement. The nuclear testing method will be ASTM D2950, *Standard Test Method for Density of*

Bituminous Concrete in Place by Nuclear Methods and the test results will be correlated to pavement cores tested by ASTM D1188, *Standard Test Method for Bulk Specific Gravity and Density of Compacted Bituminous Mixtures Using Coated Samples*. When discrepancies occur between test results, the tests results obtained from pavement cores will supersede.

Deficiencies

Remove and replace or install additional material where test results, measurements, or observations indicate that it does not comply with the specified requirements.

Warranty

The contractor or builder shall guarantee in writing the materials and workmanship of the completed pavement for a period not less than 3 years from the date of project acceptance by the Owner or Architect.

Chapter 12

Draft residential street specification

Cujo Pointe
Residential Development
Streets and Drives
Elmhurst, Illinois

Hot Mix Asphalt Paving Section

GENERAL

Description of work

Furnish and construct a hot mix asphalt (HMA) pavement for the streets of the Cujo Pointe Residential Development according to the materials, workmanship, and other applicable requirements of ASTM D3515 and other applicable requirements of the standard specifications of the Illinois Department of Transportation. The HMA pavement shall consist of a base or binder course and a surface course. The HMA pavement will also include a granular, unbound aggregate sub-base course and all required drainage structures as specified.

Related documents

Drawings and general provisions of the Contract, including General and Supplementary Conditions apply to this section. The Standard Specifications for Road and Bridge Construction, January 1, 2002 of the Illinois Department of Transportation and the 2002 Annual Book of ASTM Standards, Section Four, Construction, Volume 4.03 apply to this section.

Submittals

1 *Product data*: For each product specified. Including technical data and tested physical and performance properties where applicable.
2 *Asphalt mixture design*: An approved asphalt mixture design for the Illinois Department of Transportation Class I, Type two mixtures or an asphalt mixture design completed specifically for this work, meeting all referenced requirements and signed and sealed by an engineer registered in the State of Illinois.

3 *Shop drawings*: Indicate pavement markings, lane separations, and traffic control devices.
4 *Qualification data*: Firms or persons specified in quality control and quality assurance, ability to demonstrate their capabilities and experience. Include lists of completed projects with project names and addresses, and names and addresses of Developer or City Engineer.
5 *Material test reports*: Indicate and interpret test results for compliance of materials with the requirements specified.
6 *Material certificates*: Certificates signed by the manufacturer certifying that each shipment of material complies with the requirements.

Quality control

1 *Contractor or builder qualifications*: Engage an experienced contractor or builder who has completed HMA paving similar in quantities, materials, design, and the extent to that indicated for this Project and with a record of being successful in service performance.
2 *Producer qualifications*: Engage a firm experienced in the production of HMA similar to that indicated for this Product with a record of in service performance. It is permitted that the Producer and the Contractor or Builder be the firm. The firm must be registered to do business with the Illinois Department of Transportation and City of Chicago Department of Public Works.
3 *Testing agency qualifications*: Demonstrate to the Developer or City Engineer's satisfaction, based on the Developer or City Engineers evaluation of criteria substantially conforming to ASTM D3666, *Standard Specification for Minimum Requirements for Agencies Testing and Inspecting Road Materials*, that the independent testing agency has experience and capability to satisfactorily conduct testing indicated without delaying the work.
4 *Regulatory requirements*: Conform to applicable standards and regulations of authorities having jurisdiction for asphalt paving work on public property.
5 *Preconstruction conference*: Conduct a meeting with the Developer or City Engineer at the Project site or other suitable location to review methods and procedures relating to asphalt paving including, but not limited to, the following:

   (a) Review proposed sources of paving materials, including capabilities and location of the asphalt mixing facility that will manufacture the HMA.
   (b) Review the condition of the subgrade, including clearing, grading, and installation of drainage structures.
   (c) Review requirements for protecting paving work, including the restriction of traffic during the installation period and for the remainder of the construction period.
   (d) Review and finalize the construction schedule for paving and related work. Verify the ability of materials, personnel, and equipment required to complete the work without any delay.
   (e) Review inspection, testing, and acceptance requirements, governing regulations and proposed construction procedures.
   (f) Review forecasted weather conditions and procedures during inclement weather.

Delivery, storage, and handling

1 Deliver the pavement marking paint to the Project site in original packages with seals unbroken and bearing the manufacturers labels containing brand name and type of material, date of manufacture and directions for storage. Pavement marking paints shall not be delivered in bulk, but in individual containers no larger that 200 l.

2 Store the pavement marking paint in a clean and dry location and within the temperature range recommended by the manufacturer for storage. Protect the materials from exposure to direct sunlight.

Project conditions

1 Do not apply asphalt materials or HMA if the subgrade is wet. Do not apply any materials when it is raining or snowing.

2 Do not apply asphalt emulsion prime coats and tack coats when the ambient temperature is less than 15 °C.

3 Do not apply HMA base or binder course when the ambient temperature is less than 4 °C.

4 Do not apply the HMA surface course when the ambient temperature is less than 7 °C.

5 Do not apply the pavement marking paint when the ambient air temperature is greater than 35 °C and less than 10 °C.

Aggregates

1 *Coarse aggregate*: Sound angular crushed stone or crushed gravel complying with ASTM D692, *Standard Specification for Coarse Aggregate for Bituminous Paving Mixtures*, or the Illinois Department of Transportation Standard Specification's functional equivalent.

2 *Fine aggregate*: Sound angular natural sand or manufactured sand prepared from stone or gravel or a combination thereof, complying with ASTM D1073, Grading 1, 2, or 4, *Standard Specification for Fine Aggregate for Bituminous Paving Mixtures*, or the Illinois Department of Transportation Standard Specification's functional equivalent.

3 *Granular sub-base aggregate*: Sound angular crushed stone or crushed gravel complying with ASTM D2940, *Standard Specification for Graded Aggregate Material for Highways and Airports*, or the Illinois Department of Transportation Standard Specification's functional equivalent.

4 *Mineral filler*: Stone dust, flyash hydrated lime, or other inert material complying with ASTM D242, *Standard Specification for Mineral Filler for Bituminous Paving Mixtures*, or the Illinois Department of Transportation Standard Specification's functional equivalent.

Asphalt materials

1 The asphalt binder used in the HMA shall be a PG64-22 meeting the requirements of ASTM D 6373 *Specification for Performance Graded Asphalt Binder*. A 50/70 penetration graded asphalt binder may be substituted for the PG64-22.

2 The asphalt emulsion for the tack coats and prime coats shall be a CSS-1 meeting the requirements of ASTM D2397, *Specification for Cationic Emulsified Asphalt.*

Table 12.1 Mixture design properties

Test parameter	*Base or binder course*	*Wearing course*
Marshall design method		
Compaction blows	50	50
Air voids	3.0–5.0%	3.0–5.0%
VMA	≥13.0%	≥14.0%
Marshall stability, kN	≥5.3	≥5.3
Marshall Flow, 0.25 mm	8–16	8–16
SuperPave design method		
Gyrations, N_{design}	75	75
Air voids	3.0–5.0%	3.0–5.0%
VMA	≥13.0%	≥14.0%
VFA	65–78%	65–78%

Asphalt pavement

1 *Granular sub-base*: 300 mm of compacted sound angular crushed stone or crushed gravel complying with ASTM D2940.
2 *Hot mix asphalt for streets and drives*: 150 mm of compacted Illinois Department of Transportation Class I Type 2B binder course mixture. 50 mm of compacted Illinois Department of Transportation Class I Type 2D surface course mixture. 120 mm of a 19 mm mixture and 50 mm of a 12.5 mm dense graded mixture meeting the requirements of ASTM D3515, *Standard Specification for Hot-Mixed, Hot-Laid Bituminous Paving Mixtures* may be substituted by permission of the Developer or City Engineer.
3 *Hot mix asphalt mixture design*: The HMA shall be designed using either the Marshall or SHRP SuperPave® mixture design method. The mixture design properties shall meet the requirements of Table 12.1.

Pavement marking paint

A commercially available paint formulated specifically for pavement markings consisting of a latex water based emulsion. The color of the paints shall be white and yellow, locations as specified on the plans, and meeting the requirements of ASTM D6628, *Standard Specification for Color of Pavement Marking Materials*.

Construction

1 The subgrade should be prepared and compacted as described in the engineering geotechnical report. The subgrade should be compacted to not less than 95 percent of the standard laboratory density as described in ASTM D698, *Moisture Density Relations of Soils using a 2.5 kg Hammer and a 305 mm Drop*. The subgrade should be proof rolled with a pneumatic roller prior to the placement of the granular sub-base. Notify the Developer or City Engineer of any observed unsatisfactory subgrade conditions prior to paving.
2 Place and compact the granular sub-base to the required compacted thickness as shown on the plans. The granular sub-base should be compacted to not less than 100 percent of the standard laboratory density as described in ASTM D698.

3 Place an application of a prime coat on the granular sub-base at an application rate of 0.8 l/m^2.
4 Place the HMA base or binder courses and surface courses to the required compacted thickness. Compacted thicknesses of 200 mm or less can be placed as one lift. The placement temperatures shall be between 125 and 175 °C.
5 Begin placing the HMA along the centerline of the crown for the streets and drives with crowns and on the high side of one way slopes on parking areas. Place the HMA to grades and cross slopes as indicated on the plans. Place the HMA in consecutive strips of not less than 3 m wide, except where edge strips or confinement requires less width to be placed.
6 After the first strip has been placed and compacted, place succeeding strips and extend rolling to overlap the previous strips.
7 Complete the entire HMA base or binder course prior to the placement of the surface course. The application of a tack coat at an application rate of 0.3 l/m^2 shall be used when there are more than 30 days between the placement of the base or binder course and the surface course on top. The tack coat shall be placed between the base or binder course and the surface course.
8 Promptly correct surface irregularities in the HMA placed behind the paver. Use suitable hand tools to remove excess HMA forming high spots. Fill depressions with HMA using suitable hand tools to provide a smooth surface.

Joints

Construct construction joints to ensure a continuous bond between the adjoining pavement sections. Construct joints free of depressions with the same texture and smoothness as the mainline or other sections of the HMA course. The number of construction joints needed should be kept to a minimum. Construction joints should be constructed in the following manner:

1 Clean contact surfaces and apply a tack coat at an application rate of 0.3 l/m^2.
2 Offset longitudinal joints in successive courses to a minimum of 150 mm.
3 Offset transverse joints in successive courses to a minimum of 600 mm.
4 Construct transverse joints as either a butt or taper joint.
5 Construct longitudinal joints as a vertical joint.
6 Compact the joints as soon as possible to a density that is within 2 percent of the specified mainline density requirements.

Compaction

1 Begin compaction as soon as the placed HMA will bear the roller weight without excessive displacement. Compact the HMA with a vibrating plate compactor in areas that are inaccessible to rollers.
2 Complete compaction before the HMA cools to 85 °C.
3 *Initial rolling*: Begin initial rolling immediately after rolling joints and the outside pavement edge. Examine the pavement surface immediately after initial rolling for indicated crown, grade, and smoothness. Repair surfaces by loosening displaced HMA, filling in with new HMA and compacting to the required elevations.

4 *Intermediate rolling*: Begin intermediate rolling immediately after initial rolling with the HMA if it is still hot enough to achieve the required density. Continue rolling until the HMA has been uniformly compacted to the required density. The roller used for initial compaction may also be used for intermediate rolling.
5 *Required density*: The average density shall be a minimum of 93 percent of the reference maximum theoretical density obtained from the HMA mixture design. The maximum theoretical density of the HMA shall be measured by ASTM D2041, *Standard Test Method for Theoretical Maximum Specific Gravity and Density of Bituminous Paving Mixtures*. The average density is the average of five tests completed by the nuclear density gauge across the full paving width. The number of average density tests completed shall be one every 250 tonnes of HMA placed.
6 *Finish rolling*: Finish rolling the paved surfaces to remove roller marks.
7 *Edge shaping*: While the surface is being compacted and finished, trim edges of the pavement to the proper alignment as shown on the plans. Bevel the edges while the HMA is still hot, with the back of a rake or lute. Compact thoroughly using a hand tamper or a vibrating plate tamper.
8 *Repairs*: Remove paved areas that are defective or contaminated with foreign materials. Remove the paving course from areas affected and replace with HMA. Compact by rolling to the specified density, elevation, and smoothness.
9 *Protection*: After final rolling do not permit vehicles on the pavement until it has cooled to 30 °C or until the following day.

Pavement construction tolerances

1 The compacted thickness of the base or binder course shall be within 13 mm of the required thickness.
2 The compacted thickness of the surface course shall be within 13 mm of the required thickness.
3 The final pavement smoothness shall be determined by a 3 m straight edge applied transversally or longitudinally to the paved areas. The smoothness of the base or binder course shall be 6 mm for every 3 m. The smoothness of the surface course shall be 3 mm for every 3 m. On the crowned streets and drives, the allowable variance for the crown is 6 mm.

Pavement markings

1 The pavement shall cure 30 days prior to the application of any pavement markings.
2 The layout, colors, and placement of the pavement markings shall be verified with the Developer or City Engineer prior to application.
3 The pavement surface shall be clean and dry prior to the application of the pavement markings.
4 Apply paint with mechanical pavement painting equipment to produce pavement markings of the dimensions on the plans with uniform and straight edges. The paint should be applied at the manufacturer's recommended rates to provide a minimum wet film thickness of 0.4 mm. Pavement markings on streets and drives shall receive an application of glass spheres spread by broadcast and uniformly into the wet pavement markings at an application rate of 0.7 kg/l of pavement marking paint.

Quality assurance

1. *Testing agency*: Developer or City Engineer will engage a qualified independent testing agency to perform field inspections, tests, and test reports. The testing agency will conduct and interpret test results and will state in each report whether the tested work complies or deviates from the specified requirements.
2. Additional testing on deficient or corrected work to determine that the corrected work complies with the specified requirements will be done at the contractor's expense.
3. The acceptance of the completed HMA pavement thickness will be determined by the methods described in ASTM D3549, *Standard Test Method for Thickness or Height of Compacted Bituminous Paving Mixture Specimens*.
4. *Surface smoothness*: The finished surface of each HMA course will be tested for compliance with smoothness tolerances.
5. *In-place density*: Samples of uncompacted paving mixtures will be obtained by the testing agency according to ASTM D979, *Standard Practice for Sampling Bituminous Mixtures*. The reference maximum laboratory density will be determined by averaging the results from three samples of HMA delivered daily to the project site and tested according to ASTM D2041. The in-place density for acceptance will be determined by nuclear density test for core from every 500 tonnes of installed asphalt pavement. The nuclear testing method will be ASTM D2950, *Standard Test Method for Density of Bituminous Concrete in Place by Nuclear Methods* and the test results will be correlated to pavement cores tested by ASTM D1188, *Standard Test Method for Bulk Specific Gravity and Density of Compacted Bituminous Mixtures Using Coated Samples*. When discrepancies occur between test results, the tests results obtained from pavement cores will supersede.

Deficiencies

Remove and replace or install additional material where test results, measurements, or observations indicate that it does not comply with the specified requirements.

Appendix A

Terminology used in cross references

United States terminology	*UK, European terminology*
Asphalt, asphalt binder, asphalt cement, binder	Asphalt, binder, bitumen
Asphalt concrete	Asphalt concrete, asphalt macadam, bituminous concrete
Asphalt finisher, paver	Spreader
Asphalt mixture	Asphalt concrete, asphalt macadam, bituminous concrete
Base course	Binder course
Blacktop	Asphalt macadam, wearing course
Bucket	Tub
Cationic "C"	Cationic "K"
Chip seal	Surface dressing
Cold inplace recycling	Retread, Linear quarrying
Color	Colour
Cover aggregate	Chippings
Distributor	Surface dressing sprayer, sprayer
Drainage layer	Porous asphalt
Expressway	Motorway
Fiber	Fibre
Flushing	Fatting up
Freeway	Dual carriageway
Hydroplaning	Aquaplaning
Hot inplace recycling	Remix
Hot mix asphalt (HMA)	Asphalt concrete, asphalt macadam, bituminous concrete, dense graded mixture
Intermediate course	Binder course
Interstate	Motorway
Micro surfacing	Micro-asphalt
Novachip®	Coated chippings
Open graded friction course (OGFC)	Porous asphalt, friction course, drain asphalt, flusterasphalt, pervious macadam
Parking lot	Car park
Patching mixture	Deferred set macadam
Paver	Spreader
Pit run	Hardcore
Popcorn mixture	Porous asphalt (wearing course)
Raveling	Fretting

(Continued)

Terminology used in cross references (Continued)

United States terminology	*UK, European terminology*
Recycled asphalt pavement (RAP)	Road planings
Roadway	Carriageway
Roller	Compactor
Sand	Grit
Sand seal	Sand-carpet Pad coat
Scratch coat	Thin surface course
Seal coat	Surface dressing Veneer coat
Shoulder	Lining
Slurry seal	Slurry dressing, Schlaemme™
Speed bump	Ramp
Static roller	Deadweight roller
Stockpile mixture	Depot stock Deferred set macadam
Stone matrix asphalt, SMA	Stone mastic asphalt, SMA, Splittmastixasphalt, Mastimac™
Surface treatment	Surface dressing Veneer coat
Surface course	Wearing course
Tire	Tyre
Truck	Lorry

Appendix B

Unified soil classification

Unified soil class	*Description*	*Typical CBR*	*Permeability*
GW, GP	Gravel, crushed stone, little or no fines <0.02 mm	17	Excellent
SW, SP	Sand, sand-gravel mix, little or no fines <0.02 mm	17	Excellent
GW, GP	Gravel, crushed stone, some fines <0.02 mm	17	Good
SW, SP	Sand, sand-gravel mix, some fines <0.02 mm	17	Good
GW, GP, GM	Gravel size soil mix, medium fines <0.02 mm	8	Fair
SW, SP, SM	Sandy soils, medium fines <0.02 mm	8	Fair
GM, GW-GM, GP-GM	Silt gravel soils, high fines <0.02 mm	8	Fair to low
SM, SW-SM, SP-SM	Silt sand soils, high fines <0.02 mm	8	Fair to low
GM, GC	Clay gravel soils, high fines <0.02 mm	5	Fair to low
SM, SC	Clay sand soils, high fines <0.02 mm	5	Low to very low
SM	Very fine silt sands	5	Low
CL, CH	Clay, PI > 12	3	Very low
ML, MH	All silt soils	3	Very low
CL, CL-CM	Clay, PI < 12	3	Very low
OL	Other fine grained soils	<3	Very low
OH	Highly organic soils	Replace	Very low

Unified soil coding

Prefix	*Soil type*	*Suffix*	*Subgroup*
G	Gravel	W	Well graded
S	Sand	P	Poorly graded
M	Silt	M	Silt
C	Clay	L	Clay, liquid limit < 50%
O	Organic	H	Clay, liquid limit > 50%

Appendix C

Asphalt penetration grade cross reference

Penetration grade		
ASTM D946	*British Standard 3690*	*EN 12591 (from 1/1/2002)*
	15	
	25	20/30
	35	30/45
		35/50
40–50	50	40/60
60–70		50/70
	70	
85–100		70/100
	100	
120–150		100/150
	200	160/220
200–300		
	300	250/330
	450	

Appendix D

Standard sieve cross reference[1]

CEN Sieves, EN 933-2	*United States Sieves (SI)*	*United States Sieves, Standard*
50.0 mm	50.0 mm	2 inchs
40.0 mm	37.5 mm	1 $^{1}/_{2}$ inch
31.5 mm		
25.0 mm	25.0 mm	1 inch
20.0 mm	19.0 mm	$^{3}/_{4}$inch
16.0 mm		
12.5 mm	12.5 mm	$^{1}/_{2}$ inch
10.0 mm	9.5 mm	$^{3}/_{8}$ inch
8.0 mm		
6.3 mm		
5.0 mm	4.75 mm	No. 4
4.0 mm		
2.0 mm	2.36 mm	No. 8
1.0 mm	1.18 mm	No. 16
500 μm	600 μm	No. 30
250 μm	300 μm	No. 50
125 μm	150 μm	No. 100
63 μm	75 μm	No. 200

Note
1 Not all sieves may be specified or used in any one gradation.

Appendix E

Recommended minimum parking space requirements

Facility type	*Minimum number of parking stalls*
Retail store	1 per 35 m^2 of shopping floor area
Movie theater	1 per 3 seats
Restaurant, site down	1 per 25 m^2 of sitting space
Restaurant, fast food	1 per 40 m^2 of sitting space
Hospital	1.75 per bed
Medical office	1 per 16 m^2 of floor area
Nursing home	1 per 4 beds
General offices	1 per 40 m^2 of leased space
Motel	1.5 per room
Hotel	1 per room

Parking stall length and widths, 90°

Service letter	*Use*	*Length* (m)	*Parallel width* (m)
G	Grocery stores, others with shopping carts	5.6	3.0
A	Convenience stores, etc. with very high turnover rates	5.6	2.8
B	High turnover rates, such as general retail	5.6	2.7
C	Medium turnover rates, such as airport or residential parking	5.6	2.6
D	Low turnover rates, such as employee parking	5.6	2.5

Lane widths

Lane type	*Lane width* (m)
No parking, one way	3.1
No parking, two way	4.9
Traffic and parking lane	5.5
Traffic and curb parking lane	3.1
Fire lane	6.1

Driveway number and widths

Commercial type use	*Entry lanes*	*Exit lanes*	*Total width* (m)
Typical retail	1	1	8.0
Large volume	1	2	10.5
Very high volume	2	2	13.0

Acceptable grades for parking lots

Use type	*Grade* (%)
Routine, to drain water runoff	2
Continuous slope	≤6
Non parking automobile ramps, with pedestrian usage	≤12
Non parking automobile ramps, with no pedestrian usage	≤15

Bibliography

Alaska Department of Transportation and Public Facilities (2001) *Asphalt Surface Treatment Guide*, Fairbanks, Alaska: State of Alaska.

American Association of State Highway and Transportation Officials (1993) *AASHTO Guide for the Design of Pavement Structures*, Washington, DC: American Association of State Highway and Transportation Officials.

American Association of State Highway and Transportation Officials (1995) *A Policy on Geometric Design of Highway and Streets*, Washington, DC: American Association of State Highway and Transportation Officials.

American Association of State Highway and Transportation Officials (1997) *Segregation, Causes and Cures for Hot Mix Asphalt*, Washington, DC: American Association of State Highway and Transportation Officials.

American Association of State Highway and Transportation Officials (2000a) *AASHTO Provisional Standards, April 2000*, Washington, DC: American Association of State Highway and Transportation Officials.

American Association of State Highway and Transportation Officials (2000b) *Standard Specifications for Transportation Materials and Methods for Sampling and Testing*, Washington, DC: American Association of State Highway and Transportation Officials.

American Society for Testing and Materials (2001a) *Concrete and Aggregates*, West Conshohocken, Pennsylvania: American Society for Testing and Materials.

American Society for Testing and Materials (2001b) *Road and Paving Materials; Vehicle-Pavement Systems*, West Conshohocken, Pennsylvania: American Society of Testing Materials.

American Society for Testing and Materials (2001c) *Roofing, Waterproofing and Bituminous Materials*, West Conshohocken, Pennsylvania: American Society of Testing Materials.

American Society for Testing and Materials (2002) *Road and Paving Materials; Vehicle-Pavement Systems*, West Conshohocken, Pennsylvania: American Society of Testing Materials.

Asphalt SealCoat Manufacturers Association (2000) *Standard Specifications*. Available online at <http://www.sealcoatmfg.org/smallasma/asma.html> (visited June 16 2002).

Ball, John, III (2001) *Paving Cul-De-Sacs in Good Time, April 2001, Asphalt Contractor Magazine*, Independence, Missouri: The Asphalt Contractor.

Barber-Greene Company (1976) *Bituminous Construction Handbook*, Aurora, Illinois: Barber-Greene Company.

Blaw-Knox (1988) *Paving Manual*, Mattoon, Illinois: Blaw-Knox Construction Equipment Co.

Bohacz, R. T. (2001) *The Plain Truth about Gasoline, March 2001, Hot Rod Magazine*, New York: Hot Rod Magazine.

California Department of Transportation, Division of Maintenance (1994) *Memorandum to All District Directors: Roadway Maintenance Surface Treatment Strategies (Recommended Guidelines)*, Sacramento, California: State of California Business, Transportation and Housing Agency.

California Department of Transportation (2002) *California Test 368, Standard Method for Determining Optimum Bitumen Content for Open Graded Asphalt Concrete*, Sacramento, California: State of California Business, Transportation and Housing Agency.

Camillo, Jim (2001) *Benefits of Sealcoating a Parking Lot, Pavement Magazine, Sealcoating Reprint Collection*, Fort Atkinson, Wisconsin: Pavement Magazine.

Canadian Strategic Highway Research Program (2000) *Pavement Design and Performance: Current Issues and Research Needs, Millennium Research Brief #2, November 2000*: Ottawa, Ontario: Canadian Strategic Highway Research Program.

Caterpillar (1990a) *Asphalt Paver Manual*, Peoria, Illinois: Caterpillar Company.

Caterpillar (1990b) *Compaction Manual*, Peoria, Illinois: Caterpillar Company.

Cedarapids (1993) *Quality Paving Guide Book*, Cedar Rapids, Iowa: Cedarapids, Inc., A Raytheon Company.

Chellgren, Jon (2002) *Preventing Pavement "Birdbaths," January 2002, Pavement Magazine*, Fort Atkinson, Wisconsin: Pavement Magazine.

Chemical Lime Company (2000) Lime *in Soil, Applications, Environmental*. Available online at: <http://www.chemicallime.com/limesoil.html> (visited December 20 2000).

Childs, Mark, C. (1999) *Parking Spaces, A Design, Implementation, and Use Manual for Architects, Planners, and Engineers*, New York: McGraw Hill Companies.

Cleaver, Lisa (2001) *Making Sealer, Pavement Magazine, Sealcoating Reprint Collection*, Fort Atkinson, Wisconsin: Pavement Magazine.

Collins, R., Watson, D. and Campbell, B. (1995) *Development and Use of the Georgia Loaded Wheel Tester*, Washington, DC: Transportation Research Board.

Czarnecki, Raymond (2000) *Get Ready for a New Way to Bid, December 2000, The Asphalt Contractor Magazine*, Independence, Missouri: The Asphalt Contractor.

Deahl, Chuck (2002) *Pneumatic Know-How, April 2002, The Asphalt Contractor Magazine*, Independence, Missouri: The Asphalt Contractor.

Department of Transport (1999) *Design Manual for Roads and Bridges, Volume 7: Pavement Design and Maintenance, Bituminous Surfacing Materials and Techniques*, London: Her Majesty's Stationary Office.

Dubey, Girish (2002) *Improving Sealer Using Additives, January 2002, Pavement Magazine*, Fort Atkinson, Wisconsin: Pavement Magazine.

European Asphalt Pavement Association (1998) *Heavy Duty Surfaces: The Arguments for SMA*, Breukelen, The Netherlands: European Asphalt Pavement Association.

Federal Aviation Administration (1989) *Engineering Brief #44l*. Available online at: <http://www.faa.gov/arp/engineering/briefs/eb44.htm> (visited April 1 2002).

Forsyth, Raymond, A. (1991) *Asphalt Treated Permeable Material-Its Evolution and Application*, Lanham, Maryland: National Asphalt Pavement Association.

Harlow, G., Peters, J. and Prinz, M. (1978) *Rocks & Minerals*, Milan: Simon & Schuster.

Hyster Company (1986) *Compaction Handbook, Sixth Edition*, Kewanee, Illinois: Hyster Company, Construction Equipment Division.

International Slurry Seal Association (1990) *Design Technical Bulletins*, Washington, DC: International Slurry Seal Association.

International Slurry Seal Association (1991) *Recommended Performance Guidelines for Micro-Surfacing*, Washington, DC: International Slurry Seal Association.

Kandhal, Prithvi, Ramirez, Timothy and Ingram, Paul (2002) *Evaluation of Eight Longitudinal Joint Construction Techniques for Asphalt Pavements in Pennsylvania*, Washington, DC: Transportation Research Board.

Kirk-Othmer (1978) *Encyclopedia of Chemical Technology, Volume 3*, New York: John Wiley & Sons.

Knight, B. H. (1935) *Road Aggregates, their Use and Testing*, London: Edward Arnold and Company.

Kriech, A. J. (1994) *Understanding the SHRP Binder Specification*, Indianapolis, Indiana: Heritage Research Group.

Langer, W. and Glanzman, V. M. (1993) *Natural Aggregate, Building America's Future*, Washington, DC: United States Government Printing Office.

Lavin, P. G. (1999) *A Comparison of Liquid Antistrip Additives and Hydrated Lime Using AASHTO T-283*, Peterson Asphalt Research Conference, Laramie, Wyoming: Western Research Institute.

Lavin, P. G. and McCain, M. (1993) *Level III Asphalt Concrete Course Manual*, Mattoon, Illinois: Lake Land College.

Lindeburg, M. R. (1999) *Civil Engineering Reference Manual for the PE Exam*, Belmont, California: Professional Publications, Inc.

Local Road Research Board (1991) *Flexible Pavement Distress Manual*, St Paul, Minnesota: Minnesota Department of Transportation.

LTPP (2001), *Long Term Pavement Performance, LTPPBind2.1 download*, Available online at: <http://tfhrc.gov/pavement/ltpp/.html> (visited October 11 2001).

Miller, T., Ksaibati, K. and Farrar, M. (1995) *Utilizing the Georgia Loaded-Wheel Tester to Predict Rutting*, Washington, DC: Transportation Research Board.

Morgan, Paul and Mulder, Alan (1995) *The Shell Bitumen Industrial Handbook*, Surrey, United Kingdom: Shell Bitumen United Kingdom.

Murphy, T. R. and Bentsen, R. A. (2001) *Marshall Mix Design, Getting the Most out of your Marshall Mixes*, Norridge, Illinois: Humboldt Mfg Co.

National Asphalt Pavement Association (1986) *The Design of Hot Mix Asphalt for Heavy Duty Pavements, Quality Improvement Series 111*, Lanham, Maryland: National Asphalt Pavement Association.

National Asphalt Pavement Association (1990) *Hot Mix Asphalt Joint Construction, Quality Improvement Series 115*, Lanham, Maryland: National Asphalt Pavement Association.

National Asphalt Pavement Association (1991) Design *of Hot Mix Asphalt Pavements for Commercial, Industrial and Residential Areas, Information Series 109*, Lanham, Maryland: National Asphalt Pavement Association.

National Asphalt Pavement Association (1993) *Open-Graded Asphalt Friction Course, Information Series 115*, Lanham, Maryland: National Asphalt Pavement Association.

National Asphalt Pavement Association (1994) *Guidelines for Materials, Production, and Placement of Stone Matrix Asphalt, Information Series 118*, Lanham, Maryland: National Asphalt Pavement Association.

National Asphalt Pavement Association (1997) *Quality Control for Hot Mix Asphalt Operations, Quality Improvement Series 97*, Lanham, Maryland: National Asphalt Pavement Association.

National Asphalt Pavement Association (2001) *HMA Pavement Mix Type Selection Guide, Information Series 128*, Lanham, Maryland: National Asphalt Pavement Association.

National Asphalt Training Center (1994) *SuperPave™ Asphalt Mixture Design and Analysis, Demonstration Project 101*, Washington, DC: Federal Highway Administration.

National Center for Asphalt Technology (1996a) *Implementation Notes, Note No. 3, Spring 1996, Field Management of Hot Mix Asphalt Volumetric Properties*, Auburn, Alabama: National Center for Asphalt Technology, Auburn University.

National Center for Asphalt Technology (1996b) *Implementation Notes, Note No. 4, Fall 1996, Designing Recycled HMA with SuperPave Technology*, Auburn, Alabama: National Center for Asphalt Technology, Auburn University.

National Center for Asphalt Technology (1998a) *Asphalt Technology News, Volume 10, number 2, Fall 1998*, Auburn, Alabama: National Center for Asphalt Technology, Auburn University.

National Center for Asphalt Technology (1998b) *Implementation Notes, Note No. 5, Fall 1998, Designing Stone Matrix Asphalt (SMA) Mixtures*, Auburn, Alabama: National Center for Asphalt Technology, Auburn University.

National Center for Asphalt Technology (2001) *NCAT Develops New Generation Open-Graded Asphalt Friction Course, HMAT Magazine, Volume 6, Number 6*, Lanham, Maryland: National Asphalt Pavement Association.

National Cooperative Highway Research Program (1975) *Bituminous Emulsions for Highway Pavements*, Washington, DC: Transportation Research Board.

National Stone Association (1992) *Stone Base Construction Handbook*, Washington, DC: National Stone Association.

National Stone Association (1994a) *Design Guide for Parking Areas*, Washington, DC: National Stone Association.

National Stone Association (1994b) *NSA Flexible Pavement Design Guide for Road and Streets*, Washington, DC: National Stone Association.

Newcomb, David and Timm, David (2002) *Mechanistic Pavement Design, A Homogenized, Icy-tropic Half-What? HMAT Magazine, Volume 7, Number 1, January/February 2002*, Lanham, Maryland: National Asphalt Pavement Association.

Ng, Tony (1996) *Slurry Seal and Polymer Mixes, the Facts You Need to Know, December 1996, Better Roads Magazine*, Des Plaines, Illinois: James Informational Media, Inc.

Nicholls, J. C. (ed.) (1998) *Asphalt Surfacings*, London, United Kingdom: E & FN Spon.

Pavement Coatings Technology Center (1997) *Coal Tar Sealer: The Truth Comes to the Surface*, Reno, Nevada: University of Nevada, Reno, Department of Civil Engineering.

Plummer, Richard (1996) *Chip Seal Use-Back to the Basics, April 1996, Better Roads Magazine*, Des Plaines, Illinois: James Informational Media, Inc.

Roberts, F. L., Kandhal, P. S., Brown, E. R., Lee, D. and Kennedy, T. W. (1991) *Hot Mix Asphalt Materials, Mixture Design, and Construction*, Lanham, Maryland: NAPA Education Foundation.

Rosenberger, Carlos and Buncher, Mark (2000) *Structural Adequacy in Asphalt Intersections, August 2000, Better Roads Magazine*, Des Plaines, Illinois: James Informational Media, Inc.

Rushmoor Borough Council (1998) *Porous Asphalt*. Available online at: <http://www.rushmooor.gov.uk/hig9750.html> (visited October 22 2001).

Sanders, S.R. (1984) *Refined Petroleum Products, A Short Course*, Oklahoma City, Oklahoma: Kerr-McGee Refining Corporation.

Schmidt, Paul (1995) *Physical Maintenance: Repairing Asphalt Pavement Problems*. Available online at HTTP: <http://www.facilities.com/NS/NS3mh56.html> (visited August 29 2001).

Stick, Greg (2001) Asphalt *Patches that Last, October 22 2001, Construction Digest Magazine, West Edition, Volume 76, Number 20*, Norcross, Georgia: Cahners Business Information.

Strategic Highway Research Program (1990) *Distress Identification Manual for the Long-Term Pavement Performance Studies*, Washington, DC: National Research Council.

Strategic Highway Research Program (1993) *Materials and Procedures for the Repair of Potholes in Asphalt-Surfaced Pavements, Manual of Practice*, Washington, DC: National Research Council.

Strategic Highway Research Program (1994) *Asphalt Pavement Repair Manuals of Practice*, Washington, DC: National Research Council.

Summers, C. J. (2001) *The Idiots Guide to Highways and Maintenance, Modified Bitumen and Bituminous Materials*. Available online at: <http://www.highwaysmaintenance.com/polybitxt.htm> (visited October 22 2001).

Tappeiner, Walter, J. (1993) *Open-Graded Asphalt Friction Course*, Lanham, Maryland: National Asphalt Pavement Association.

Terrel, Ronald (1990) *Pavement Preparation Prior to Overlaying with HMA*, Lanham, Maryland: National Asphalt Pavement Association.

The Asphalt Contractor (2000) *Putting the breaks on Skidding, Skid Test Evaluates Aggregates, December, 2000, The Asphalt Contractor Magazine*, Independence, Missouri: The Asphalt Contractor.

The Asphalt Contractor (2001) *Establishing Perfect Rolling Patterns for Bonus Quality Mats, November 2001, The Asphalt Contractor Magazine*, Independence, Missouri: The Asphalt Contractor.

The Asphalt Institute (1973) *Model Specifications for Small Paving Jobs, CL-2*, College Park, Maryland: The Asphalt Institute.

The Asphalt Institute (1975) *Differences between Petroleum Asphalt, Coal Tar Pitch and Road Tar*, College Park, Maryland: The Asphalt Institute.

The Asphalt Institute (1981a) *How to Design Asphalt Pavements for Streets, IS-96*, College Park, Maryland: The Asphalt Institute.

The Asphalt Institute (1981b) *Asphalt Hot Mix Recycling*, College Park, Maryland: The Asphalt Institute.

The Asphalt Institute (1982) *Asphalt Surface Treatments-Construction Techniques, ES-12*, College Park, Maryland: The Asphalt Institute.

The Asphalt Institute (1983) *Asphalt in Pavement Maintenance*, Lexington, Kentucky: The Asphalt Institute.

The Asphalt Institute (1984) *Full-Depth Asphalt Pavements for Parking Lots, Service Stations and Driveways, IS-91*, College Park, Maryland: The Asphalt Institute.

The Asphalt Institute (1985) *Asphalt Pavement for Athletics and Recreation*, College Park, Maryland: The Asphalt Institute.

The Asphalt Institute (1986a) *Asphalt Plant Manual*, College Park, Maryland: The Asphalt Institute.

The Asphalt Institute (1986b) *Soils Manual for the Design of Asphalt Pavement Structures*, College Park, Maryland: The Asphalt Institute.

The Asphalt Institute (1987) *Asphalt Paving Manual*, College Park, Maryland: The Asphalt Institute.

The Asphalt Institute (1988) *The Asphalt Handbook*, Lexington, Kentucky: The Asphalt Institute.

The Asphalt Institute (1989) *Large Aggregate Asphalt Mixes, TB-5*, Lexington, Kentucky: The Asphalt Institute.

The Asphalt Institute (1990) *Design of Hot Asphalt Mixtures, ES-3*, Lexington, Kentucky: The Asphalt Institute.

The Asphalt Institute (1992) *Factors Affecting Compaction, ES-9*, Lexington, Kentucky: The Asphalt Institute.

The Asphalt Institute (1993) *Mix Design Methods for Asphalt Concrete and Other Hot-Mix Types*, Lexington, Kentucky: The Asphalt Institute.

The Asphalt Institute (1997) *Performance Graded Asphalt Binder Specification and Testing*, Lexington, Kentucky: The Asphalt Institute.

The Asphalt Institute (1998a) *A Basic Asphalt Emulsion Manual, Second Edition*, Lexington, Kentucky: The Asphalt Institute.

The Asphalt Institute (1998b) *Construction of Hot Mix Asphalt Pavements, Second Edition*, Lexington, Kentucky: The Asphalt Institute.

The Asphalt Institute (1999) *Thickness Design*, Lexington, Kentucky: The Asphalt Institute.

The Asphalt Institute (2001) *SuperPave Mix Design, Third Edition*, Lexington, Kentucky: The Asphalt Institute.

Transportation Information Center (1989) *Asphalt-Paser Manual, Pavement Surface Evaluation and Rating*, Madison, Wisconsin: University of Wisconsin.

Transportation Research Board (TRB) (1994) Highway *Capacity Manual, Special Report 209*, Washington, DC: United States Government Printing Office.

Transportation Research Board (TRB) (2000) *Hot-Mix Asphalt Paving Handbook 2000*, Washington, DC: Transportation Research Board, National Research Council.

Transportation Research Board (TRB) (2001) *Recommended Use of Reclaimed Asphalt Pavement in the SuperPave Mix Design Method: Technician's Manual, NCHRP Report 452*, Washington, DC: Transportation Research Board, National Research Council.

United States Army Corps of Engineers (2001) *Unified Facilities Criteria, Standard Practice Manual for Flexible Pavements*, Washington, DC: United States Government Printing Office.

United States Department of Agriculture, Forest Service (1999) *Asphalt Seal Coat Treatments, Publication No. 1201-SDTDC, April 1999*, Washington, DC: United States Government Printing Office.

United States Department of Transportation, Federal Highway Administration (1990) *Technical Advisory 5040.31, Open Graded Friction Courses* Available online at: <http://www.fhwa.dot.gov/legsregs/directives/techadvs/t504031.htm> (visited April 8 2002).

United States Department of Transportation, Federal Highway Administration (1994) *State of the Practice Design, Construction and Performance of Micro-surfacings, FHWA-SA-94-051*, Washington, DC: United States Government Printing Office.

United States Department of Transportation, Federal Highway Administration (1997) *Pavement Recycling Guidelines for State and Local Governments, Publication No. FHWA-SA-98-042*, Washington, DC: United States Government Printing Office.

United States Department of Transportation, Federal Highway Administration (2000) *Recycled Materials in European Highway Environments, Uses, Technologies, and Policies, October 2000*, Washington, DC: United States Government Printing Office.

University of New Hampshire T2 Center (1997) *Crack Sealing*. Available online at: <http://www.t2.unh.edu/summer97/pg7.html> (visited May 26 2002).

University of Southern Queensland, Australia (2000) Maintenance *of Roads and Streets*. Online. Available online at: <http://www.usq.edu/users/ayers/munsermod16.htm> (visited April 8 2002).

Wallace, Jai (2000) *Joint Construction Wedges its Way onto Roadways, October 2000, Asphalt Contractor Magazine*, Independence, Missouri: The Asphalt Contractor.

Wardlaw, Kenneth and Shuler, Scott (1992) *Polymer Modified Asphalt Binders, STP 1108*, Philadelphia, Pennsylvania: American Society of Testing Materials.

Whiteoak, D. (1991) *The Shell Bitumen Handbook*, Surrey, United Kingdom: Shell Bitumen United Kingdom.

Wright, P. H. and Paquette, R. J. (1987) *Highway Engineering*, New York: John Wiley & Sons.

Ziegler Chemical and Mineral Corporation (2001) *What is Gilsonite?* Available online at: <http://www.zieglerchemical.com/gilsonit.htm> (visited November 28 2001).

Index